AF901948

Biosemiotics

Volume 26

Series Editors
Kalevi Kull, Department of Semiotics
University of Tartu
Tartu, Estonia

Alexei Sharov, Lab Genetics, Rm 10C222, Ste 100
National Inst on Aging
Baltimore, MD, USA

Claus Emmeche, Department of Science Education
University of Copenhagen
Kobenhavn K, Denmark

Donald F. Favareau, University Scholars Programme
National University of Singapore
Singapore, Singapore

Combining research approaches from biology, semiotics, philosophy and linguistics, the field of biosemiotics studies semiotic processes as they occur in and among living systems. This has important implications and applications for issues ranging from natural selection to animal behaviour and human psychology, leaving biosemiotics at the cutting edge of the research on the fundamentals of life.

The Springer book series *Biosemiotics* draws together contributions from leading scholars in international biosemiotics, producing an unparalleled series that will appeal to all those interested in the origins and evolution of life, including molecular and evolutionary biologists, ecologists, anthropologists, psychologists, philosophers and historians of science, linguists, semioticians and researchers in artificial life, information theory and communication technology.

Kalevi Kull, Professor in biosemiotics, University of Tartu
Alexei Sharov, National Institute of Aging, Baltimore

Daniel Carlos Mayer-Foulkes

The Nature of Living Being

From Distinguishing Distinctions to Ethics

Springer

Daniel Carlos Mayer-Foulkes
Living Leadership
San Diego, CA, USA

ISSN 1875-4651 ISSN 1875-466X (electronic)
Biosemiotics
ISBN 978-3-031-24788-0 ISBN 978-3-031-24789-7 (eBook)
https://doi.org/10.1007/978-3-031-24789-7

© The Editor(s) (if applicable) and The Author(s), under exclusive license to Springer Nature Switzerland AG 2023
This work is subject to copyright. All rights are solely and exclusively licensed by the Publisher, whether the whole or part of the material is concerned, specifically the rights of translation, reprinting, reuse of illustrations, recitation, broadcasting, reproduction on microfilms or in any other physical way, and transmission or information storage and retrieval, electronic adaptation, computer software, or by similar or dissimilar methodology now known or hereafter developed.
The use of general descriptive names, registered names, trademarks, service marks, etc. in this publication does not imply, even in the absence of a specific statement, that such names are exempt from the relevant protective laws and regulations and therefore free for general use.
The publisher, the authors, and the editors are safe to assume that the advice and information in this book are believed to be true and accurate at the date of publication. Neither the publisher nor the authors or the editors give a warranty, expressed or implied, with respect to the material contained herein or for any errors or omissions that may have been made. The publisher remains neutral with regard to jurisdictional claims in published maps and institutional affiliations.

This Springer imprint is published by the registered company Springer Nature Switzerland AG
The registered company address is: Gewerbestrasse 11, 6330 Cham, Switzerland

A monk asked Jōshū, "Has a dog the Buddha nature?"
Jōshū answered "Mu."

—*Mumonkan*[1]

[1] Wumen Huihai. "Mumokan, Case 1." In *Two Zen Classics*: *Mumonkan (The Gateless Gate) and Hekiganroku (The Blue Cliff Records)*.

To Vicky, my beloved and best possible partner in the adventure of joining that is marriage,
to our sons Moy and Gabriel and their wives Tammy and Jacky,
and to our marvelous and dear grandchildren
Ezra, Elian, Alex, Daniela, Arie, and Michelle.
May this project of a lifetime contribute to a luminous time of life.

Introduction

The remote origin of this project was a decade living at The Monkey Sanctuary, preservation center for the South American woolly monkey. Time and again, we were confronted with the analytical problem of the anthropomorphic fallacy, the *a priori* judgment that claims about the resemblance of some aspect of the animal with the human are fallacious or, at best, suspect of being human projections. The more so when a trait is remarkable precisely because of such a resemblance. By trying to conform to "the scientific," this results in the general tendency to reduce the animal to the physiological and ecological, leaving aside the psychological, the cognitive, the social, and, needless to, say, the emotional. The richness of the individuality of individuals is thus suppressed. Thus, formal study of the animal tends to be in a reductionist and a mechanistic frame that consigns what is most characteristic of individuals to "anecdotal accounts" in publications aimed at the general public.

Many years later, I understood that this was an epistemological problem of the first order. At the same time, the efficacy of the approach suggested, not only how to solve—or rather dissolve—the problem of the anthropomorphic fallacy, but it radicalized the inquiry, extending it from zoology to biology as a whole. Aliveness itself became the topic of investigation. Finally, this led to an understanding of the deep roots in living being of our ethical nature. In very broad terms, then, this investigation establishes what shall be referred to as the proper content of distinction and logic and then the proper content of distinguishing and relationship. This in turn leads to the origin of ethics.

The endeavor took shape around a single fundamental question: What is the minimum *qua* cognitive activity that any living being must be capable of? One answer stood out head and shoulders above all others: the capacity to distinguish. This insight turned out to be decisive. It is from this answer that all argumentation in this treatise proceeds. It first examines what a distinction is and what it is to distinguish; the noun and the verb; the activity and what results from it. It then probes what kind of thing can distinguish. It is found that distinction-distinguishing is the appropriate way inquire into aliveness. Distinction-distinguishing obtains in physiology as much as it does in psychology, fusing botany and zoology into a single continuous-discontinuous field. Thus, from distinction-distinguishing follows a re-foundation of the notion of the biological. Here, then, are the Abstracts of the chapters that follow.

Chapter 1, Distinction-Distinguishing, begins by showing that living beings are distinctions whose most fundamental activity is distinguishing. This is what makes them unique and how inquiry into aliveness must proceed. All aspects of living being involve physical and chemical conjugate forms, but the highest, supervening conjugate form to understand them, is distinction-distinguishing. Organisms distinguish by establishing A/Ā, A=A. They do so to remain themselves. Primary sources in this exposition are introduced: Spencer-Brown's *Laws of Form*, Bernard Lonergan's *Insight*, and a decade's experience living with monkeys. The second part of this chapter provides an account of one of many nights I slept in the monkey territory looking after the recently orphaned Laura. Sharing the monkeys lives, being there with them, even for them, at critical times, with the stirring of emotion this involves, illustrates what is lost to primatology, and to zoology generally, so long as the problem of anthropomorphism persists. Such intimate knowledge of animals must have some bearing on a zoology that is proportional to its object of study.

Chapter 2, Distinguishing Distinctions examines definitions of distinction, along with problems of circularity. It is shown that the commonsense notion of distinction conceived abstractly as a circle is not sustainable Elementary distinguishing is to differentiate differences that make a difference from differences that do not. Given the nature of concreteness, arbitrariness is required for equivalences. This is "forness", the ground of repetition. This is how a bacterium distinguishes and what allows for equivalences, or A=A. The notion of bet structure (ב) is introduced to account for the possibility A=A, or repetition. Repetition is inherent to environments, given their cyclical nature. Aliveness is to establish equivalences between previous and current events through anticipation. Thus, anticipation is intrinsic to distinguishing and thereby foundational of aliveness. Nestings of distinctions lead to ever higher-levels of distinction.

Chapter 3, The Emergence of Living Being, tests the notion of distinguishing distinctions. Biogenesis is examined in terms of the question, How and when could ב emerge? Living beings self-maintain and are therefore [subjects]. (Square brackets are introduced to indicate a generalization that postpones indispensable further discussion.) The notions of autopoiesis and homeostasis obtain for an entity that is an in itself. Living beings make sense of the world in terms of their own nature. In contrast to anthropomorphism, biomorphism is to ascribe features of aliveness to the nonliving. When inquiring into the advent of life, we must guard against our own biomorphism just at the point where it matters most. Aliveness involves two layers, 'ב'and ב. Narrow Darwinism addresses random natural selection but not selection as carried out by living cells with their distinguishing.

Chapter 4, Empirical Evidence for ב, presents three bodies of empirical findings as evidence of the foundational role of ב. First, a recently proposed scenario for biogenesis through emergent probability in volcanic pools. Second, the ubiquity of circadian clocks. These are also prototypical instances of two-component regulatory systems. Third, Earliest Indication (EI), a general model for the unfolding of anticipation. A prototypical Integrator-Operator, EI exemplifies a general heuristic structure for inquiry into aliveness. Through EI organisms systematically apprehend events or circumstances that regularly precede differences that make a difference to them. Anticipation is a selective and ultimately a cognitive advantage. Nested elementary anticipations yield ever more complex anticipations.

Chapter 5, Formal Definition of Distinction, takes up the definition of distinction by means of two axioms in *Laws of Form* (*LoF*) by George Spencer-Brown. This formalism demonstrates a primary non-numerical arithmetic that grounds Boolean algebra (logic). However, the two axioms use repetition, which is not justified as required by the formalism. This taking for granted of repetition, or ב, renders LoF a phenomenology of aliveness. In turn this points to the relation between life and mathematics. The mathematical ideal of knowledge, so successful in physics, could not in principle obtain in biology. Concretely, oscillation, which antecedes the pure discrete values of logic, underlies distinguishing, including the process of human judging. This fundamental isomorphism of all living beings must radically transform the method of biology and will contribute to the claim that living being is inherently [ethical]. Organisms are always already in interrelationships, they are not abstractly independent from each other.

Chapter 6, The Mathematical Ideal and the Elision of the Subject, discusses contemporary science as bound to the mathematical ideal of knowledge, the key to which is the elision of the knowing subject. Can biology, as inquiry into [subjects], be bound by, and to this ideal? In terms of the primary non-numerical mathematics that models the foundation of logic in LoF, a differentiation and contrast must be made between the proper and borrowed contents of @@L (logic), and @@A (aliveness). Differences that make a difference are values. In formalisms values are named. But values pertain to Liminalism and names to Substantialism. As a highly differentiated instance of EI, naming too is an Integrator-Operator. Formalisms express insights. Understood as systems of distinctions, the Platonic and phenomenological interpretations of mathematical formalism converge. This convergence speaks to the nature of [subjects] and the relation I/World.

Chapter 7, Four Kinds of Things, begins by clarifying the notion of 'thing'. On the side of the knowing subject, a thing as a unity-identity-whole is a kind of insight. On the side of the object, only living beings are things for themselves. In terms of distinction-distinguishing there are four kinds of things. The non-living are things only insofar as knowing subjects distinguish them. Distinction-distinguishing is inherent to living being, products, and byproducts, as objects. Much may be understood from these differentiations. A key blunder is to conflate one kind of thing with another, for instance, to conceive products as nonliving or organisms as mechanisms. Living being emerged in plurality, joinings of distinguishing distinctions for themselves in relationship. Additional empirical evidence of living being as joinings is provided by environments, endosymbiosis, major evolutionary transitions and social evolution, and sexual reproduction.

Chapter 8, [Subjects], shows the convergence between organisms as objects and organisms as [subjects]; how [subjects] are built, interrelate, and how they construct the world. [Subjects] are organized, belong to organizations, and organize the world. On the side of the object, organisms are built and function hierarchically as identity-unity-wholes. This is how [subjects] make sense. Skin is a primary instance of the function of identity. The originary skin—membranes—are so central to the functioning of cells that they may be dubbed 'mem-brains'. The organization of [body subjects] as identity-unity-wholes and our making sense of them as unity-identity-wholes, sets the foundation for inquiry into [subjects]. This brings up the question, What is it to understand when it comes to organization?

Chapter 9, Living Being, starts from the point that living being takes place in plurality, and plurality involves boundaries. Microbial partnerships are examples of joinings. There are three kinds of wholes, two of which are joinings: semi-autonomous organic wholes made up of semi-autonomous organic wholes (WW); social wholes, not alive but made up of organic wholes (WW); and mechanical wholes made of parts (wp). There are key differences and deep isomorphism between the two kinds of joinings. Social wholes are products. Organic wholes make up a larger living whole, a plurality of living wholes with identity. Paradox is inherent to joinings because they are made up of relationships. The intelligibility of relationship brings a shift from substantiality to liminality. Hymenomorphism, contrasted to Aristotle's hylomorphism, underlies the intelligibility of wholes made up of relationships and leads to an overarching conception of [ethics].

Chapter 10, The Paradoxical Nature of Aliveness, delves into relationships. Our experience of relationships provides us with direct access to the intelligibility of joinings in terms of paradox. This includes the paradoxes of dependency, boundaries, regression, and repetition. Being isomorphic, social and organic wholes may be understood in terms of each other. Social life unfolds anticipation, transparency/opaqueness, and thereby our ethical nature. From this follows the notion of [ethics] as the inherent character of aliveness. As a founding feature of @@A (aliveness), [ethics] is the foundation of @@L (logic). This dovetails with Levinas' "ethics as first philosophy." Our own ethical nature is a recent specific instance of [ethics]. That being so, our ethical character does not alienate us but rather grounds us intimately in the world. Renewing old forms of inquiry reenchants the world, as we so long for. Reintegrating them, we bring ourselves home.

A Glossary of coined terms is provided. Their full meaning derives from their use, particularly those whose meaning unfolds as the argument proceeds.

This book has been over thirty years in the making. Keeping abreast with the fields to which distinction pertains is an impossible task. Indeed, narrowing the scope of this investigation to distinction had the perverse paradoxical effect of multiplying the fields that are relevant to it.

It has been said that a long piece of writing is never completed, only finally abandoned. Its ending, like any other distinction, is irreducibly arbitrary and fuzzy.

I was once confronted with the question, Is the re-founding of biology worth a life? My response remains a decisive Yes.

Acknowledgments

This adventure began back in 1972 when, as a student at Atlantic College in South Wales, Mr. Jenkins, my house-master and Marine Studies professor, suggested spending a project week at *The Monkey Sanctuary*. I met then its founders Leonard and June Williams and Sue Rickard and their coworkers, Lorna and John Tucker. Living at the Sanctuary from 1979 to 1989, I worked with Simon Rickard, Kathy Day, Geri Skeens, Peter Maxey, Rachel Hevesy, and many others. Len embraced me as a son and took it upon himself to teach me British culture through endless conversation and trenchant humor. He reminisced about his life; discussed philosophy, primatology, ethology, evolution, and history; and broke my bones. We played chess, and he taught me jazz guitar and, not least, how to write. Sue taught me to look after the monkey colony, and from her, I took over as curator the last years of my decade there.

Before living at the sanctuary, Jessy, the monkey matriarch, had greeted me with expressive snuffles. Her family followed suit and so did the rest of the monkey group. Once I lived there, Emma, Jessy's second daughter, welcomed me as a nanny. Over the years, I babysat her babies Lucy, Charlie, and Laura. Laura grew up and one day gave birth to her own Lucy a few inches away from my nose. When I left the sanctuary, I had known four generations of the Jessy clan.

Upon my return to Mexico in 1989, I found my bearings in Tepoztlán, Morelos, at my parents Robert and Vera Mayer' country house. For six months, I revised my notes and wrote up everything I could recall about the sanctuary. In time the project took shape as an inquiry in the philosophy of science. Eventually, I found my way to the Philosophy Department at Universidad Iberoamericana. Learning about my project, Paco Galán, then the director and now an old friend, pointed out that I needed foundations. He introduced me to the work of Bernard Lonergan, to Phil MacShane, and later suggested I apply for the Lonergan Fellowship in Boston College. Also at Ibero, Luis Vergara always cheered on this project. He was a member of my three synods and generously discussed and encouraged the development of some key ideas. In Boston, I spent a year at the Friday seminars with Fred Lawrence and his colleagues—among them Pat Byrne who would one day become director of the dissertation that was the first version of this book. Later I met Mark

Morelli and his group in LA when I read papers at the yearly Loyola Marymount Lonergan conferences. I remember fondly a dinner I once shared with four students—unfortunately I do not recall their names—who spontaneously launched into a stress-test of the notion of four kinds of things, the topic of the paper I had read that eventually became Chap. 5. During those years, I also found my way to Biosemiotics. Applying at the very last minute for my first conference in Braga, Portugal, as I walked in the door Marcello Barbieri invited me to read a paper. I was also warmly welcomed by Kalevi Kull, Jesper Hoffmeyer, and Donald Favareu. By then living in San Diego, I had met Terry Monroe, director of the Leadership Institute at the University of San Diego, and was for many years immersed in Group Relations teaching leadership experientially.

Over the decades, eight dogs and two cats have beckoned me back to life outdoors, insistently announcing it is time for their daily walks. They have rescued me from abstraction, and their delightful company has brought endless insights: Toc and Tepe in Tepoztlán; Alpha, then Max and Jeronimo in Mexico City and later in Boston; Sancho in San Diego; and more recently the lovely female cats Itayetzi and Paulina who gracefully don their leashes.

I am particularly indebted to Pat Byrne who, enchanted by tales of monkeys, accepted being the director of my PhD dissertation. Little did he know that he was taking on a task measured in decades. It is difficult to imagine his burden being carried out in a more exemplary manner, with limitless grace, patience, and diligence well beyond the call of duty. I am eternally grateful.

Having Don Favareu and Kalevi Kull zoom in for my defense was a wonderful treat, and their follow-up invitation to publish with them at Springer was a blessing. I thank the series committee for welcoming it, and two anonymous readers for excellent incisive feedback. Together with Tarren Andrews from Dissertation Editors, these comments made for a significantly improved final version. Needless to say, all errors are my own responsibility.

Throughout my life, I have enjoyed the company and the deeply insightful conversation of my brilliant brothers David and Benjamin, on all topics imaginable. It would be impossible to even gage their effect on me, on my life, and on this undertaking.

Nothing would have been possible without Vicky. At the beginning, in a variation of the Arabian nights, at bedtime over two years she asked me to tell her stories of the primates at the sanctuary, both monkey and human. So, this project was baked into our relationship from the start. Over thirty years, I have foisted it on her and on our marriage. It is impossible to express my gratitude for her endurance, which has been nothing less than heroic.

Copyright Permissions

Permission is gratefully acknowledged to the following publishers for excerpts cited:

Dao De Jing: A Philosophical Translation by Laozi, translated and with commentary by Roger Ames and David L. Hall, copyright © 2003 by Roger T. Ames and David L. Hall. Used by permission of Ballantine Books, an imprint of Random House, a division of Penguin Random House LLC.

William, Large, 2015, *Levinas' Totality and Infinity*. London: Bloomsbury, Academic, an imprint of Bloomsbury Publishing Plc.

Bruce H. Lipton, PhD, 2016. *The Biology of Belief: Unleashing the Power of Consciousness, Matter & Miracles*. www.brucelipton.com.

Bernard Lonergan, 1992. *Insight: A Study of Human Understanding*, Collected Works of Bernard Lonergan, vol. 3. Toronto: University of Toronto Press.

Daniel Mayer, 2022 *Divyadaan* 33/1, "McShane on the Transformation of Botany".

Carlo Rovelli, *Seven Brief Lessons on Physics*, translated by Simon Carnell and Erica Segre, copyright © 2014 by Carlo Rovelli. Translation © 2015 by Simon Carnell and Erica Segre. Used by permission of Riverhead, an imprint of Penguin Publishing. Group, a division of Penguin Random House LLC. All rights reserved.

Contents

Chapter 1
Distinction-Distinguishing

The Proper Conjugate Form

We. We are. We are alive. This aliveness is the precondition for any reflection on life, and in turn the precondition for biology. The purpose of this book is at once extremely simple and extremely complex: to define the precise sense in which living beings are fundamentally different from the nonliving. Without such a clear definition, inquiry into the nature of life cannot begin. Furthermore, this fundamental difference must be expressed in the appropriate terms—*the proper conjugate form* as will be explained below. If the nature of life is *aliveness,* then it is in living terms that inquiry into life must proceed. This is so for fundamental reasons.

Physical or chemical terms—or *conjugate forms*—are appropriate for inquiry into physical and chemical processes. So long as questions about the nature of life are asked and answered in these terms, aliveness itself will be bypassed altogether. Beyond the indispensable physics or chemistry, this book argues that the nature of living being is *to distinguish.* Moreover, it is proposed that living beings are themselves distinctions. They self-regulate to be so: this is how they remain themselves and distinct from the environment. A living being is a distinction (noun) that distinguishes (verb): it is a *distinguishing distinction.* This underscores the biological autonomy or *functional closure* of living cells and aggregates of living cells. The notion of distinguishing distinction is not meant to substitute equivalent biological notions. However, it will soon become clear why it makes sense to think about the nature of living being in these terms. If the fundamental difference between living being and the nonliving is, as I propose, that living beings distinguish whereas the nonliving does not, then the proper way to inquire into living being is not in terms of physical, chemical, or even logical terms, but in 'distinctional' terms. For, as we shall later see, at the heart of inquiry into the distinctional is the questioning of our

© The Author(s), under exclusive license to Springer Nature Switzerland AG 2023
D. C. Mayer-Foulkes, *The Nature of Living Being*, Biosemiotics 26,
https://doi.org/10.1007/978-3-031-24789-7_1

notion of boundaries themselves and their very nature. Altogether, this is what makes the distinctional the *proper conjugate form* to inquire into aliveness.[1]

Let us start this inquiry and further explain the notion of *proper conjugate form* by considering what all living beings must do.

The root of a plant distinguishes between nutrient (A) and non-nutrient (Ā). It distinguishes between A and Ā or, what is the same, establishes A, Ā, and /, the boundary between them. It establishes A/Ā. A major topic of this chapter is to delve into what this *establishing* entails. The establishing of A/Ā is ubiquitous throughout life: a white blood cell distinguishes between infectious (A) and non-infectious (Ā), or A/Ā; a bacterium distinguishes between a viable pH level and an unviable pH level (A/Ā); for a virus, there is host and non-host (A/Ā). Myriad instances illustrate distinguishing. We shall see why this is an active process inherent to aliveness, to living, to being alive, to being a living being.

A major subtopic is how distinguishing or establishing A/Ā could have occurred in the first place. How could distinguishing first come about? This is the topic of the advent of life. It will be argued that distinguishing emerged with life. Thus, inquiry into life in terms of distinguishing offers a new approach to inquiry into life's emergence.

While certain nonliving processes present features that are like distinguishing, I shall later show that the distinguishing of living beings is inherently anticipative, that it involves homogeneity, or A=A, by establishing similarity and difference. Inherent to A=A, then, is anticipation and a degree of arbitrariness utterly uncharacteristic of the nonliving. As we shall see, for this reason alone, the method of the hard sciences is utterly incompatible with the object of study of the life sciences. For the root of a plant, given a multiplicity of nutrients and a multiplicity of non-nutrients, a nutrient is a nutrient, and a non-nutrient is a non-nutrient. In both cases a multiplicity is arbitrarily boiled down to nutrients (A=A) and non-nutrients (A=A). Thus, A=A refers to any instance in which a multiplicity is treated as homogenous, 'the same.' For a white blood cell, infectious is infectious (A=A) and non-infectious is non-infectious (A=A). For bacteria, optimal and sub-optimal concentrations of glucose or pH levels are treated as the same (A=A). A bee seeks nectar among a bunch of flowers; a squirrel selects among many nuts to eat; a child chooses between pieces of candy; a concertgoer selects a seat. In all cases, multiple optimal instances are regarded as the same (A=A) just as the sub-optimal are. Throughout, an arbitrary homogeneity allows multiplicities to exist, multiple instances of 'the same.'

---

[1] So, this is not an inquiry into the history or sociology of inquiry into life or the philosophy of biology. For this, see for an example José Luis Gonzáles Recio, *Teorías de la Vida* (Madrid: Editorial Síntesis, 2004). A presentation of his about the theories from the Greeks to our day is available at http://www.gonzlezrecio.com/bionomos/textos/Publicaciones/JLGRecio/Libros/lib1. Several authors' treatment of various topics in theoretical biology and philosophy of biology converges significantly with sections of this essay. A few will be discussed, but many more must be omitted. Notable instances include Hans Jonas (*The Phenomenon of Life*), Hans Blumberg (*Teoría del Mundo de la Vida*), Helmuth Plessner (*Levels of Organic Life and the Human*), and Robert Rosen (*Anticipatory Systems* and *Life Itself*).

The reader may recognize 'A=A' as *identity*, one of the fundamental laws of logic.[2] This points to something not immediately obvious to be discussed in Chap. 4: distinction is the foundation of logic. If living beings are distinguishing distinctions, it follows that this has deep implications for logic and its foundations, shall we say the relation between the nature of logic and the logic of nature. Regardless, the terms 'distinction' and 'distinguishing' are here used in their usual sense.

Our own distinguishing (A/Ā), which calls for grasping sameness (A=A), is always at work in our thinking. Above it is written "the same is true of a white blood cell" and "the same is true of a bee." We understand at once, both that white blood cells are distinguished from bees (A/Ā), and that they are the same (A=A) insofar as both distinguish. Thus, white blood cells, bees, and humans formally do the same when they distinguish: A/Ā, A=A. Therefore, distinguishing is the proper conjugate form to understand living being. For the meaning of *conjugate form* and hence *proper conjugate form*, here is Bernard Lonergan's use of the terms in his four-step account of scientific method:

> At the root of classical method [of inquiry] there are two heuristic principles. The first is that similars are understood similarly, that a difference of understanding presupposes a significant difference of data. The second is that the similarities relevant to explanation lie not in the relations of things to our senses but in their relations to one another. Next, when the heuristic principles are applied, there result classifications by sensible similarity, then correlations, and finally the verification of correlation and of systems of correlations … . What is verified accurately is not this or that particular proposition but the general and abstract proposition on which ranges of ranges of particular propositions converge. Accordingly, there is a fundamental heuristic structure that leads to the determination of conjugates, that is, of terms defied implicitly by their empirically verified and explanatory relations. Such terms as related are known by understanding, and so they are form. Let us name them conjugate forms.[3]

Notice that, throughout, Lonergan's account is in terms of similarities (A=A) that imply differences (A/Ā). Among the terms he uses to indicate similarity are *classification, correlation,* and *systems of correlations*. The term *verification* is an instance of A=A because verification is *similarity displaced in time*. Verification, or ratification, is accomplished through repetition, whether carried out by bacteria or by a vast team of scientists running a multibillion-dollar particle accelerator. If, for science, the verifiable is the repeatable and to verify is to repeat, for living being in general, to repeat is to verify.

If the claim of this investigation is correct, then scientific inquiry is living activity *par excellence*; it is to verify highly rigorously by repetition.

For animals, repetition is affirmation and thereby verification. Try lifting a spider onto a piece of paper. After the second or third attempt it resolutely avoids being touched. This is no far-fetched instance of anthropomorphism. Vindication has recently come from a team of scientists who trained Kim, a spider, to jump. Hoping

---

[2] According to different points of view, the laws of logic are: (1) the law of contradiction, (2) the law of the excluded middle, (3) the principle of identity, and (4) the principle of sufficient reason.

[3] Lonergan (1992, 460–461).

to study her prowess at jumping, food was not a very practical incentive because spiders eat only once a week. So, Kim was repeatedly moved manually from one platform to another. After a few days she started to jump on of her own accord to be transported back to her nest after her day's work.[4]

If all living processes involve distinguishing, this primary characteristic must be at the foundation of the heuristics for inquiry into living being. Are living beings primarily organized around distinguishing?[5] If they are, this must be the foundational, general, and abstract proposition to be verified. If verified, it follows that distinguishing is indeed the proper conjugate form for inquiry into living being. Classifications must be based on it as well as correlations and the verification of correlations and of systems of correlations.

In relation to conjugate forms, Lonergan pointed to successive genera of kinds of things. To each correspond distinct and autonomous empirical sciences: physics, chemistry, and physiology. Regularities, investigated empirically and correlations verified, result in implicitly defined sets of conjugates. Thus, physical correlations are verified that result in physical conjugates, chemical correlations are verified that result in chemical conjugates, and physiological correlations are verified that result in physiological conjugates.[6]

However, as Lonergan pointed out, there is no logical progression from one set of conjugates to the next, else empirical sciences would not be distinct and autonomous. Thus, physical correlations may be verified in a living being (e.g., a body that falls and rises according to the laws of physics). Chemical correlations may be verified (e.g., the body that falls and rises again is warm for reasons accounted for in chemical terms). Physiological correlations may be verified (e.g., in this warm body the heart is pumping blood at a very high rate). Psychological correlations may be verified (e.g., the heart of a human subject is pumping wildly because, having just won the lottery, she is leaping up and down with excitement). None of the correlations at one level can account for further intelligible relations at another. "Successive higher viewpoints," as Lonergan referred to them, take in what is irrelevant for lower ones.[7] The conjugate forms that make up a higher viewpoint account for data that occur systematically yet is merely coincidental for a lower viewpoint. From the point of view of the physics that explains the rising and falling movements of a body, a small variation in temperature is merely coincidental. From the point of view of chemistry, the rapid beat of a heart pumping blood is merely coincidental. From the point of view of physiology, the cause of excitement that has brought

[4] http://www.dailymail.co.uk/sciencetech/article-5688065/Scientists-train-spider-jump-demand--unlock-arachnids-hunting-secrets.html (Retrieved May 29, 2018).

[5] The double meaning of this question is significant. It may be taken as asking about the organisms themselves, or about the field of inquiry into them. If the answer to the first sense is in the affirmative, it must have profound implications for the second.

[6] Lonergan, *Insight*, 464. Chapter 7 treats the notion of thing and kinds of things.

[7] Lonergan, 465.

about a rapid heartbeat is merely coincidental. This is why, though all viewpoints from lowest to highest must be integrated, there is no logical progression from one set of conjugates to the next.

If everything that is verified systematically must be included, then the proper conjugate form must be the highest viewpoint possible—else there is data left out. This would be the viewpoint that allows for the verification of correlations of correlations of all intelligible relations that occur systematically leaving out only what occurs coincidentally. Clearly, for a science of living being, the shared aliveness of living beings cannot be coincidental. But it will be so if living being is only inquired into in physical, chemical, or physiological terms. Therefore, there must be a conjugate form that is proper to aliveness.

This study argues that this form is distinguishing. Distinguishing is unintelligible in physical, chemical, or physiological terms that are limited to the physical or chemical. As shall be shown in detail, the arbitrariness and anticipativeness of the distinguishing, or the distinctional, is the foundation of the *qua cognitional*, that is, of knowing as such, starting from the most rudimentary level possible. Thus, the distinctional is to that which distinguishes as the cognitional is to that which cogitates. Both the manner of distinguishing, which varies immensely, and what is distinguished remain coincidental at this fundamental level. But it shall be argued at length, that distinction-distinguishing *per se* is what living being both *is* and *does*, what being alive is all about. And therefore, that it must be the highest viewpoint to understand life.

In the chapters that follow, this will be demonstrated by a thorough inquiry into distinction. Before entering into the argument, what I wish to show here is, first, that what is being argued for is intuitive and, second, to provide an initial overview of what it implies.

Consider an example from a newspaper heading: "Body of knowledge: on average, women's hair is about half the diameter of men's hair." This single brute piece of data pertains to human beings. Women's hair is on average about half the diameter of men's hair. Why? One must work up through various conjugate forms to arrive at an answer.[8]

Since one must start somewhere, let's start with the molecular design that results in hairs of different strengths, the physical and chemical conjugate forms. Having verified the physics and chemistry of hair, one may move up to the next viewpoint. At the level of physiology, questions may be asked about the process of hair growth: stages of growth, their sequence and duration, lengths of hair, why and how hair decays, and so on. In the next higher viewpoint, psychology, one may frame questions about hair preferences (e.g., what kind of hair do humans like?). This, it turns out, involves the tastes of males and females. Thus, one gradually approaches one of the reasons for hair and kinds of hair in the first place. Eventually two questions may be framed: (a) Of the psychological traits of males and females that have

[8] "Health," *Union Tribune*, Thursday, April 21, 2015.

resulted from natural selection and sexual selection, which of the physiological mechanisms that underlie mating patterns, promiscuity, perception of beauty, and so forth, affect hair diameter? and (b) Which of these tend to lead to women's hair being, on average, about half the diameter of men's hair?

Implicit throughout is the further underlying question: What kind of hair do *I* like—a real subject, whether male of female? Now clearly, while my own taste in hair is, on the one hand, merely coincidental to the account, on the other hand, it is significant, at least statistically, because such preferences are crucial. Indispensable as each higher viewpoint is, none fully makes sense until the highest viewpoint is reached, in this case human beings—real particular actual living subjects with real particular actual hair preferences. All preferences involve distinctions, and these are very concrete—how does *this* hair of a human woman look and smell and feel compared to *that* one (A/Ā)? Distinguishing is the conjugate form by which all systematic correlations of correlations of living beings are actually verified.

Clearly, hair preferences pertain to a very high level of the cognitional. But, in fact, as shall be demonstrated, distinguishing takes place throughout what we name "physiology." Notice how it is implicit in the following instance of recent research that points to a link between red meat and cancer. The sugar molecule Neu5Gc (a sialic acid) is found in most mammals but not in humans. Because it is foreign to human biology, ingesting this molecule results in an immune response that likely increases lifetime cancer risk. Researchers state the role played by sialic acids such as Neu5Gc in vertebrates as follows: "They bristle from cell surfaces, and *the exact structure helps the immune system determine self from foreign*."[9] In other words, sialic acids allow the immune system *to distinguish*. *How* this is done is at the molecular level: "Structures with human sialic acids are ignored by a person's immune system. Some pathogens take advantage of this fact by coating themselves with human-like sialic acids. Other pathogens specifically target chains of sugar molecules called glycans that contain Neu5Gc."[10] But this function of the immune system can only be understood at the highest conjugate level. At this level it is further understood that the loss of Neu5Gc is linked to human evolution because it may be central to the divergence between species. Such diversion is itself a distinction—at the species level. Most other animals, including chimpanzees, the closest human relative, make the Neu5Gc molecule. But sometime after diverging from chimps, humans lost the ability to do so. According to Varki and colleague Pascal Gagneaux this loss may bear on later steps in human evolution by reducing fertility between individuals who kept the molecule and those who did not. "Moreover, the loss itself may have been a response to infection with malaria or some other disease that may have been targeting the nonhuman sialic acid, according to Varki. Humans are immune to the malaria parasite that infects chimps."[11]

Again, this research considers *how* distinctions are made at the molecular level, *how* the immune system distinguishes between self and foreign, and even *how*

[9] Fikes, *Red Meat*, A1 on research by Dr. Ajit Varki and colleagues at the University of San Diego.

[10] Fikes, A1.

[11] Fikes, A1

species differentiate. This is the physiological *how* limited to the physical and chemical. However, throughout, the *what* is distinguishing. Distinguishing is the highest viewpoint. "Fertility" is an abstract concept; choosing one's mate involves all manner of distinctions, including the allure of hair. Not only is the *what* only understood at this higher level; it *must* be so understood because only at this level are the results systematic and not merely coincidental. Moreover, because only the highest level makes it intelligible, it takes the distinguishing of researchers to make it intelligible—as well as the distinguishing of the author of the article, and our own reading of it.

A key point that follows from the foregoing is that research into life is research into distinguishing. The lower-level physical and chemical conjugate forms are indispensable, just as much as the higher-level conjugate form that makes them intelligible. It takes living beings to make sense of living being. Herein lies the potential complication of the fallacy of anthropomorphism that merits scrupulous attention. Living beings that we are, we cannot help but bring our own aliveness to any inquiry. Chapter 3 shows just how crucial our framing of the data is to inquiry into the advent of life.

The previous examples show that the full intelligibility of even the most rudimentary living processes requires the appropriate conjugate form. What discipline is best suited to its study? The obvious answer is biology. However, mainstream biology remains a reductionist science. It reduces its problems to the lower-level conjugate forms—for instance, the highly complex chemistry of genetics. Why does it do this? In short, because the highest viewpoint—such as the kind of hair you or I like—involves *interpretation*. This is why, recently, the growing field of *biosemiotics* explicitly centers on "the production and interpretation of signs and codes in the biological realm."[12]

In biosemiotic inquiry into living being, interpretation is front and center. Questions about aliveness are no longer framed in physical and chemical terms but in terms if signs, semiosis. The research agenda of biosemiotics is to build an understanding of organic life in terms of signs, which necessarily involve interpretation—of the organism in question and of the researcher. Why and how chemical reactions conform signs is, from the point of view of chemistry, merely coincidental. However, for biosemiotics, the sign constitutes the higher-level conjugate form because inquiry into living being can be systematized according to heuristic principles that result in classifications, then correlations, and finally the verification of correlations and systems of correlations in terms of signs. As a new field of study, biosemiotics seeks a new hermeneutics to understand aliveness. A very young discipline, it is still in the process of establishing its foundations. Insofar as distinction-distinguishing is the key condition for the possibility of signs, it is the foundation for the science of living beings, a new biology, with biosemiotics as one of its subdisciplines.

Again, living being inquires into living being. So long as aliveness is not understood in its own terms, with the proper conjugate form, it is not possible to realize

---

[12] "Biosemiotics."

what gratuitous attributions we may be making to the nonliving that properly belong to us, *qua* living being. And vice versa, we may also not realize what gratuitous attributions that pertain to the nonliving we may be making to ourselves. Without the proper conjugate form, we will fail, not only to properly distinguish between living being and the nonliving, but further, we shall fail to grasp why there is a need in the first place to define a proper conjugate form for inquiry into living being. This conjugate form is required for inquiry into life, both on the side of the object and on the side of the subject who inquires. In Chap. 7 it will be shown that deep underlying questions turn on the issue of things and kinds of things. What is a thing? What kinds of things are there? What is the effect of treating one kind of thing as another kind of thing? This, no less, is how biology must be re-founded to be radically differentiated from physics.

Starting from the proposition that distinguishing is the nature of living being, this examination frames its inquiry into aliveness as an inquiry into distinction and distinguishing. Conceivably, it could be shown that distinction and distinguishing, though important, are *not* fundamental to aliveness. Still, a key point of the overall argument will succeed if it is shown that inquiry into the nature of living being calls for its own proper conjugate form.

The proposition that distinguishing is the proper conjugate form to inquire into aliveness soon gives rise to complexity. Interestingly, the unusual character of this complexity itself weighs in favor of the argument that distinguishing is the proper conjugate form. The problem, to put it bluntly, is that everything and anything that pertains to living beings falls under the purview of distinguishing.

Is this so? Is it truly the case that everything and anything that pertains to living beings falls under the purview of distinguishing? To be alive at all, living beings must self-regulate, and self-regulating is only possible in *functional closure*—that is, remaining an autonomous or relatively closed system by opening and closing of membranes in the process of relating to the rest of the world. Just how living beings manage this functional closure, what it requires and what it implies, will be discussed at length. The basic point, however, is that this is how organisms remain themselves. That is, *distinct*. Remaining distinct is the foundational condition for the possibility of life. This is both the condition and the purpose of distinguishing. It is to remain alive, and therefore distinct, that living beings distinguish between nutrient and non-nutrient, toxic and non-toxic, threat and not-threat, and so forth. It is in this strict and concrete sense that living beings are distinctions that distinguish, or distinguishing distinctions. In Chap. 2 it is shown that this is the primordial form of distinguishing from which all others follow. Thus, for this reason alone, everything and anything pertaining to living being involves distinguishing.

That organisms self-maintain as distinct, and that they do so by distinguishing, is the root of the isomorphism between all living beings. This fundamental isomorphism underlies all inquiry and becomes at once increasingly important and increasingly problematic as organisms become more complex. Growing complexity is expressed in additional, increasingly nested, and therefore increasingly complex distinguishing activity. All manner of relevant topics arise that would merit discussion must be left aside. Among these, the centrality of distinguishing makes increasingly evident the need to acknowledge the isomorphism between the object of biological inquiry and the biologist who inquires—the aliveness shared by living beings and the subject inquiring into them. This is the general context in which the topic of interpretation must be discussed.

There is a strong precedent for the notion of distinguishing distinctions and the cluster of issues that surround it. Aristotle, for one, in *On the Soul*, sets up his inquiry into living being very similarly. To begin with, he too points to what I have referred to as conjugate forms. He notes that a living being is only very partially intelligible as a body because this involves only lower conjugate forms. Its full intelligibility is only possible when understood as *entelecheia*—as being-at-work-staying-itself or "holding together actively as a whole"—a beautiful term coined by Aristotle.

> *Being-at-work-staying-itself* (ἐντελέχεια *entelecheia*) fuses the idea of completeness with that of continuity or persistence. Aristotle invents the word by combining ἐντελής *enteles*, (complete, full-grown) with ἔχειν *echhein*, (= ἕξις *hexis*, to be a certain way by the continuing effort of holding on in that condition), while at the same time punning on ἐνδελέχεια *endelecheia* (persistence) by inserting τέλος *telos* (completion). This is a three-ring circus of a word, at the heart of everything in Aristotle's thinking, including the definition of motion. Its power to carry meaning depends on the working together of all the things Aristotle has packed into it. Some commentators explain it as meaning being-at-an-end, which misses the point entirely, and it is usually translated as 'actuality,' a word that refers to anything, however trivial, incidental, transient, or static, that happens to be the case, so that everything is lost in translation, just at the spot where understanding could begin. A number of Aristotle's most definitive explanations fall with the meaning conveyed by *entelecheia*.[13]

This is a very helpful term that I shall make use of. In *entelecheia* a living being makes distinctions to remain itself, a distinguishing distinction. Crucially, Aristotle pointed out that *entelecheia* takes place at various levels. This is true both synchronically, in the constitution of organisms now at all levels and diachronically, as they unfold. Throughout this exploration, I will use the term *unfolding* to refer to the full set of the diachronic as it pertains to living being, including evolution, development, history, and biography.

Unfolding, as it occurs in living beings, contrasts sharply with the transformations of other kinds of things. This, again, is why the theme of kinds of things is central to this investigation. Chap. 7 treats the fallacies that result from ignoring the fundamental differences between kinds of things. Again, recognizing these differences is crucial if biology is to cease attempting to emulate physics and become a discipline that rigorously investigates aliveness. Beyond the arbitrary and anticipative nature of distinguishing, being alive necessarily involves relationships with other organisms, all semi-autonomous wholes made up of smaller wholes, such as semiautonomous cells. This calls for reflecting on organization, which is to say on relationships whose very nature, as we shall see, is paradoxical. These paradoxes, at the heart of aliveness, are beyond the bounds of classical logic and thereby the hard sciences.

---

[13] *On the Soul*, trans. Sachs, 2004, 189.

Inquiry

To inquire into aliveness is to inquire into the process of becoming. Becoming is to cease to be what you are and to begin to be what you are not. This paradoxical process always already involves distinguishing and therefore anticipation. Anticipation is the foundation of inquiry. Thus, aliveness, as such, is inquiry, and prefigures what we habitually refer to by this term. With the emergence of life, species of organisms test the possibilities of the environment in what might be thought of as an incipient form of distinguishing or 'inquiry.' In time, individual organisms begin to do so. Eventually, this unfolds into the full-fledged inquiry that in humans takes the form of questions.

Many shades of differentiation may be found between kinds of distinctions drawn by organisms, but the fundamental isomorphism between them, from the most compact to the most unfolded, allows them all to be compared. All these forms may be thought of as a *lineage* of distinctions, which again points to isomorphism as the general heuristic structure for inquiry into distinction by drawing parallels between them, no matter how compact or differentiated they may be.

That biology must be inquiry into inquiry brings us to our own lineage. For one thing, such inquiry into inquiry/ understanding/ knowledge is a central concern of the philosophical tradition. These are a few titles that state it explicitly: *An Essay Concerning Human Understanding* (Locke); *New Essays Concerning Human Understanding* (Leibniz); *Treatise Concerning the Principles of Human Knowledge* (Berkeley); *An Enquiry Concerning Human Understanding* (Hume); *Critique of Pure Reason*, *Critique of Practical Reason,* and *Critique of Judgment* (Kant); *Foundations of the Science of Knowledge* (Fichte), and so forth. Thus, there is a deep continuity between distinction and human inquiry, and specifically between the present inquiry into the distinctional and the general field of inquiry into human inquiry, if only because all human inquiry begins with distinctions. To take a single example, Kant's inquiry in *Critique of Pure Reason* begins by distinguishing a distinct set of cognitions. He points to "the boundaries of possible experience"—a distinction—and goes on to make his key distinction between things as objects of experience and as things in themselves. Again, the Introduction begins by presenting the crucial distinction between analytic and synthetic judgments that animates Kant's project as a whole.[14]

Drawing distinctions is the norm in human knowing. Recall Genesis, which begins with a long series of separations and namings, two fundamental features of human distinguishing, as we shall later see. This is only one of many creation myths that sets up its narrative in terms of initial distinctions. Like all living beings, this is how humans construe the world, how we construct ourselves and how we materially participate in the construction of the world. As part of our inquiry, then, we can, and must, examine our own distinguishing.

---

[14] Kant, 112.

This brings me to the primary sources that have been used for this exposition. The first, *Laws of Form*, a mathematical text that could just as well have been entitled *Laws of Distinction*, inquires into form as such, as given to any observer that draws distinctions. It shows that a possible interpretation of it is as the foundation of logic.[15] The second source, *Insight, A Study of Human Understanding*, explicitly situates itself among the titles listed previously but with a difference; uniquely, it invites us to carry out a rigorous exercise in 'self-appropriation;' that is, to attentively advert to our own cognitional operations. It serves as a 'flying manual' along various legs of this long journey.

The third source, to which we now turn, was my own experience over a decade living and working with monkeys. The deep and overarching conclusion of this experience was the certain knowledge that the deep bonds of friendship among the monkeys themselves, and between them and us, where rooted in a profound and subtle understanding of each other. Far from dismissing this as the illusory effect of anthropomorphism, it was this deeply moving experience, sedimented day by day over many years, that gave birth and has sustained this project—the conviction that our understanding of the living beings around us, far from being illusory, rests on our single shared nature.

Monkeys

My own encounter with the set of problems taken up in this study began during a decade living and working at The Monkey Sanctuary in Cornwall, United Kingdom, from 1979 to 1989. This center for conservation, founded by Leonard Williams in 1964, was for 35 years home to a natural colony of woolly monkeys and the first place where this beautiful species from the Amazon survived and bred outside the wild. Living in conditions of semi-captivity in a large territory consisting of interconnected large outdoor enclosures and indoor areas, the group hovered around 20 individuals over many years, eventually including some belonging to the seventh and eighth generations.[16]

Having visited a few weeks at a time as a student volunteer starting in 1973, in 1979 I became a full-time keeper and eventually, years later, curator. With each species living in its own quarters, we could observe each other virtually at all times. Daily, we visited the monkeys in their territory during cleaning routines, and very often they visited us in our part of the house. During the summer months, visitors met the monkeys at close quarters in the sanctuary grounds, including mothers with their babies and, on occasion, even babies born the night before. We asked ourselves

[15] George Spencer-Brown, first published by George Allen and Unwin, London, 1969; Bernard Lonergan, first published by Longmans, Green & Co., London, 1957.

[16] The Sanctuary runs to this day, though its purpose has changed to rescuing individuals of various species of monkeys. See http://www.monkeysanctuary.org. For a history of the Sanctuary see http://www.monkeysanctuary.org/files/2010/08/History.pdf. This PDF file features a picture of a keeper in lively conversation with the monkeys during a cleaning routine, this author over 30 years ago!

endless questions about the monkeys—practical day to day concerns about their territory, cleaning routines, and their diet; issues about their development, their individual characters and relationships, and the health of the group as a whole; and theoretical questions about the nature of their social and emotive life, their intelligence, communication, abilities, learning, creativity, and so forth. Our conversations ranged over primatology, ethology, anthropology, and philosophy. Williams' method of animal husbandry long predated today's norms and allowed him to reach original insights into the primate mind. Father and teacher of the celebrated guitarist John Williams, he maintained an ongoing correspondence with renowned personalities such as art historian Kenneth Clark, author John Fowles, naturalist Gavin Maxwell, geneticist C.H. Waddington, and ethologist Konrad Lorenz. Most importantly, we simply delighted in our friendships with these beautiful and wonderful animals.

Many conversations with visitors about life with the monkeys was thwarted by the straitjacket of the anthropomorphic fallacy, the charge that behavioral traits attributed to animals—whether these are gestures, display of emotions, demonstrations of friendship and trust—merely amount to human projections. Significantly, the issue hardly ever came up during the very busy summer months with the general public. They had no impediments listening to our detailed accounts of the lives of the monkeys. They understood perfectly well, even if, to our chagrin, they often did compare our superb Amazon wooly monkeys to their pet dog, cat, or parrot. Vexing as this was, however, they shared our principal assumption: people and animals—certainly animals such as apes, dolphins, dogs, or cats—can be friends, and such friendships involve deep and subtle knowledge of each other. It was the highly educated visitors of the off-peak season that contested the moving experience of watching a natural colony of monkeys at close quarters. Even as they were able to touch them, rather than open their hearts, they would often adopt a stance of dry 'objectivity' and ply us with 'scientific' questions routinely framed in terms of numbers—length of gestation, weight at birth, age at puberty, life expectancy, frequency of estrus, frequency of breeding, ratio of males to females, and so on—as if the essential character of monkeys lurked in such quantitative data. Often the next question was impatiently asked before the previous one could be answered, conveying that they found these dry bits of data as devoid of meaning as we did. The deeper the commitment to scientific 'doctrine' the more rigid they wished the boundary to be when relating to animals as objects of study. As Patrick Byrne astutely observed:

> Whether humans can be "friends" with monkeys or other animals depends a great deal upon exactly what one means by "friend." Aristotle seems to me to be unsurpassed. Affection is essential to friendship, but so is mutual good will, and mutual knowledge of good will. The challenge would be which non-human animals truly do have good will toward humans, and which ones are truly capable of knowing of human affection and good will toward them. But one thing is clear—by blocking out affection, the "objective" visitors are missing the objective goodness and beauty of the monkeys, which can only be known by using what feelings can reveal.[17]

[17] Patrick Byrne, personal communication.

I will never forget a conversation I had years afterwards with a graduate primatologist at Harvard who had spent time in the field researching the diet of gorillas. She did this by carefully and regularly collecting samples of feces. One day, as she diligently went about her task, she came face to face with a gorilla. Excited at the very idea of this, I implored her to tell me more. But there was little she could add, only that she patiently waited for the gorilla to depart. Wondering alone through the dense foliage of the tropical forest, she was more excited to find her cherished samples of shit than meeting face to face with a wild gorilla! To this day, it saddens me to think that anyone would have missed this opportunity. But I am most shocked to recall the reason; she missed it for the sake of the science of primatology!

Monkeys as Philosophical Object

"He who understands baboons would do more towards metaphysics than Locke,"[18] wrote Darwin. Monkeys, along with apes, straddle a highly controversial, even strategic place in our cultural imagination. They are the human-like animals or the animal-like humans. Cities in India inhabited by macaques attest that our common kinship was assumed many centuries before Darwin. It can be no surprise that "apes and monkeys are among the most prominent inhabitants of the misty borderlands between beast and human in the Western cultural imagination," said Raymond Corbey, leaving "their mark on the fundamental concepts and discussion of human nature and human origins … . [Thus], apes have been, and still are, 'good to think [with],' a phrase coined by Claude Lévi-Strauss."[19] In a lucid and carefully documented argument, Corbey showed that, even as far back as Plato:

> …the history of the anthropological disciplines to a considerable degree has been an alternation of humanizing and bestializing moves with respect to both apes and humans, a persistent quest for unambiguousness and human purity, and an ongoing rebuff of whatever has threatened to contaminate that purity.[20]

He traces the controversies around the extremely sensitive ape-human boundary in the vocabularies of different disciplines at various times, offering this thesis "for its heuristic value, for further corroboration and refinement or, alternatively, for partial or complete refutation, with sound argument."[21] The grand Christian narrative, inherited from Aristotle of immutable essences and nature as God's harmonious, hierarchical creation, is still subtly at work in the nineteenth century idea that persists even today—that as high as humans have ascended, the brutish nature of our apish ancestors is still, quite literally, within us. Twentieth-century paleoanthropology, preoccupied with discovering the "natural" facts of human evolution, interprets

[18] Darwin, "Notebook M," 539.

[19] Corbey, *Metaphysics of Apes*, 5–6.

[20] Corbey, 1.

[21] Corbey, 1.

these in terms of human unicity and the concomitant dichotomizing of the archeological record into the still beastly and the already "fully human."[22]

Since the mid-seventeenth century, the animal-human boundary has been policed by revising and stating, ever more subtly, the case for our uniqueness—only humans have thumbs, make tools, use language, grasp symbols, create culture, and so forth. This, which is a battle over the distinction between the human and the nonhuman, takes place across various disciplines—natural history, biology, paleoanthropology, ethnology, primatology, and occasionally psychology and philosophy—including what today are considered the 'four fields' of American anthropology: cultural anthropology (ethnology); biological and physical anthropology (including paleoanthropology and primatology); archeology; and linguistics.[23] Most significantly, perhaps, what lies at the heart of the controversy is the nature of boundaries themselves, which is grounded on the nature of distinction itself. An epigraph by Mary Douglas heads Corbey's chapter 1, *Ambiguous Apes*: "It is part of our human condition to long for hard lines and clear boundaries."[24]

In short, aside from potential friendships with them, monkeys and apes are interesting in themselves and highly significant as we reflect on ourselves and our place in nature. At the heart of this highly charged controversy is the notion of distinction itself. Hard and clear distinctions keep ideas on this topic abstract and removed from the concrete world, yet "nowhere in the study of the animal world … is the emotional involvement of the human species so great as with its own ancestry and next of kin."[25] So, monkeys are splendid philosophical objects.

A philosophical object is anything we care to think about.[26] The objects of academic philosophers are typically tables, chairs, and pieces of chalk—the objects at hand where philosophical inquiry officially takes place. More sophisticated objects are triangles, pi, mathematical theorems, mathematics as such, language, and the findings of quantum mechanics, sometimes even music and art. Few of us reflect much about some of these objects, though we all have first-hand experience of language, music, and, though unaware of it, the quantum physics that allow our phones to operate. We mostly gain deeper access to them by reading and through the internet these days. But suppose our object of philosophical reflection are monkeys. A lucky few may be familiar with them, but what of those who are not? How to overcome this hurdle? What is it to be 'familiar' with monkeys? What is it like to inhabit the monkey-human boundary? I did not personally seek them out as philosophical object, but 10 years living and working with them brought a growing number of questions. It turns out in retrospective that they are indeed splendid objects for philosophical reflection. Darwin was right, perhaps more than he might have suspected, as this inquiry shows. Its overall intent might be conveyed by a series of transformation of his words:

---

[22] Corbey, 2.

[23] Corbey, 4.

[24] Mary Douglas, *Purity and Danger*, cited in Corbey, *Metaphysics of Apes*, 5.

[25] Corbey, *Metaphysics of Apes*, 6.

[26] "By an *object*, I mean anything that we can think, *i.e.* anything we can talk about." Charles Sanders Peirce, "Reflections," *MS 966.*

He who understands baboons would do more toward metaphysics than Locke.

↓

He who understands how baboons understand would do more
toward metaphysics than Locke.

↓

He who understands how he understands baboons would do more
toward metaphysics than Locke.

↓

She who understands how she understands how baboons understand
would do more toward metaphysics than Locke.[27]

How does one get to understand baboons? I can only speak from my own experience, and I should like to assist the reader in obtaining at least a glimmer of what that was like.

Laura, First Night Sleeping with Her in Room 1[28]

[Laura was suddenly orphaned during a flu epidemic. Her mother Emma, full of vitality and the most dominant female in the group at the time, had died suddenly. Laura was still nursing for comfort then. Emma was Jessy's second daughter. For two decades, Jessy was the matriarch of the group and head of the largest clan in the colony, a family who we referred to as the aristocrats because, like her, they tended to be extraordinarily intelligent and refined. Jessy gave birth to Poly, Emma, Bella, Kim, Django, Jimmy, and Anna. Emma, highly intelligent and wily, was herself the mother of Jojo, Lucy, Millie, Charlie and Laura.

At the time of the events recounted below, both Jessy and Poly, Emma's beloved mother and older sister, had long passed away. Kate, Poly's only daughter and Emma's niece was there, a wonderful mischievous female. With her mother and sister gone, Emma became the most dominant female in the group because of her very large family. Years later Charlie, her son, would become the leader of the group. Laura would become a mother herself. As the Emma family's official babysitter, I was at her side, only inches away.]

---

[27]The most outstanding contributions to field primatology, and to primate language studies, have been by women in research projects lasting over whole lifetimes: Jane Goodall, Dian Fossey, Biruté Galdikas, Jeanne Altmann, Shirley Strum, Linda Fedigan, Barbara Smuts, Sue Savage-Rumbaugh, Marina Cords, Alison Jolly, Francine "Penny" Patterson, (Roger and) Deborah Fouts, to name a few.

[28]With a few corrections and editions, the following sections are reproduced from my notes during my time at the Monkey Sanctuary.

The following events took place in Room 1. What had once been the spacious windowed living room of Murrayton, a large Edwardian house, had been adapted as the monkey's main indoor area. Decked with specially designed furnishing, large wood beams, ropes and daily fresh wood shavings, the roof of the large ante-cage provided an ample cozy area were most of the colony chose to spend the night.

It is 8:30 in the evening and the monkeys are starting to settle for the night. The territory is divided. The front monkeys are in Room 1, and the rest of the colony, including the males, is in the Back Enclosure and in the Gym. I had better make my way into Room 1 now so as not to keep them up. I am going to be sleeping on the top area with them and must get organized without upsetting them. It will take them a while to get used to my sleeping gear. The dark gray blanket is likely to be far more intimidating to them than the light blue Styrofoam camping pad.

Quietly as possible, I let myself into the ante-cage, silently closing the door behind me, and *eeolking*. This is their reassuring sound. They know my voice and it tells them what mood to expect. A few upside-down faces peer curiously from the ample platform overhead where I shall be sleeping. But there is no mad scramble onto the beams. A young monkey unfamiliar with someone coming into the room this late could easily precipitate this. The tightly rolled blanket and sleeping pad are tucked innocently under my arm. I want them to see it while I stand in the ante-cage, before I enter the territory proper. A lumpy shape would be disastrous. They immediately spot such deadly creatures from the underworld, and, when they do, their hair stands on end, their eyes open wide, they purse their lips and start with their *yuk-yuk*. I greet them casually as if I just happened to stroll by. But they eye the intruder. It is late and we all know I am doing something unusual.

Bella glances at me with suspicion. She may or may not be as wary as some of the others, but she does manage to look far more suspicious than anyone else. She sits and eyes me by the door on her haunches with arms crossed in front of her. She does these little stooping movements with her head while her big moon-face shrivels into an uncontroversial scowl. Anticipating the worst, I look at her directly and cheerfully coo "Be-lla!" in the brightest sweetest possible voice. This evokes a half-hearted *eeolk* from her. There is some hope that her grimace will soften. I desperately need to nip in the bud the slightest dark impulse she might have to start *yuk-yuking*. That would be catastrophic. Once this little sound starts and catches on with the others it puts a terminal stigma on an object. Suspicion piles upon suspicion and this can swiftly reach such a fever pitch that you'd think someone is being sucked to death by a giant poisonous toad. And anyway, it is clear to them that you are out of your mind to be anywhere near that thing and something hideous is surely going to happen to you; you are being so utterly reckless. Their concern for our safety will often turn to *nyook-nyooking*, the aggressive cheer of the mob from a safe distance. Excitedly, they verbally assail an offending object on your behalf and are delighted when you vanquish it. They heartily express their admiration and approval with loud and resonant *eeeolks!* No doubt they find our inveterate foolishness appalling, but they cannot help admiring our courage. But if we bring an offensive object into their territory things can become fraught very quickly indeed.

So far so good. No *yuk-yuking*. I keep a close eye on Bella's lips. She has a humorous way of pursing them, just as we do, when she is on the verge of expressing apprehension. This preliminary to a *yuk-yuking* must be caught in the nick of time. It must be said: at no time does one appreciate Sarah more than on these occasions. Admittedly, she is not the brightest of monkeys, bless her soul, and, truth be told, I am not excessively fond of her. She is just not as pally as the monkeys from the Jessy clan are, and we have had our disagreements. But she has a lovely full bright and rich *eeolk!*, and, for the most part, she sees no reason why all the geniuses from the Jessy clan should be so scared. Often, while everyone shivers and shudders and *yuk-yuks* inconsolably, Sarah can be quite chipper. When one desperately longs for one single opposing *eeolk!*, it is Sarah who obliges. One merry *eeolk!* from her can fill a pause in a torrent of *yuk-yuking*, momentarily abating the flood and marking the turning of the tide. The end of torment of many a *yuk-yuking* session—as one stands in the ante-cage in a bathrobe at 4:00 o'clock in the morning, cold, sleepless, forlorn and at the end of one's tether having tried everything—has been announced by one of Sarah's bright *eeolks!* Yes Sarah, you tell them, they are all silly buggers, there is nothing there you twits, let it go you lot, come on, let's all go to sleep, that's right, thank you Sarah, ol' girl, thank you!

Sarah *eeolks* authoritatively and Bella pulls a face. She looks away as if saying, "I hate new things but, hard as I try, I cannot see what could be nasty about that horrible thing under his arm." They all start to settle, and I plan my next move—opening the inner door and getting in. I must get up on the top to see where Laura is. But it would be rash to take the bedding up there now. Although initial suspicion has been allayed somewhat, the alarm could still be raised. I must take things very slowly. Even leaving the hideous roll in the ante-cage could be fatal. For now, its harmlessness is guaranteed by being tucked under my arm, but on its own it could present quite a different proposition, much like a dog one is dubious about but willing to tolerate until, without warning, its owner lets off its leash. The monkeys make no bones about, even if Roman law pales by comparison—*everything is guilty*, period. And only fools think otherwise. I nip out of the room to leave my precious roll outside well out of sight.

Back in the ante-cage, I quietly unlock the enclosure door, let myself in, and swiftly climb to the top, just as I do every morning during the cleaning routine. This is their favorite place, the heart of the monkey territory. Sitting there, they can reach up and touch the ceiling, and we humans can crawl comfortably on all fours. Kate and Bella sit by the radiator and Sarah is already leaning against it at the other end, close to the shut hatch that opens the corridor to the Back Enclosure. On the mantelpiece—Monkey Room 1 was once the living room of an Edwardian Mansion—Charlie and Sophie still pick at little bits of food. It was a very late fruit-bowl, and Millie and Laura sit contented and close to each other on one of the beams. Where will they decide to settle, I wonder?

Little Laura looks at my hands obviously wondering if I have something else for her to taste and reject today. It has been only the first day of her course of antibiotics, and already I have an ominous notion of what lies ahead. Still, injecting the tiniest amounts of bitter white powder into all kinds of tasty morsels for today's dose, is done. She looks at me from the beam and then over to Sarah by the radiator. Clearly, she would like to approach as she always does, but she is wary of walking past Sarah, even though she is already curled up and looking the picture of peace. I realize that the best chance of Laura being warm tonight is by organizing for her to sleep by the radiator, where Sarah is now, which is where she has always slept until Emma's untimely death earlier this week. This is a favorite place and where the Jessy clan traditionally sleeps. Emma and her children, including her three youngest—Millie, Charlie, and Laura—have simply taken it for granted. It is heartbreaking to think that Emma's death represents nothing less than the collapse of the Jessy clan. This is the monkey family that made the Sanctuary.

There is no option but to remove Sarah, and then to sleep by Laura to ensure she is warm and not chased away during the night. I can lie along the radiator between Laura and the rest of the room, and hopefully Millie will join us and also enjoy the benefit of my protection. That will also be the most comfortable for me. I must catch up on my sleep as this terrifying epidemic is still making its way through the colony. We have already lost several monkeys. Although on the face of it my own comfort is not a priority, it is essential to establish a routine that will work well as we have decided that it is indispensable for me to sleep with Laura for the next 2 weeks. Her very survival may be at stake—and what little is left of the Jessy clan.

The Top, as we call it, is a lovely place to sleep; with a quick wipe, the polished surface is perfectly clean. It is warm and all the cozier because most of the monkeys in the room snuggle to sleep together up there. Of all the areas in the territory, this is the one that best lends itself for a cuddle. There are no wood shavings here and the smooth fiberglass surface warmly invites one to lie down and curl up for a comfortable long snuffle with them.[29] The idea of sleeping with them up there is thrilling, even at such a difficult time as this.

On hands and knees, I approach Sarah. Her two half-closed eyes look out at me from a ball of fur soaking up the warmth of the radiator against her back. "Sarah, you are going to have to move," I say, tapping her gently as though awakening her. Her face disappears as she curls up more tightly. Placing my face gently against hers I talk to her. She can detect a certain lack of intimacy in my tone of voice as I prod her a little with my nose and even push with my face. There is no better way to do it; a face this close is by far the most appeasing and therefore the most inhibiting of gestures. And, yet, albeit gently, I am physically removing her from the spot. How else to say, "I am sorry, but I am afraid you are going to have to move out of the way"? I am almost acting as someone who is being overbearing—not unfriendly, but rude. None of the monkeys like this. It is like politely requesting someone to do something they dislike doing. After a few moments of this, she gets up and walks away and by the edge of the top she *eeolks* into the room. Sophie, her daughter, and a few others, respond. Then she makes her way to the other radiator. It is not as

[29] "Snuffling is a highly emotional response that belongs to appeasement, greeting and friendship behaviour. It begins usually with the gesture of covering the mouth with the hand and leading finally into an intimate huddle or embrace. In a snuffling situation infants respond in their own way with appealing little sounds of *tuff-tuff* and much head-shaking." Williams, *The Woolly Monkey*, 23.

warm, but she is quite used to it since this is where her own family habitually sleeps. She makes a couple of attempts to return but I wave her off saying "no." She gives up, *eeolks!* and settles for the rest of the night. I am always impressed by how well they take things in stride.

Laura looks impressed. Fancy getting rid of Sarah as easily as that! Soon she comes over and sits by me, no doubt wondering what I am doing up here at this time of night. Since no answer is forthcoming, save for my greeting her and grooming her affectionately, she settles down. However, she does not lie with her back to the heater as Sarah did, but perpendicular to it, along the far wall. So, she will not lie along me but close to my head when I finally settle down, which is just fine. In the meantime, Millie has disappeared into the far hatch at the other end of the room, the one leading to the Vertical. Lately, she has taken to sleeping there when the hatch is closed, like tonight. Laura, for her part, having always slept by the radiator with Emma, is all too happy to be in her familiar place. The lack of closeness between the two orphan sisters on this very first night is surprising in retrospect, given that they grew to be inseparable. No doubt Millie slept with Emma when baby Charlie was born, and when Laura was born it was Charlie who still regularly slept with mum and baby Laura. However, the next few nights Millie did not sleep with us, even as over the weeks that followed, she gradually learnt to be a mother to Laura, looking after her as her own daughter. As Laura's constant and best loved companion, she was always available to groom her, play with her, sleep with her and protect her, and on occasion even carried her on her back as mothers do with offspring who are already independent. But tonight, Millie sleeps alone at the other end of the room.

As her closest sibling, Charlie has spent the most time near Laura. Having babysat him too, he is delighted to see me. He approaches, curious to find me up here at this hour. He and I have snuffled and played together up here ever since he can remember. The three of us sit together on the top. Emma first chose me as babysitter not long after I arrived at the Sanctuary when Lucy, her eldest daughter after Jojo, first started venturing away from her. This was the dream older sister that Millie adored, and together they doted on Charlie when he was born. Laura never met Lucy because she died suddenly of the flu before she was born. I will never forget the awe in Charlie's expression when he saw her dead body, his hair standing on end. One could not affirm that he realized she was dead, but he palpably sensed she was not alive as he had always known her. It is mulling over incidents such as this during many years that we realize just how much we learn about the monkeys merely because we live in such close quarters with them.

Of Jessy's many children, Emma was the most prolific, though not anywhere near as prolific as Jessy herself. Many years after tonight, Charlie will become the leader of the group by displacing Django, his uncle, and Laura's baby will be named Lucy, after her beautiful sister. I greet Charlie playfully, but he hardly responds. He has other things in mind, what with playing with Sophie and the late fruit-bowl.

Laura lies down. It is time to fetch my bedding. Better to get that over with, a very ticklish step before everyone is really settled. I must avoid anyone taking fright. With Laura lying down I will have to be very quick. It would be awful if Sarah came over and bullied her away. To avoid trouble, I must leave my shoes and all other spooky objects outside, well out of sight—not even in the ante-cage. Step by step, I mentally rehearse what I am about to do, a habit one acquires when the slightest misstep could raise the alarm and cause endless unnecessary trouble. Moving calmly away from Laura, I glide past Sarah giving her a knowing look to convey that I very definitely know what I am doing and what she is doing. Her glance satisfies me that she knows, too. I nimbly climb down from the top looking at her, and, when I can just barely see them both, I quietly say "Sarah!" in a clipped tone of voice as I wag my finger at her. Then I am gone.

As soon as I touch the floor, I quietly open the door. Just outside the enclosure door, and in a single sustained movement, I take off sweater, shirt, belt and shoes, empty pockets (everything except for a few raisins that could prove to be essential), decide to put on the sweater again—it may make a pillow—and then *eeolk!* as I quietly open the outer door to leave everything out of sight and pick up my bedding. There I dim the light (but only slightly because of the underworld monster I am about to carry in), shut the outer door again, and with another *eeolk* I let myself back in the enclosure. The latch must be properly closed; Kate knows how to open it if it isn't, and that is the last thing I need tonight. I am back on the top as if I had never left. Everything is the same. Not surprising; it has only been 30 seconds. It is the little look one or two of them give you on an occasion like this that makes you wonder whether you haven't gone completely mad. I giggle quietly to myself and imagine them patiently exclaiming, "What *are* you up to? It's been a long day and some of us are trying to get some sleep."

As long as I can, I keep the bedding under my arm and then discretely place it down and sit on it. I am hedging my bets that this will make it look the most innocuous and I am now in deep study of something I am picking at on the ceiling. The whole thing has been done so quickly and nonchalantly, and I hope no one has noticed anything. Then I hear a quiet, rather touching little *yuk-yuk* behind me. Bella has noticed. She sounds a little embarrassed. I am utterly concentrated on the ceiling. She turns quiet, obviously wondering what I could possibly be doing. I *eeolk*, lamely attempting to mimic Sarah, and she and Kate *eeolk* back. I could kiss them. Bella sits up and *yuk-yuks* a moment longer, but, observing my intent stare at the ceiling, she quiets down. Now Kate sits up and looks at what Bella is looking at. Like Bella, she is curious about what I seem to have found overhead. I must not turn to look at them so I can only catch glimpses of their mystified expressions. At the far end of the room a hatch opens slightly and Millie peers out: "Did I hear something?" Laura is sitting up now. Charlie and Sophie sit on the beams looking in my direction. The room is dead still. I am Michelangelo assessing progress on the Sistine Chapel. I rub my nose and talk to myself, oblivious of the seven pairs of eyes scrutinizing me. This is the moment of truth.

I can almost hear them thinking: "He could not possibly be sitting on that thing *and* foraging so successfully if it were dangerous." To allay any suspicion that I might have made the horrifying mistake of sitting on the thing without even noticing, I allow my hand to casually brush against the roll, making it clear that I take it absolutely for granted. What could be more reassuring? The pad is not the problem. Its electric blue color is so synthetic that even a monkey knows it could not possibly be dangerous. Unfortunately, the blanket is a smooth dark gray. After a few minutes, I can breathe. The monster is now in the territory. It will require a good half hour to slowly unroll the pad and then get under the blanket. This will create the very lumps I am so scrupulously attempting to convince them are nowhere to be seen. For now, I sit on everything and keep it flat. For the better part of the next hour monkeys intermittently peer at me or come over to take a look. It goes without saying that no one touches anything. Once or twice, Millie holds open the flap of the far hatch to take a look. Sarah *eeolks* brightly, bless her, and Bella makes a second feeble attempt at *yuk-yuking,* but she cannot quite get into the swing of it. All this time I am straining every muscle in my body to appear to do one thing while doing another, like a mime artist who conveys walking while staying in place.

Outwardly, I scrutinize this and that, not a care in the world, while inwardly I am intent on doing everything as slowly and delicately as possible. By and by, somehow, after not having done much of anything I lie close to Laura on a blue pad and rest my head on a very flat grey pillow. It is a bit like persuading someone of the reasonableness of a conclusion by showing them that they did not object to any of the tiny steps into which you broke down an argument. Neither a reasonable debating opponent nor a monkey can get themselves to object to the whole if they find no legitimate objection to the parts. I have done the whole thing so painfully slowly that, hard as they try to interest themselves in what I am doing, they have become bored. Their eyes start to blink and soon they are dozing off to sleep.

While I pursue this new artform, Sophie and Charlie potter away on the floor below. Occasionally, Charlie looks in on Laura and gives me the odd curious look, as though wondering how I am getting on. He knows of my eccentricities and, though he couldn't possibly be more tolerant, I must be an unending source of wonder for him. Every so often he sits with arms folded and at a respectable distance, gazing quietly at me.

The worse is yet to come; having very gradually unfolded the blanket, I must get under it. Suddenly Kate comes to the rescue. She saunters over and almost mischievously curls up on the blanket, almost as if to say: "I wish you'd said this was a blanket; I love blankets! What a lovely place to sleep." I embrace her and she *eeolks* happily. This dispels any remaining suspicion of the blanket. But I can no longer get under it. As I cuddle up to Kate for warmth and close my eyes, I cannot help but smile at this mixed blessing. But for the creaking of the ropes as Charlie and Sophie swing, everything is quiet now.

My muscles ache as I go over the curious contortions, I have put myself through over the last hour just to get a blanket in the monkey territory. The thought of Kate now lying on it as I shiver with cold starts me chuckling, and then the thought of me chuckling at being cold is suddenly too amusing. I start laughing, and the thought of me laughing alone on the Top of the monkey room because I am cold is too much. Quietly as I can, I heave with unbearable uncontrollable laughter, a relief after the terrible anxiety we have all been through the past 2 weeks. Many *eeolk*s around me reassure me that everything is fine.

A while later Kate gets up to pee. Quickly, I get under the blanket. With luck, when she returns, Kate herself will camouflage the lumps. But when she returns, she looks skeptical. It is not as welcoming as it was a moment ago. I do my best to cajole her by snuffling emotionally. But there is nothing doing; she has noticed the movement of one of the lumps. I delicately move the blanket aside to show her my leg. She is quite prepared to believe it is my leg, but still, she does not like it. Monkeys have a way about them that conveys *nothing personal, but this is just not my scene*. They do precisely as they wish with no awkwardness or trace of apology. She moves over by the heater and lies down there. Keeping as still as I can, and the lumps as lifeless as possible, I close my eyes.

Laura is settled and so is everyone else. Everybody, that is, except for Sophie and Charlie, who swing gaily together around the room as if on a huge horizontal merry-go-round while we all listen to the familiar creaking of the center rope. They do so utterly silently, we can only hear their breathing and how they alight every so often on the beams or on the floor shavings below. This seems to go on for at least an hour.

As I doze off, I muse about the tolerance they have for each other. Even if they could conceive a desire to, no one could ask Charlie and Sophie to end their game. "'Come on, time to go to sleep" is not the kind of thing monkeys can say to each other. That this possibility is foreclosed to them no doubt explains their complete lack of irritability over this sort of thing. There is no such a thing as obedience amongst them. Except for physically asserting themselves, no monkey tells another what to do. With this, of course, comes the implied threat of getting close and the corollary threat by merely conveying the intention of approaching. So, if no one bodily interferes, monkeys do exactly as they please. Should I, for some reason, try to stop Charlie and Sophie, I could only do so by forcefully relieving them of the rope, a show of crude strength. When there is a need, this can be done in several ways. One way would be to do so physically, completely un-emotionally by avoiding looking at them or saying anything. This would cause surprise and the resulting state of the rope—if it were tied, for instance—would elicit puzzlement, a problem to be resolved. Another is what I did to remove Sarah from the radiator, which amounts to a very detached, almost mindless, way of being rude. This causes surprise, but it is usually the most effective. The third would be to convey a state of mind through sounds, looks, and gestures, and therefore act as an overbearing dominant monkey. This would likely result in anger and a bitter sense of unfairness along lines typically evoked by a bully.

In the two latter cases, however, they would persist at trying to get hold of the rope and resume their activity. In the first case, they would do so by attempting to solve the unusual problem of untying a knot; in the second, they would insist for a bit, just as Sarah did; in the third, they would do so while looking over their shoulder to ensure the bully remains at a safe distance. It would then be the inscrutable nature of the knot or our own reiteration, or persistent threats of further interference by the bully that would stop them swinging. The limitation imposed on possible threats by the non-abstract nature of their gestures makes these remain just that—threats.

The monkeys' response to such threats is not obedience since it is merely reiteration of the threat that settles the point. At most, they internalize the possibility that the threat may recur. Monkeys simply do not obey in the sense we mean the term, for instance, with a child or with a dog. There is no such thing as disobedience in monkey society. We exact obedience from a child or from a dog by repetition. That is, by consistently iterating hard and fast rules and making consequences follow from causes with such predictable regularity that they come to be taken as given. The point of repetition is, precisely, to establish the recurrence of a prohibition as necessary, and not contingent on mood, time of day, quantitative or qualitative considerations, and so forth. "No" means no, period, like a "fact of nature." Even if they could conceive it, which they obviously cannot, no monkey could be bothered with such consistency.

Just as with a dog, if anything resembling obedience takes place among them in relation to limits we set, it is because of us, not because of them. In a few instances such as electric cables when they could get injured, we establish hard limits. However, as for Charlie and Sophie swinging into the night, I am a guest in the monkey's territory. We establish the rules in our territory; they do in theirs. Nor would I want them to stop; it is a unique opportunity to listen to them. And it soothes us all to sleep.

Tonight, Sophie is full of beans well after everyone else is asleep. Charlie arrives and settles next to Laura for the night. Sophie continues to swing. A short while later she lands close to Charlie, gives him a yearning look, and scampers away looking back at him, encouraging him to follow. He cannot help himself. As they swing together, the ropes creak, and every so often Sarah *eeolks!* either with reassurance or in approval. Charlie returns and starts to settle, only to have Sophie seek him out again, her face lit up. Off they go for another round. The creaking comes and goes, forever it seems, and I open my eyes time and again just for one more look at Sophie's enchanting face.

During the night, I wake up to realize that the rope has long gone quiet. Close by there is a ball of fur. It must be Charlie finally asleep. But where is Laura? I look behind me and there she is, sound asleep by the radiator where I hoped she would be all along. My bones ache: I am stiff; and the night rushes past as I lie awake stranded. But Laura is warm, and one can always drift and dream and make up for lost sleep some other time.

The Problem of Anthropomorphism

Accounts such as the foregoing are largely absent from primatology textbooks. For one thing, many primatologists are lucky to catch a fleeting glimpse of their object of study. At the Sanctuary, we had many hours of direct contact with the whole of the monkey colony every day. At all times, we could see them across a wire mesh throughout their territory, even a few inches away. And, of course, they visited with us and us with them. Indeed, they were as interested in us as we were in them.

Significantly, though they are mostly absent from the scientific literature, one may read accounts such as the foregoing in books directed at the general public. But the epistemological problem is not taken up in either. For one thing, readers of such books need no convincing since most have probably enjoyed deep and lasting friendships with their pets. The same is true of most primatologist. No doubt that is why they became primatologists in the first place. But, for primatology the discipline, an anecdote is simply not scientific data. If referred to, it is likely to be accompanied with a cautionary note. It would be highly exceptional to find a statement indicating that this is a deep problem in metaphysics with deep roots in both epistemology and ontology.

The point, however, is that hardly anything that we learned about monkeys at the Sanctuary would qualify as 'serious' primatology. For one thing, this led Leonard Williams to argue for the complementarity between research in labs, field studies, and studies of primates living in artificial but natural social settings.[30] His books, rich in anecdote and descriptions of life with the monkeys, refer to studies "in interliving."[31]

Intuitively, Williams and his colleagues took Darwin's cue seriously. To understand monkeys, and to understand how one understands their understanding, is not only indispensable to look after them, but also to do something about metaphysics. How else could one succeed at maneuvers in the monkey territory such as the one just described? How do you go about understanding monkeys and have monkeys understand you, such that you can understand their understanding and yours? Interliving is to learn about an object that happens to be a subject, which in turn inevitably leads to learning about oneself. Two subjects thereby become objects, and yet these two subjects enter a relationship with each other; thereby, this relationship itself becomes an object of inquiry. This is a We—a fundamental point I shall be working towards. As all science inevitably leads us to a better understanding of ourselves, the deliberate understanding of ourselves to better understand, is understanding of an altogether different kind. Much of my understanding of the monkeys derived from adverting to what I discovered myself doing in various situations, such as the one described. However, perhaps the deepest source of knowledge about the monkeys at the Sanctuary, and the least verifiable in a standard scientific sense, are myriad glimpses over many years into their psyche through tiny incidents such as those just described, which hardly even qualify as anecdotes. I shall recount more of these in the chapters that follow.

In this era of internet and with everyone holding a powerful video camera, we are privy, like never before, to endless glimpses into the lives of animals. A duck feeds the goldfish, a bear unlocks the door of a house to find food, a dog brings home his new deer friend, five or six cows lick a cat affectionately, a frog holds his arm around his mate while it rains, a crow urgently prods a hedgehog to get off the road, a cat dives headfirst into a bin, and so forth. These incidents are at our fingertips simply by searching with the right phrase. Listen to David Attenborough's account

[30] Williams, *Challenge to Survival,* 3.

[31] Williams, *The Wooly Monkey*, 8. His other books include *Samba and The Monkey Mind* and *Man and Monkey*. These were preceded by *The Dancing Chimpanzee: A Study of the Origins of Primitive Music.* His magnum opus, *Challenge to Survival: A Philosophy of Evolution*, includes additional superb anecdotes of life with the monkeys. Originally, all were published by Andre Deutsch, London. There are some second and revised editions by Allison and Busby, London.

of a weaver bird sowing a nest.[32] No doubt his account elicits the charge of anthropomorphism. And yet, we can only make sense of this, and all the other scenes, because we know how we would do it ourselves. This self-knowledge is what allows us to understand what these animals are doing and, watching, we learn something about ourselves. Clearly, sewing is sewing, regardless of who is doing it. This doing is sustained by a knowing, and any such knowing takes us to the foundations of all knowing, indeed to the nature of knowing as such, to how any living being makes sense of the world. The foundation of all such knowing is distinguishing.

Distinguishing, knowing, and discipline long predate the splitting of the arts from the sciences. Given that divide, one may contrast between a "scientific approach" to monkeys and our own "artistic approach." The founders of the Monkey Sanctuary were musicians. They took up the challenge of understanding monkeys in their own terms by building a way of life with them. They did so with the rigor of musicianship. Decades later, I would refer to this as the contrast between the "romantic approach" and the "classic approach". The latter aims at theory while the former aims at deep understanding within oneself. Of course, and already mentioned, this "leads to a unique understanding which, taken far enough, is just as inaccessible to others as the nonhuman primates themselves. Only those who have spent years with them will understand such accounts."[33] Each approach come with its own drawbacks.

Again, regardless of the approach, at the heart of all lies the problem of interpretation. One may attempt to dodge the problem by following the method of the hard sciences, but this leaves out the most significant data. Or, one can take up the challenge of interpretation. As I wrote elsewhere, interpretation is an issue that will have to be confronted head on by primatology and zoology. But it has been there all along under a different heading—the problem of anthropomorphism:

> The problem which is raised by anthropomorphism is not some incidental problem; it belongs to the groundswell of the centuries; it is something which has been building up over the past four or five centuries. Because it is a first-class problem, it is not something that is going to be solved in any simple fashion.[34]

The remote origin of this research was the challenge we faced at the Sanctuary when talking to visitors about the monkeys. However, the problem of anthropomorphism is not merely a problem for the philosophy of science and for biology; it is a

[32] "BBC Homemaking: Weaver Bird."

[33] Daniel Mayer, "Interiority," 146–147.

[34] Mayer, "Interiority," 156. The present investigation indirectly addresses problems of interpretation in biology generally. The paper cited addresses these problems in primatology although in a very different idiom. It may be read as a summary reflection of the initial stage of this project, which was about the concrete experience of living and working with monkeys. As written in terms of Lonergan's cognitional philosophy, it represents a second stage, with programmatic conclusions that prefigure the present examination, the fourth stage. An intervening third stage was the encounter with the field of *biosemiotics*. Key themes in the present essay where first presented in several *Gatherings in Biosemiotics* (see http://www.biosemiotics.org).

symptom of a deep cultural malaise. Immersed in a world of our own mechanical and mass-produced products, we are ever less in touch with aliveness, both our own and that of the few living beings around us. Just how deep the problem runs, is a central topic of this exploration. Methodologically, a key to the problem is to become intimate with the nature of living being through our encounter with others and with ourselves. We are alive and, whether consciously aware of it or not, we cannot help but understand the aliveness of other living beings in terms of our own aliveness. As we inquire into aliveness as such, we may as well do so explicitly and as rigorously as possible. Bernard Lonergan referred to inquiry into ourselves as *self-appropriation* and its practice as an exercise in *interiority*. Thus, to inquire into aliveness we must self-appropriate our distinguishing. As I will argue, distinguishing is nothing less than the core of living being. Thus, while there is much to learn about monkeys by observing them, there is also much to learn about them by observing ourselves in their company as they make sense of the world, even asking what we ourselves would do in their place.

We cannot get *over* the problem of anthropomorphism; we must get *under* it, so to speak, by digging down to the foundations that make it a problem in the first place. The problem is not solved; rather, it must be dissolved. But, how to do that? Thinking along these lines during my days at the Sanctuary, I one day asked myself what the minimal possible *qua cognitional* activity is that all living beings must share. By this I had in mind any activity of knowing, no matter how rudimentary. Eventually, the answer came—*distinguishing*. To be alive is to distinguish. Thus, common to all living beings is *distinctional activity*. All draw distinctions, no matter how simple or complex. If our understanding of aliveness is founded on this, the problem of anthropomorphism can be dissolved. In short, the problem is dissolved once biology ceases to leave out the highest conjugate form of living being and comes to terms with the problem of interpretation. This will radically differentiate biology from the hard sciences and its methods. With this, the far larger general underlying question may be addressed: How to include the knowing of living beings in our knowing of them?

Sadly, addressing the root of these issues calls for a fair amount of abstract inquiry. Only limited space can be devoted to wonderful and heartwarming glimpses into the lives of monkeys. The poet David White talked about the conversational nature of reality and referred to beautiful questions. The beautiful question I started from was: Why are the wonderful relationships we have with the monkeys, and what we learn from them, fundamentally not admissible as science? What does it say about science? Paradoxically, so long as proper foundations are lacking, the more suggestive and thrilling these glimpses into aliveness are, the more they elicit the charge of anthropomorphism, thereby reinforcing the stance that human knowing is in some sense unique and that comparisons to it are unjustified. Not without regret, then, we must largely leave stories about the monkeys aside. Doing so, I recall Boswell's anecdote about a schoolmate of the great English lexicographer who once said to him: "You are a philosopher, Dr. Johnson. I have tried too in my time to be a philosopher; but I don't know how; cheerfulness was always breaking in."[35] I will endeavor to be a philosopher yet bring in some cheer to the following chapters by including at least some stories about the monkeys.

[35] Boswell, *Life of Samuel Johnson*, 294.

Chapter 2
Distinguishing Distinctions

Now why a thing is itself is not a quest after anything . . .
Aristotle. (Aristotle, *Metaphysics*, 151)

Advertencia al lector:
Esta página reproduce la anterior. Otra, *es la misma. Pero para evitar que el lector no tome en cuenta esta segunda primera página atribuyéndola, por ejemplo, a un error de re-lectura, he tenido que incluir esta advertencia, que no figura en la primera página. Para ser la misma, hace falta que sea otra.* Vincent Descombes. (*Lo Mismo y lo Otro: cuarenta y cinco años de filosofía francesa,* emphasis in original. *["Notice to the reader:* This page reproduces the previous one. *Other,* it is *the same*. But to prevent the reader from *not taking into account* this second first page, attributing it, for example, to a re- reading error, I had to include this warning, which does not appear on the first page. To be *the same*, it must be *other*." (trans. Daniel Mayer)])

What we're doing now in medicine is what Einstein did for physics.

He brought time to physics. We're bringing time to biology.
Fred Turek, circadian scientist. (Turek, "Circadian Surprise." An updated written version of this interview on the occasion of the Nobel Prize in medicine for their work on circadian clocks earned by Jeffrey C. Hall, Michael Rosbash, and Michael W. Young is available at http://www.npr.org/sections/health-shots/2017/10/02/555054483/how-messing-with-our-body-clocks-can-raise-alarms-with-health*)*

Distinction: A/Ā, A=A

Distinctions are of vastly different orders and different kinds. As humans, we take pride in the subtlety of the distinctions we draw in all fields of endeavor. Language is all about subtle distinctions, whether what is said is about math, the physical sciences, or the arts. Thus, the Pythagorean theorem elegantly distinguishes a set of

© The Author(s), under exclusive license to Springer Nature Switzerland AG 2023

D. C. Mayer-Foulkes, *The Nature of Living Being*, Biosemiotics 26,
https://doi.org/10.1007/978-3-031-24789-7_2

geometrical objects and relations; the periodic table of the elements assembles a vast array of distinctions; painting, poetry, dance, and music depend on distinctions. We draw subtle non-verbal distinctions, and, even when we caress each other, we play with exceedingly faint and evanescent distinctions.

And yet, vast as the range of distinctions is, there is no formal difference between what a T cell accomplishes when it discriminates between pathogen and non-pathogen and what is accomplished by a group of astronomers when, after many years of discussion, they arrive at a definition that distinguishes between planet and non-planet.[1]

Let us start from what has already been pointed out. All distinctions involve A/Ā: something (A), what that something is not (Ā), and/, the boundary between them. Distinction also involves A=A. How so? Inherent to any distinction A/Ā is the homogeneity of A and of Ā on one side of / and on the other. That is, for a distinction to obtain, A must be homogenous in respect to the distinction, else a further distinction is introduced.[2] Likewise, Ā must be homogenous, insofar as Ā is that which is not A, else again, a further distinction is introduced. Therefore, homogeneity, or A=A, must obtain for both A and Ā for the distinction A/Ā to obtain. Of course, further distinctions may always be drawn within a given A or Ā. In these distinctions too A=A obtains, else they do not constitute distinctions.

The homogeneity of both A and Ā together imply—and are implied by—the homogeneity of /, the boundary between them. That is, A=A as it obtains in A and in Ā implies and is implied by the A=A of /. To make explicit the homogeneity of distinctions, the full expression for A/Ā may be taken to be is A/Ā, A=A.

Though A/Ā, A=A might be taken as a definition for distinction, how else might distinction be defined? Let us start by noting that one thing is the act of distinguishing and another is the resulting distinction. For shorthand we may refer to the *act* of distinguishing as the *verb* and to distinction, its *result*, as the *noun*. The same applies to defining and definition. A moment's reflection, and a quick look at a dictionary, will confirm that a definition is a kind of distinction. Therefore, a definition, the noun, is a kind of distinction, and defining, the verb, is a kind of distinguishing. A definition is a formal and, therefore, particularly precise way of stating a distinction. Being distinctions, definitions are constituted as A/Ā, A=A. Their purpose is to precisely indicate a boundary (A/Ā) and the condition of homogeneity to be fulfilled (A=A). Whatever meets certain criteria falls within the definition.

[1] "IAU Definition of Planet." After many years of heated debate, the International Astronomical Union voted on, and passed, a final definition of planet on the 24th of August 2006. Pluto, now designated a dwarf planet, was recognized as the prototype of a new sub-category of Trans-Neptunian objects.

[2] Koko the gorilla provides a comical illustration of this. One afternoon Koko was playing by herself with some white towels, signing *red.* She was asked, "You know better, Koko. What color is it?" She insisted, signing *red* three times, each sign larger than the preceding. Then, with a grin, she picked up a minute speck of red lint clinging to the towel, held it up, and signed *red.* Days later she repeated exactly the same routine. See Patterson and Linden, Education of Koko, 81.

Explanatory disciplines make sense of the world by starting from, or arriving at, rigorous definitions. The present is a case in point. But it involves a performative circularity—when what one does and what one is referring to are the same, such as when one talks about talking or writes about writing.[3] Of the two inseparable aspects of distinction, A/Ā and A=A, start with the definition of A=A. A definition is a proposition of the form A=X, a statement that X stands for something that A is equivalent to. Therefore, the definition of A=A would take the form (A=A)=X. But then this definition, indeed, definition as such, is an instance of A=A. Clearly, the same applies to the definition of A/Ā since, in both cases, the boundary is expressed by the differentiation between what they are and what they are not, or A/Ā. This circularity is unavoidable.

This circularity points to the knowing subject without whom there is no defining/definition, or more generally, no distinction/distinguishing. We may wish to abstract them, but distinctions originate with subjects. Below, I consider distinctions accomplished by a product, a household thermostat, and argue these distinctions are due to the living beings that produce them.[4] This also obtains for any instance of artificial intelligence (AI), for example, a device capable of arriving at accurate medical diagnoses. There is much to learn about living being by examining products, which are radically different from them, and also radically different kinds of things to the nonliving such as the sun, the moon, or stars. It is impossible to overestimate the importance of these differentiations. Moreover, if robotics and AI pose an epistemological problem, it is only because they point to a profound underlying ontological shift. As Yuval Noah Harari pointed out, for the first time in its 4 hundred-billion-year history, robotics and AI are allowing life to break out of its organic limitations.[5] In the not-too-distant future this will allow life to reach other parts of our galaxy and beyond, thereby overcoming limitations imposed on organic forms by conditions on our planet. Aside from the emergence of life, we are witnesses, and even participants, in the most radical shift to have occurred in the history of our planet.[6]

To show why distinguishing pertains to living beings, we must dig into the nature of distinctions and of distinguishing. Thus, a key topic of this analysis is the

[3] There are also performative contradictions. To say "I believe I am not expressing myself clearly" is plainly clear. This gives rise to performative paradoxes such as the liar's paradox. When one says, "I am lying," is this lying or telling the truth? We shall see in Chap. 5 that these issues lie at the very heart of the nature of distinction, and therefore the very nature of living being.

[4] Other authors, Luhmann for instance, identify other kinds of systems that distinguish, for instance social systems. The issue is taken up in the section below "Do only living beings distinguish?", discussed more generally in Chap. 7 *Four Kinds of Things* and followed up in the discussion about wholes in Chap. 9 *Living Being*. For now, it can be simply said that it is only because living beings distinguish, that products of theirs, or social systems made up of them, may also be said to distinguish.

[5] Harari, *The Future of Humanity*.

[6] To see a talk on the future of humanity go to https://youtu.be/XOmQqBX6Dn4

obscuring or elision of the subject that draws distinctions. This abstraction is the one that products such as a thermostat appear falsely to confirm. These are discussed below. Ironically, a key challenge to understanding living beings comes from the method of the exact sciences, precisely because of their mechanical framework. This framework not only collapses the differentiation between products and the nonliving. It also elides the aliveness of subjects by treating organisms as physical objects. Thus, distinction, the noun, is abstracted from distinguishing, the verb.

This brings us back to the highest conjugate form to understand aliveness—a living being is a distinguishing (verb) distinction (noun). We remain a distinction (noun) by distinguishing (verb); we can distinguish (verb) because we are distinctions (noun). This performative circularity is inherent to living being. Inevitably, then, it underlies this examination. The circularity is the foundational link between a subject's distinguishing and the distinctions that result. This is why an inquiry into the nature of living being may be carried out as an inquiry into distinction and distinguishing. Because being a distinguishing distinction is the foundation of living process, of aliveness as such, it is the very nature, *the form,* of aliveness.

Being distinguishing distinctions, living beings are *subjects*. Even the simplest organism is a subject in this most rudimentary sense. A subject is a unique distinguishing distinction in *entelecheia*, being-at-work-staying-itself, holding together actively as a whole. A subject is a unique living being that makes distinctions to remain itself. Subjects, then, display that mode of being whose fundamental nature is inquiry. Self-maintaining by distinguishing, it is inherent to the nature of subjects to anticipate. These anticipations of subjects are *for* them. Thus, events, things, etc. are always already *for* subjects.

Subjectivity is our own experience of subjecthood, or subjectness. This opens a vastly complex and subtle discussion of levels or qualitative degrees of subjectness and what constitutes them. This discussion is indispensable and must be central to biology. But it can only be touched on in this essay. To sidestep it, reference will henceforth be made to the full range of 'subjectness' of subjects by bracketing. So, all organisms are [subjects] because they are *qua* subject; they are, of themselves, subjects and to some degree subjective.

As was said in Chap. 1, distinction underlies signification. In the case of bacteria, say, the presence or absence of attractants and repellants is *for* an organism, for a living being. Thus, insofar as the distinctional is foundational to aliveness—to distinguishing distinctions—it is, thereby, also foundational to signifying. Reciprocally, and for the same reason, signifying is foundational to aliveness. Thus, long predating what we understand as 'knowing'—human or otherwise—the concrete foundation of aliveness and signification is the *entelecheia* of living cells, the most fundamental distinguishing distinctions. Being distinct—or distinct being—by means of distinguishing is the *form* of cells. At this most primary level, living/knowing/signifying converge. We shall return to this topic and its ties to logic and mathematics, which are founded on distinction. Thus, our being and our knowing—ontology and epistemology—converge: the distinctional is the foundation of living, knowing, sense-making, signification.

The vast scope of distinction arises from its foundational character. It is embodied in the most rudimentary organisms as open/closed, nourishing/non-nourishing, toxic/non-toxic, light/dark, start/stop, on/off, here/there, this/that, up/down, is/is-not, better/worse, acid/non-acid, alkaline/non-alkaline, hot/cold, quick/slow, wet/dry, and so forth. As complexity increases, distinguishing unfolds into endless subtle differentiations—always embodied—such as I/not-I, you/me, yours/mine, friend/foe, us/them, now/then, big/small, positive/negative, something/nothing, alive/dead, yes/no, one/two, and thing/not-thing. Billions of years of evolution lead to our own sophisticated distinctions mediated by language—true/false, beautiful/ugly, good/evil, sense/nonsense, beginning/ending—as well the subtle and dynamic shades of differences expressed in processes such as growing, developing, maturing, and learning. These are processes that we conceptualize in terms of distinction, and which pertain to the very nature of living being. They do so because the *entelecheia* of organisms, their maintaining themselves as such. Aliveness is continual dynamic distinguishing activity. That being so, what discontinuities characterize the different levels of complexity of living beings? The distinctional must therefore be capable of accounting for both continuities and discontinuities between organisms, including the very diverse kinds of distinctions of which they are capable.

Notwithstanding the performative circularity involved, let us go back to the definition of distinction. We begin with etymology and go on to consider some definitions, their adequacies, and inadequacies.

Definitions of Distinction

The etymology of 'distinction' is very suggestive. To distinguish comes from the Latin *distinguere* (*dis*, apart; *stinguere*, to prick)—to separate by means of pricked marks.[7] The Old English *sticca* (rod, twig, spoon) from Proto-Germanic *stikkon* (pierce, prick)[8] is closely related to the Greek *stigma*, a mark made by a pointed instrument. These, and many other similar words in various languages, derive from the ancient Indo-European root *steig*, meaning "to prick" or "sharp."[9] The Latin word encompasses two crucial notions: *marking*, as derived from the ancient Indo-European root, and *separating in two*, through the Latin prefix *dis* (apart, asunder). Inherent to separating is the notion of two, or twoness; thus, the Latin root derives from the Greek *dis* (twice).

In short, the etymology of the word suggests that distinguishing involves marking, separating, and twoness. Our inquiry into A/Ā, A=A will show why twoness,

---

[7] Partridge, *Origins, XX*, 667.

[8] Online Etymology Dictionary.

[9] Edward and Pastor, *Diccionario Etimológico*, 170, trans. Daniel Mayer.

inherent to both separating and marking—or boundaries and indication—has its source at the very foundation of living being. As distinguishing distinctions, living beings set themselves apart, separate, and doing so by distinguishing, they indicate, signify. The etymology of the term reveals a profound intuitive understanding of the nature of distinguishing.

Consider the etymology from a different language group. The seminal Hebrew term *lehavdil* literally means *to separate* or *to distinguish.* In the Torah (Jewish Bible), creation is the making of distinctions: God separates between the light and the darkness, the water that was under the space and water that was above the space, dry land and seas, plants and animals, day and night, and so on. Creation is the act of separating and then naming, indicating.[10] *Lehavdil*, is the key term used in the prayer that ends the Sabbath and ushers in the new week. According to this prayer, distinguishing constitutes the possibility of wisdom, understanding, judging, redemption, salvation, by allowing the separation between holy and secular, light and darkness, Israel and other peoples, the Sabbath and workdays, good and evil.

Thus, the kernel of these two etymologies is the same: separating and indicating; differentiation and duality. Unspoken yet implied in both is the homogeneity on either side of what has been separated.

In this review, I shall examine three definitions of distinction used in *Laws of Form* by George Spencer-Brown (hereafter *LoF*). There is a weighty reason for this choice. *LoF* constructs a formal mathematical system starting from distinction. And this formal system, in one of its possible interpretations, provides the foundation for logic. Recall that logic, as we commonly know it, is an algebra—Boolean algebra. To state it precisely, *LoF* defines distinction by means of two axioms, and constructs upon them a simple arithmetic. This arithmetic is the hitherto unknown arithmetic of distinction that underlies Boolean algebra. That being so, *LoF* could just as well have been entitled *Laws of Distinction*. The author sometimes refers to what it presents as the *calculus of indications*. In short, the mathematical definition of distinction in *LoF* has a lot going for it. But, in fact, *LoF* provides two other definitions of distinction, both of which are also of great interest. We will start this chapter with the first, later touch on the third, and treat the second one expressed formally in terms of two axioms, in Chap. 4.

Since *LoF* is a mathematical text, it is well worth recalling a remark by Kant:

> [In] philosophy one must not imitate mathematics in putting the definitions first, unless perhaps as a mere experiment. For since they are analyses of given concepts, these concepts, though perhaps only still confused, come first, and the incomplete exposition precedes the complete one, so that we can often infer much from some marks that we have drawn from an as yet uncompleted analysis before we have arrived at the complete exposition, i.e., at a definition; in a word, it follows that in philosophy the definition, as distinctness made precise, must conclude rather than begin the work.[11]

[10] Freedman, 7.

[11] Kant, *Critique*, A731/B759, 639. Saying "definition, as distinctness made precise" Kant concurs that a definition is a kind of distinction. He adds a footnote himself: "Philosophy is swarming with mistaken definitions, especially those that actually contain elements for definition but are not yet complete. If one would not know what to do with a concept until one had defined it, then all phi-

In short, mathematics is presented by starting from precise definitions; philosophy presents tentative definitions to experiment with them to arrive at more precise definitions. We have already noted, both the performative circularity of defining distinction, and the largely unacknowledged significance of A=A. Kant's remark points precisely to how we have been proceeding and shall continue to proceed. Namely, we shall experiment with the definitions of distinction in *LoF*. We do to assess whether the foundation of distinction itself must be sought in the aliveness of living beings. Or to be precise, to assess whether the axioms of *LoF* require the distinctional. It this is so, then the distinctional precedes the mathematical axioms upon which logic is built. This would make the distinctional the condition for the possibility arithmetic of logic, let alone Boolean algebra. And, if this is so, then, historically, the distinctional must have preceded the logical. This is true in two senses, one obvious and one not so obvious. Life went on for billions of years before the discipline of Logic appeared with humans. Less obvious is that aliveness tends to be logical because organisms are distinctional—distinctions that distinguish. I have already begun to argue that the distinctional must be the foundation of biology if this discipline is to inquire into aliveness. Experimenting with the definitions of distinction in *LoF* allows us to assess whether the distinctional is the condition for the possibility of its axioms. If so, if the distinctional does indeed turn out to be the condition for the possibility of logic, then, at the very least, we will be able to affirm that the distinctional is not itself bound by logic. Logic is built on the distinctional, not the other way around. Therefore, strictly, the distinctional, which is the biological, can be—indeed, is likely to be, at least in some sense—nonlogical. It is certainly prelogical. Thus, we shall be performing experiments with the definitions of distinction in *LoF* to assess whether we stand on firm ground when we say that the biological need not be logical. Two major points would follow from this: first, it would show why aliveness cannot but remain unintelligible so long as it is inquired into in physical and chemical conjugate forms, since these are founded on logic. Second, this inevitably leads to the question: how then is aliveness to be made intelligible? In what sense is the distinctional not logical? But we are getting ahead of ourselves. Let us turn to the definitions of distinction in *LoF*.

---

losophizing would be in a bad way. But since, however far the elements (of the analysis) reach, a good and secure use can always be made of them, even imperfect definitions, i.e., propositions that are not really definitions but are true and thus approximations to them, can be used with great advantage. In mathematics definitions belong *ad esse* [to the being], in philosophy *ad melius esse* [to the improvement of being]. Attaining them is fine, but often very difficult. Jurists are still searching for a definition for their concept of right."

Abstract Definition: Distinction as Perfect Continence

An elementary way to think of distinction is in terms of closure. This is often represented graphically as a circle and may be stated as follows: *distinction is perfect continence.*[12] This is the first definition of distinction in *LoF*. However, though highly intuitive, we shall soon see that this definition turns out to be unsatisfactory. Spencer-Brown often appealed to etymology, so we can do so ourselves here to flesh out what he means. *Continence*, like *contain* and *contents*, is derived from the Latin *continere*—to hold (*tenere*) together (*con-*)—thus it means to hold within, to contain, to retain.[13] Hence, *perfect continence* is to hold together thoroughly.[14] An archaic meaning of continence is "something that serves as a container or boundary."[15] Thus, *perfect continence* would be an absolute boundary. Spencer-Brown notes: "Although it says somewhat more, all the reader needs to take with him from chapter 1 are the definition of distinction as a form of closure and the two axioms which rest with this definition."[16] This is how it is usually represented (Fig. 2.1).

This image of perfect closure purports to show three terms: an inside, an outside, and the limit between them as indicated by the circular line. As Spencer-Brown noted, "[A] distinction is drawn by arranging a boundary with separate sides so that a point on one side cannot reach the other side without crossing the boundary. For example, in a plane space a circle draws a distinction."[17]

We may do worse than ask the foolish question—why is there a need for continence? Why is the following not a distinction? (Fig. 2.2).

Obviously, a linear distinction would be inadequate to keep a herd of cows from wandering away. For that, it would be a good idea to join the two ends together. Why? Well, this is just how the world is made, even if a vague sense remains that such a line segment is a distinction of sorts. After all, though not an enclosure, the

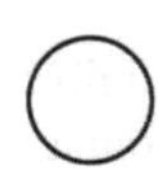

Fig. 2.1 Distinction as perfect continence

Fig. 2.2 Linear distinction

[12] Spencer-Brown, *Laws of Form*, 1. Except when noted, all references are to the Cognizer Limited Edition.

[13] Partridge, *Origins*, 117.

[14] Partridge, 485.

[15] Dictionary.com, "continent."

[16] Spencer-Brown, *Laws of Form*, 77. In the 2011 Bohmeier Verlag edition, he notes: "There have been so many jokes about this that I had better point out that I was employing 'continence' in its original sense of 'containment,' not in its later usage as 'sexual abstinence.'" Spencer-Brown, *Gesetze der Form*, 1997, 1.

[17] Spencer-Brown, *Laws of Form*, 1.

Fig. 2.3 Closure as territory

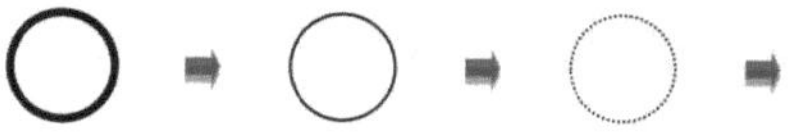

Fig. 2.4 Distinction as circumference

Great Wall of China did succeed at keeping the barbarians out for centuries. I shall return to this point later.

One may think of closure in terms of territories: an inside territory, an outside one, and the boundary between them (Fig. 2.3).

Taken literally, however, this boundary is itself a territory, a kind of 'body,' since one could stick a pin in it.[18] This is so unless we abstract and take this image to *represent* a circumference by regarding it as a line with length but not width (Fig. 2.4).

We see the first three but must suppose the boundary to be like the fourth. But this is a significant departure. What a moment ago was an instance of a distinction (indeed, its paradigmatic example) has now ceased to be a distinction and has become a *representation of a distinction*. Because the drawing itself has not changed, clearly there is a difference between what we *look at* and what we *understand*. A circle *represents* a geometric circumference by relying on our capacity to abstract—that is, our capacity to differentiate between a representation and what it represents. While the representation is by means of a 'body,' what is represented is not a 'body' at all. In short, representation heavily relies on us as knowing subjects, though we tend to forget this. We retain the distinction and forget about the distinguishing activity that brought it forth. Forgetting that we *as subjects* do the sequence shown above, and then, forgetting that we forgot, we twice elide the knowing subject. An elision is not an elision unless we forget—elide—the first elision. This double elision makes boundaries at once both excessively abstract and excessively concrete—too abstract, insofar as it is independent of a subject, and too concrete, insofar as it is conceived as a 'body.' This attributes false concreteness to the boundary by regarding it as 'a body' that is already out there now.

A circle drawn on a page represents a boundary, just as the perimeter wall of a garden or a fence around a field do. However, even these 'bodies,' material as they are, represent boundaries only by convention—graphical and pedagogical for a circle and legal, political and economic for a wall or a fence. Lacking these, a circle, a wall and a fence are just that, not boundaries at all. In our daily lives we take boundaries and distinctions to be 'bodies,' concrete physical objects. We simply assume that A and Ā do not depend on a subject—the island is the island, and the sea is the sea; the water is the water, and the tub is the tub; and so forth. We take for granted that these distinctions do not depend on us and that they are real 'bodies'

[18] This use of the term 'body' will be discussed at length in Chap. 7.

Fig. 2.5 Territory on a plane

Fig. 2.6 Boundary under a microscope

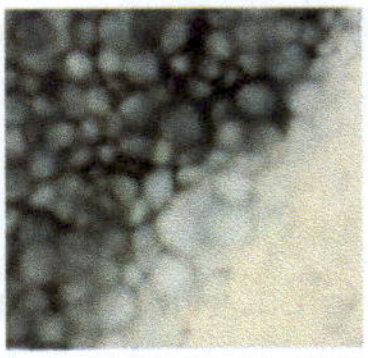

Fig. 2.7 Closer look at a boundary under a microscope

already out there now.[19] For common sense, the very idea that these distinctions in some way depend on us smacks of solipsism. But distinctions result from distinguishing—a synthetic act that involves both a subject and an object. It shall later be shown that inherent to such acts is anticipation. This is an inquiry into this synthetic act. To understand why this is the cardinal feature of living beings we must go by stages.

Perhaps a better image for distinction is not a circumference but a circle. If a circumference is an imaginary line, a circle is a territory on a plane:

The territory in Fig. 2.5 is like an island or a country on a map. But how does this boundary look under a microscope? Roughly, it looks like this (Fig. 2.6).

One may still clearly discern a 'this,' and a 'not-this,' and the fuzzy boundary between them. This fuzziness calls on a subject to determine where precisely the boundary is. In fact, if you look closely at a concrete boundary, it is nothing like a straight line. It is jagged, more like a shoreline. And the closer you look at a shoreline, the more jagged it is. This leads to the *coastline paradox*: the length of a coastline depends on the proximity at which it is measured; the closer you get to it the longer the coastline is.[20] Needless to say, measurement points directly to a knowing subject. As we get closer the fuzziness increases, to the point where either there is no boundary, or there is because we arbitrarily determine where it is[21] (Fig. 2.7).

[19] The differentiation between these two notions—the real as already out there now and the real as the intelligently understood and rationally affirmed—is central to Bernard Lonergan's *Insight* and will be discussed at length in Chap. 7.

[20] First observed by Lewis Fry Richardson, Benoit Mandelbrot took up this paradox in his inquiry into fractals. See a good illustration of this in https://en.wikipedia.org/wiki/Coastline_paradox

[21] Regarding this arbitrariness, the reader may want to see an animation of going ever deeper into a fractal found at https://giphy.com/gifs/fractal-konczakowski-coastline-1qk3wLOO91HaRRP8yZ.

This makes the point obvious: arbitrariness is inherent to any and all concrete acts of distinguishing. Below we shall discuss how an *E. coli* bacterium distinguishes during chemotaxis. As we shall see, organisms successfully manage boundaries despite the dilemmas around their fuzziness and do so even when—or precisely because—they do not constitute an enclosure. My aim is to show how organisms do this, even if—or precisely because—the closer one gets to them, the more boundaries turn out to be not only fuzzy but also in motion. This is even more true for conceptual boundaries, if only because what is understood by a given concept is always already dependent on context.

Though begrudgingly, our common sense must concede that, while it may be convenient to regard distinctions as sharply delineated and black-and-white, concretely, their clarity and distinctness is the outcome of a process that involves us. Overcoming fuzziness and movement is inherent to distinguishing. No matter how rudimentary, a distinction involves *the arbitrary establishing of differences that make a difference and differences that do not* (henceforth, I may sometimes abbreviate this as *diff-diff*).[22] Some differences make a difference to an organism but not others, both at the individual and at the species level. Gregory Bateson first suggested the idea of "differences which make a difference" back in 1970. Indeed, he suggested that "the word 'idea,' in its most elementary sense, is synonymous with 'difference'." Citing Kant's assertion in *Critique of Judgment* "that the most elementary aesthetic act is the selection of a fact" Bateson suggested this can be modified to refer to *differences*. So, of the infinite facts—differences—presented by a given object, an organism selects those that make a difference to it. Indeed, organisms are designed to function by selecting that which makes a difference to them and filter out the rest. Bateson points out that these differences are what we mean by information. So, he proposes that the elementary unit of information is a *difference which makes a difference*.[23] Along similar lines, in the field of contemporary biosemiotics, Jesper Hoffmeyer proposes that the basic unit of life is the sign, not the molecule.[24] For the reasons already presented, this inquiry proposes distinction as the basic unit of life.

Differences that make a difference also converge with the notion of *Umwelt*, "the interactive unity of the organism and the world sensed by it."[25] Sometimes translated as "self-centered world,"[26] this term was first introduced by Jakob von Uexküll. His

[22]This notion is due to Gregory Bateson. He had stated something similar before, but it became famous with the Glossary of his last book, *Mind and Nature, A Necessary Unity* (1979), with the definition *information is any difference that makes a difference*. I do not here take up the discussion of the similarities and differences between the notions of information and signification.

[23]Bateson, Gregory, *Steps to an Ecology of Mind*, p459.

[24]Hoffmeyer, J. (1996). *Signs of meaning in the universe*. Indiana University Press.

[25]Favareau, *Readings in Biosemiotics*, 41.

[26]Jakob von Uexküll (1864–1944) coined the term *umwelt* ("self-centered world") and pointed out that organisms sharing the same environment may have different *umwelten* (see Kull, "Umwelt," 348–349.)

point is that, in a single shared environment, different species inhabit different *Umwelten*. My garden is vastly different for me than for a lizard, a bee, an ant, a dog, etc. Some differences make a difference to one and some to the other. In each case, the fuzziness and shades of gray of the boundaries between the differences that make a difference and those that do not involve each of us as subjects. These boundaries point to the dynamic relation between subject and object. From this dynamic relation result the *synthetic acts* that constitute distinctions. Distinction, signs, *Umwelt*, differences that make a difference, are all synthetic acts. On one side is the irreducible and endlessly changing concreteness of the environment, call it [object]. To this corresponds a [subject] *for* whom some differences make a difference but not others. On the side of the [subject], then, because the differences of the [object] are concrete, specific, and, by nature, contingent, the synthetic act of distinguishing necessarily involves a degree of arbitrariness. Where exactly is the shoreline? Is it where the waves break? Is it at the highest point the waves ever reach? Or at the average point they reach? This kind of fuzziness is inherent to any distinction ever made by living beings. In a nutshell, in their synthetic acts, [subjects] overcome the fuzziness inherent to the concreteness of [objects] through arbitrariness. This is A=A, treating differences that make a difference as 'the same'.

The foregoing points to the foundation if we are to understand how organisms are designed and how they relate to their environment. The arbitrariness inherent to A=A pertains to an utterly different horizon than cause-effect. In this sense, organisms are only full intelligible in terms of event-response, rather than cause-effect. Event-response pertains to the distinctional as the highest conjugate form, as discussed above. It is on a level that exceeds physical and chemical cause-effect. For the nonliving, only cause-effect obtains. Not so for living being. Cause-effect certainly obtains for us, but beyond it so does event-response. Event-response is inherent to aliveness, involving as it does, differences that make or do not make a difference. Or, to state it simply, living beings are [subjects] with agency. This is the form of aliveness that must be the foundation of biological inquiry.

But let us go back to fuzziness. Perfect continence may not stand up to scrutiny. Yet the fact remains that our commonsense notion of distinction is—to recall Descartes' expression—in clear and distinct binary terms: either/or, black/white. This undoubtedly works in practice—not a moot point. Might it be that clarity and distinctness, and the fuzziness of concreteness, are both inherent to the distinctional? Perhaps the distinctional is, precisely, the interplay between fuzziness and distinct clarity? Can distinguishing be thought of as the leap from analog to digital, so that distinction, the noun, is binary in the digital sense? Perhaps this is the reason why, once drawn, distinctions are not easily dissolved? In any case, this interplay could only follow from distinctions not being 'real' as already out there now, but real insofar as they pertain to [subjects] making sense of the world in terms of differences that make a difference to them. The paradoxical concatenation of fuzziness and clear distinctness, then, would be at the heart of the distinctional activity of [subjects] and, moreover, point to something fundamental as to what it is to be a [subject].

That being so, let us turn to examine how, through arbitrariness, bacteria overcome the fuzziness inherent to their concrete environment. Their distinctional activity is highly representative of the synthetic acts of all organisms as they self-regulate. The intelligibility of these synthetic acts, beyond cause-effect, must be in terms of event-response. Indeed, these acts is what makes them [subjects].

Chemotaxis

An elementary instance of distinguishing occurs in chemotaxis, the orientation or movement of organisms in response to certain chemicals in the environment. Somatic cells, some single-cell organisms, some multicellular organisms, and bacteria, display chemotaxis. Various models have been proposed to explain exactly how, through movement, these organisms maintain essential variables in viable boundaries by maximizing and minimizing their exposure to them—temperature, pH level, and so forth. We are not concerned here with the physiology of chemotaxis and its complex chemistry. We are interested in it insofar as it involves distinguishing. Chemotaxis is representative of organic process in general because any such processes involve the monitoring and response to variables. This, after all, is what is meant by self-regulation. Organisms thrive by maintaining the variables that affect them within viable boundaries. How do they do this? How do they maintain boundaries? Again, physics and chemistry answer indispensable *how* questions. But, given the synthetic nature of boundaries which result from the play [subject]/[object], our focus is on the *what*, the distinctional as the highest conjugate form whereby organisms set boundaries.

During chemotaxis, *E. coli* bacteria respond to chemical variation gradients in water such as pH or nutrients like sugar, by seeking what they like and keeping away from what they do not like. The do so by means of two complimentary kinds of movement: (a) clockwise rotating of flagella that result in swimming in a straight line or directed motion, also known as 'running,' and (b) brief periods of counterclockwise rotation of flagella that produce 'tumbling' and result in random new orientations. A bacterium compares current and past concentration of attractants or repellants and, depending on the result, either runs or tumbles. It runs when the concentration of some repellant is higher than viable, and it tumbles when it is not; the opposite is true with attractants. This general strategy of *adaptive gradient climbing* is so called because it is suitable to a wide range of concentrations of chemicals. Another tactic combines running and tumbling to perform a random walk until relevant concentrations are acceptably high, at which point the bacterium simply tumbles in-place.[27] The bacterium does this by making comparisons. How so?

[27] Egbert et al., "Metabolism-Based Chemotaxis."

> [During chemotaxis, the *E. coli's* flagella] . . . respond to a change in chemical gradient that are regulated by a two-component regulatory system. This two-component regulatory system is comprised of sensor proteins and transmitter proteins that signal to flagella to reverse rotation. In the absence of attractant, the system progresses, causing the flagella to rotate clockwise and the bacteria tumbles. In the presence of attractant, the two-component regulatory system is inhibited, causing the flagella to rotate counterclockwise. Thus, the bacterium swims in a straight line.[28]

As shown in Fig. 2.8, the motor system M makes flagella rotate in one direction or the other, resulting in either running (displacement) or tumbling (staying in place). The sensor system S detects outside conditions causing M either to continue to rotate as it is or to change direction of rotation. This is a two-component regulatory system: S controls M and M controls S. Using the musical notation for *repeat*, the system operates in an ongoing loop as follows:

> ||: Conditions detected remain the same? M continues to rotate as it is. S detects a difference that makes a difference? M reverses direction of rotation :||

The system oscillates between two states, running and tumbling. Concretely, in the case of *E. coli* bacteria, S is more complex because it detects one of four conditions—presence or absence of attractants and presence or absence of repellants. A chief function of S is to partially control two ongoing and simultaneous reactions that result in the production of transmitter proteins thus causing flagella to rotate in one direction or the other. One reaction stimulates and the other inhibits the production of transmitter proteins. According to what S detects, one of these two reactions predominates, causing proteins to form or not, thereby determining the motion of the flagella. The transmitter proteins that lock into flagella are quickly removed so that others must take their place. Thus, the detected environmental conditions determine the rate at which these proteins succeed each other, accelerating or slowing the rotation of flagella. Additionally, the lifetime of other proteins at work in the system is very short, increasing its sensitivity to environmental conditions.[29]

What, then, is the *what* of *E.coli's* distinctional activity? There are two fundamental elements. First, *E.coli's* distinctional activity takes the form of a two-element

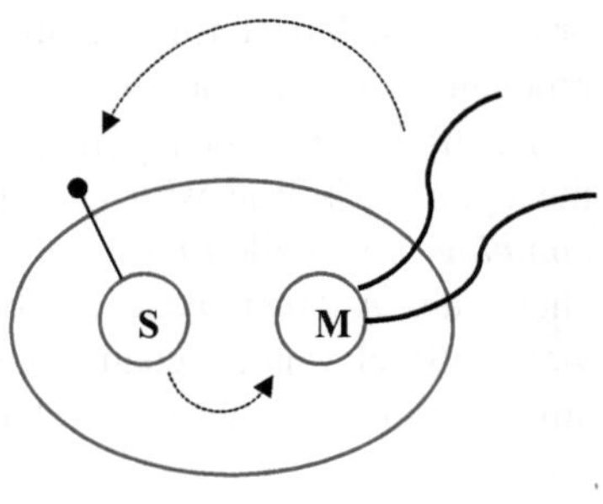

Fig. 2.8 Sensor and motor system for flagella rotation

[28] "Bacterial Chemotaxis in Plain English."

[29] For a clear animated info-graphic of this biochemical process, see https://www.youtube.com/watch?v=h4lv7cBYVug

control system. Such systems are ubiquitous in nature. Two-element control systems allow processes to start and stop. It is what makes processes and systems oscillate between binary states: on/off, start/stop, slow/fast, etc. I shall have much more to say about oscillation later. Of course, there is endless variety in *how* two-element control systems work, but this is the common underlying *what.*

The key second element in *E.coli's* distinctional activity is what I shall refer to as *neglect of concrete residue*. This is what makes A=A possible. The concentrations of attractant and repellants in the environment vary. Switching between running and tumbling occurs when differences make a difference. When differences are slight, they are disregarded, there is no switching, and the bacterium continues to either tumble or run. In distinctional terms, the differences that make a difference result in the switching/between A and Ā. On either side there is A=A or homogeneity—differences that do not make a difference. Given that, inevitably, there is some variety in the environment, A=A is maintained by *E.coli* by neglect of concrete residue. This is *how E.coli* goes about it: the rate of succession of transmitter proteins locking onto flagella and being removed remains stable. If the tipping point between tumbling and running is not reached, there is no switch. In either state, a multiplicity of concentrations, differences that do not make a difference, are regarded as 'the same', A=A. So, A=A is indispensable in two-component regulatory systems. It is how variation in the environment is neglected, and this makes possible steady states of A or Ā: sufficient glucose, of insufficient glucose; toxic or not toxic, dangerous pH or acceptable pH. Indeed, the same neglect makes /, the switching point, be 'the same'—the switching does not shift, it remains constant. In short, A=A too is inherent to process. A=A is what allows them to be binary: on/off, start/stop, slow/fast, etc.

Two-element control systems and neglect of concrete residue require each other. Both are indispensable to process, as such. And process is inherently distinctional activity. Both together—two-element control systems and neglect of concrete residue—as distinctional activity, are at work in all organisms, from the lowliest to the most complex since all involve process. Complex organisms are made up of billions of cells. Distinctional activity is carried out by each and every one of these cells; by the processes carried out by aggregates cells' and by the processes carried out by all functional units made up of aggregates of cells all the way of to the whole organism. This is why it was stated above that distinction-distinguishing obtains in physiology as much as it does in psychology. Therefore, botany and zoology are a single continuous-discontinuous field, and why a re-foundation of the notion of the biological follows from distinction-distinguishing.

If distinctional is A/Ā, A=A, the *what* by which organisms carry it out are two-component regulatory systems and neglect of concrete residue. Before continuing our examination of the distinctional, two points are worth mentioning.

First, I mentioned out the outset that we would consider three definitions of distinction in *LoF*. Borrowing terms from Buddhist philosophy, Spencer-Brown mentions a second definition of distinction: "called by Sakyamuni the links of conditioned coproduction, called by me [Spencer-Brown] the calculus of indications."[30] *Conditioned coproduction* is synonymous of two-component regulatory system, as well as *negative feedback loops* from systems theory.[31] Synonymous images are *oscillation* and *zigzagging*. Regardless how we refer to them, A=A is what makes them all possible in the first place. Otherwise, there would be an infinity of sates, rather than just two. The radical implication of this is that process is inherently related to living beings. As I shall argue later, even nonliving processes are such only because living beings construe them as such by A=A. So, just like abstract distinctions depend on the living subjects, so does process.

The second point is an invitation to consider just how alive living being is. Recall Bateson's suggestion that a difference which makes a difference is the elementary unit of information, followed by his suggestion that the word 'idea,' in its most elementary sense, is synonymous with 'difference', a unit of information, a sign in biosemiotics. While the distinctional activity of an *E. coli* bacteria is utterly non-conceptual, it is, in this sense, 'ideational' or informational or semiosic. We are at the most elementary level biologically and the most primitive level logically. The A=A or arbitrariness of homogeneity of organisms is what makes process and the distinctional possible. A=A or arbitrariness are quite simply beyond the horizon of the causality of physics and chemistry. Organisms inhabit a world of event-response superimposed on the world of cause-effect. In other words, they have agency. This is their fundamental nature as [subjects]. The living world as a whole is about differences that make a difference, and therefore ideational or semiosic. Regardless of the terms used to state it, this is what Bateson meant by the title *Steps to an Ecology of Mind*. The living world is a vast ecology of the mind-like.

Do Only Living Beings Distinguish?

If distinguishing pertains to minds, an obvious and fundamental objection must be acknowledged. Do only living beings distinguish? Are living beings truly set apart by distinguishing? The objection is important because it goes to the very heart of what is claimed in this treatise. If things other than living beings distinguish, what becomes of that claim? In what follows, I shall start by arguing as strongly as possible against this claim. I do so to find out, what, if any qualifications must be made, to maintain the fundamental claim that only living beings distinguish. This exercise will bring us to a deeper understanding of the nature of living being and its relation to other kinds of things.

Take an automatic household heating system—does it not distinguish? The system is automatic precisely because it turns itself on and off. Moreover, just like an

[30] Spencer-Brown, *Laws of Form*, viii. Above it was pointed out that it could just as well have been named "the calculus of distinctions."

[31] See, for instance, Macy, *Mutual Causality in Buddhism*.

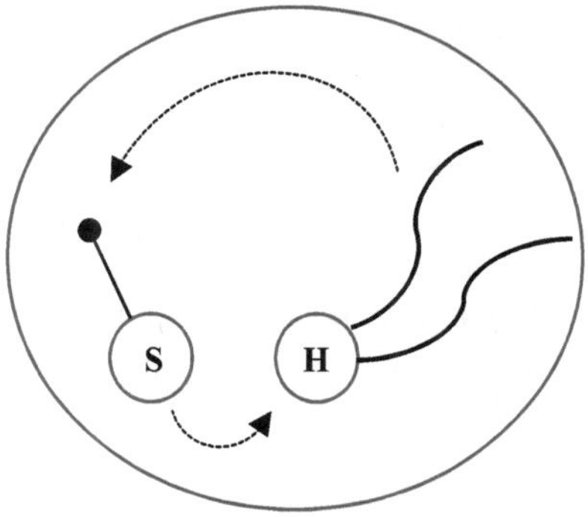

Fig. 2.9 Automatic heating system

E. coli bacterium, it is a simple mechanical two-component regulating system comprised of a heater and a thermostat in an enclosed environment. The thermostat is used to set the temperature, and based on this setting, the system turns the heater on and off. Say the thermostat is set at 72 °F. Then anything above this point is *hot* and anything below it is *cold*. Below 72 °F the system switches on and above 72 °F it switches off. So, the room temperature hovers around 72 °F. Is this not to distinguish between hot and cold, just as the bacterium distinguishes between high pH and low pH? In fact, a figure for how the heating system operates is almost identical to that of chemotaxis (Fig. 2.9).

Aside from switching the name of the active component from M for motor to H for heater, the only other difference is the placement of the outer boundary. Whereas bacteria move through their environment, a heating system alters the environment it is within. If nothing else, this points to the versatility of two-component regulating systems. The thermostat's operation is identical to *E.coli's*:

> **||:** Conditions detected remain the same? H continues to perform as it is. S detects a difference that makes a difference? H switches to its alternative state**:||**

Is there, then, no difference between the distinguishing in chemotaxis and that of an automatic heating system? What is the difference? Let us take a step back to see what is at stake here. First, if there is no difference between the distinguishing of an organism and a mechanical device, then (a) distinguishing cannot be the cardinal feature that sets living beings apart from other kinds of things. Therefore, (b) inquiry into the distinguishing of living beings is, at least at this very basic level, no different from inquiry into mechanical devices. However, the opposite may also be true: (c) the similarity between the distinguishing of living beings and mechanical devices calls for a clarification of the difference between them, which (d) would explain why, despite this difference, this similarity has led to accounts of living beings in mechanical terms. This second possibility calls for a differentiation that shows why the similarity obtains, yet also why, more generally, it is erroneous to account for living distinguishing in mechanical terms. This is a tall order. But it amounts to a more precise statement of my general thesis: if organisms are understood in physical and chemical terms, then the highest conjugate form—their aliveness—is overlooked. If they can so be understood, then there is reason to doubt the validity and value of such a higher conjugate form.

First consider a constraint that both systems have in common. As a mechanical device, the automatic heating system invites reflection on precision. This points to a fundamental insight about the nature of concreteness. A thermostat is a rudimentary thermometer; when the temperature is above the set point it switches off, when below it, it switches on. If a thermostat were excessively sensitive, no sooner had it switched to off, it would switch to on again, and then back to off, and so on. This problem is real, and why *hysteresis* is deliberately built into thermostats, a buffer to protect a system from excessive switching. Hysteresis slows down the feedback loop, the oscillating or zigzagging that is the hallmark of two-component regulatory systems.

In fact, recalling the previous section, by making the system less sensitive, the purpose of hysteresis is to enlarge the amount of concrete residue it neglects. Hypothetically, ever more precise measurements could be made of the actual temperatures at which the switching on and off occur. The differences detected by a very sensitive thermometer would produce a zigzagging line on a graph. The straight line that one could draw along the zigzag would represent the temperature at which the thermostat is set. Given what the system is *for*—in this case, people in a room who are not sensitive to very slight variations in the temperature—these differences do not make a difference. The system is functioning properly when people in the room experience the temperature as even, and not fluctuating between hot and cold. So, the system operates by *neglecting a concrete residue* that is negligible to its users. Again, this concrete residue is defined as differences that do not make a difference.

The foregoing brings us to the fundamental similarity and a fundamental difference between a living system and a mechanical one. In both systems a two-component regulatory system is at work that neglects concrete residue. So, there is an oscillation, a process, a scheme of recurrence. Insofar as it is a distinction, this scheme of recurrence establishes the two states A and $\bar{A}$. The key question, then, is this: for whom are these two states and the boundary between them, a distinction? This is the very point that was raised in relation to a circle drawn on a flat plain. As was shown, defining distinction as perfect closure makes it appear as if a distinction is real in the sense of a 'body' already out there now. The same applies to the scheme of recurrence between two states. A recurrence is not real as out there now, independently of a subject. It takes a subject for a recurrence to be a recurrence. This will be discussed at length in the section that follows.

There is no question that there is a striking similarity between the scheme of recurrence of both systems. And that this seems to pose a serious challenge to the proposition that only living beings are distinctional. The similarity forces us to delve more deeply into the distinctional and recalls Kant's remark that philosophy presents tentative definitions to experiment with them to arrive at more precise definitions. The notion of the distinctional is too valuable to concede so easily. For one thing, it is humans who have designed products that, at least on the face of it, appear to distinguish. Indeed, the example of the thermostat is fruitful because it poses a challenge that calls for further avenues of exploration.

So, what are the differences between the *E.coli* bacterium and the automatic heating system? Let us consider the question of who the distinction is *for*. What is an *E. coli* bacterium *for*? In the case of an *E. coli* bacterium, differences that make a difference and differences that do not (diff-diff) are *for itself*. Not so in the case of the automatic heating system. These diff-diff are for living beings, humans who set up the system or, say, plants in the case of a greenhouse. In this strict sense, the heating system is an extension of humans and may be thought of as a prosthetic. That being so, its capacity to regulate the temperature automatically, pertains to living being—specifically, the human beings that produced it. Certainly, the switching on and off plays no part in regulating itself. Rather, this is all to do with the self-regulation of humans. As a prosthetic of humans, the system is just another *how* whose proper intelligibility, like other *hows*, requires the highest conjugate form that pertains to living being. To put it plainly, the *what* of the automatic heating system is only fully intelligible as a human product. And this is a second key difference. Taken together, the first and second difference may be stated as follows. In the case of the *E.coli* bacterium, the scheme of recurrence or oscillation between the two states A and $\bar{A}$ is *for* itself. It plays a key part in its self-regulation. One may examine the *how*, but the *what* that makes it fully intelligible is the distinctional activity of the bacterium. This full intelligibility is independent of human knowing. *E.coli* carries out these processes of self-regulation regardless of human knowing. Thus, while the full intelligibility of the *E.coli* bacterium requires humans epistemologically, it has nothing to do with them ontologically.

In the case of the automatic heating system, the scheme of recurrence or oscillation between the two states A and $\bar{A}$ is *not* for itself and plays no part in its self-regulation. The account of its functioning is in fact an account of the *how*. The *what* that makes it fully intelligible is human distinctional activity. Without that activity, it is not fully intelligible. Its full intelligibility calls on humans to reflect on their process of knowing, given that this knowing is intrinsic to how automatic heating systems work. Moreover, in stark contrast to bacteria, humans are indispensable to how they came to be in the first place. So, the full intelligibility of the automatic heating system fully requires humans epistemologically as well as ontologically. The foregoing may be summarized simply: *E.coli* is a living being and an automatic heating system is a product. *E.coli* bacteria distinguish regardless of humans, even if it takes a human to recognize this. If it is tempting to say that automatic heating systems distinguish, it is only because they are products designed, produced, and maintained by humans to do so. Moreover, if their oscillating processes are construed as distinguishing, this is not done by the system itself. It is indispensable that humans to construe them as such.

With this we can return to the options raised above and decide between them. There is a difference between living distinguishing and the oscillation of a mechanical device. So, (a) distinguishing can be the cardinal feature that sets off living beings from other kinds of things, and, therefore, (b) inquiry into the distinguishing of living beings is different from inquiry into mechanical devices. Therefore, (c) the similarity between the distinguishing of living beings and mechanical devices has been clarified. This is so even if, despite the difference between them, (d) the similarity between them has led to accounts of living beings in mechanical terms.

In fact, there is a much deeper difference between the distinguishing activity of *E.coli* and an automatic heating system. We shall soon see that, at the highest level of intelligibility, distinguishing and distinctions—the distinctional—is inherently anticipative. This is inherent to the *what* and not the *how* of both systems. Like any organism, the bacterium's self-regulating is anticipative. The automatic heating system is anticipative only insofar as it is serves human self-regulating—which, like any other organism, is anticipative. Humans are anticipative in the production, management, and maintenance of automatic heating systems. These systems in themselves are not anticipative. An organism's anticipativeness is the outcome of its own previous anticipations and those of its ancestors. This lineage of anticipativeness is itself part of the highest-level conjugate form by which living beings are made intelligible. The lineage of the distinguishing of the thermostat is, through and through, within the human lineage of anticipativeness. Paradoxically, at one level this anticipativeness sets organisms radically apart from any product, yet it is this anticipativeness that brought products into existence and will keep doing so. That anticipativeness is not in our horizon when we compare living beings to products, points to our lack of understanding of living being. It is only this lack of understanding that has allowed us to compare them to mechanical products.

To conclude, yes, only living beings distinguish. They do so by means of various *hows*, and these may involve products, including, as we shall see, tools and automatic devices. One can, of course, wonder whether a product might eventually be produced that must be deemed alive, given its distinctional capacity in the proper sense. This is a fascinating but separate question. However, reflection on it requires clarity on the notion of the distinctional, and how, based on it, one may differentiate between living beings and other kinds of things. We now have initial glimpses into the radical difference between living being and products. As we shall see, there is an equally radical difference between products and the nonliving. Products produced by a living being are nothing like the Sun, the Moon, or the Milky Way. The point is crucial because products and the nonliving are often casually thrown together as inanimate objects. In a physics class, a piece of chalk can stand in for an apple or for a celestial body when discussing gravity and parabolic curves. But, in our topic, the highest conjugate form is essential. This is why it is indispensable to keep well in mind the difference between different kinds of things. What is to be understood by *things*, and the kinds of things there are, is discussed at length in Chap. 7. In addition to the nonliving, living beings, and products, there is a fourth: byproducts. These are also a kind of thing, namely, the outcomes of living process. These things are not alive—but neither are they nonliving nor products—though they, too, obviously fall under the domain of the living. This is the reason why I cite the Sun, the Moon, and the Milky Way as instances of the nonliving. We are so fully immersed in a world of products and by-products that we are hardly ever in touch with the nonliving.

Products pertain, at least partially, to the domain of living being. This is why there are at once key similarities and differences between these two kinds of things. It is the reason why accounts for living distinguishing in mechanical terms are at once likely and inherently flawed. The foregoing discussion suggests a more precise reading of my thesis, and an extension of it. Organisms cannot be understood merely in physical and chemical terms because this overlooks the highest conjugate form of living beings. The same is true of products—an account of them is not possible in mere physical and chemical terms. While they share in the highest conjugate form, still living beings and their products are radically different. What there can be no doubt about is that—recalling Bateson—our ecology of mind includes products.

This close discussion of automatic heating systems has laid open several topics indispensable to understand aliveness. These will be taken up in sections and chapters that follow. For one thing, it has underscored the importance of differentiating between kinds of things and pointed to the anticipativeness inherent to living distinctional activity. Also, perhaps not as obviously, it has shown that reference to the *how* and the *what* of systems brings up the issue of systems as wholes, as well as their organization. We shall later see that the distinctional grows in layers, is inherently recursive, and that this leads to paradox.

We are witnesses to singular moment in time. Add together cutting-edge discoveries in genetic engineering, AI and robotics, and we realize that planet Earth is undergoing its most radical revolution since the advent of life. We are entering an inflection point that will increasingly blur the differentiation between living being and products. Products such as automatic heating systems are early precursors of this revolution and this blurring. The topic of this exposition is the distinguishing of living beings. That is, the previous most significant ontological inflection point of planet Earth, when life emerged. My aim is to show just how radical this revolution was, and that an elegant account of it can be provided in terms of the emergence of distinguishing distinctions.

Establishing Bet Structure – ב

And it was evening, and it was morning, first day. …
…and it was evening, and it was morning, second day. …
And it was evening, and it was morning, third day. …
And it was evening, and it was morning, fourth day. …
And it was evening, and it was morning, fifth day. …
And it was evening, and it was morning, the sixth day. …
And God completed on the seventh day the work He had done.

– Genesis[32]

The distinctional activity of *E. coli* clearly does not fit the definition of distinction as perfect continence. Far from it. A bacterium's distinctions are in no sense like a

[32] Alter, *Genesis,* 3–7.

'body' already out there now. Rather, their distinguishing is a scheme of recurrence, an oscillation, founded on repetition. As has been said, only arbitrary neglect of concrete residue makes repetition possible, else there would be no such thing as 'the same.' This is A=A. We therefore arrive at a more precise definition of distinction: *a scheme of recurrence that establishes A and $\bar{A}$ through arbitrary neglect of concrete residue for self-regulation.*

How is A=A arrived at? How does repetition, with its inherent arbitrariness, come about? I shall now show why, inherent to A=A is anticipation, the root of the anticipativeness that radically differentiates the distinctional activity of organisms.

Consider an example from day-to-day life. Some of us are privileged enough to shower daily. We follow a routine—undress, turn on the hot tap, turn on the cold tap, adjust the temperature of the water, step in, get soaked, rinse our hair, hum a little, shampoo our hair and soap our body, rinse, turn off the taps, grab a towel, get dry. There is a curious timeless quality about routines that blend multiple single occurrences into one. Unless something quite extraordinary took place, we do not recall individual instances of such events. Rather, they are recalled as one. This is because each event and each of its components is linked to its counterpart 'across' time, so to speak, somewhat like the successive images glimpsed between parallel mirrors. Adjusting the temperature today is the same as it was adjusting it yesterday and as it will be to adjust it tomorrow; washing is the same today, yesterday, and tomorrow as is soaping, rinsing, drying, and so on. Clearly, there is a sameness/difference or A=A, A=$\bar{A}$ between them. It can feel like crossing a portal to find oneself back in a place long known.

In fact, there is a double relation between events and their components. This is the condition for their sameness and difference. Let this double relation be called *establishing bet structure*, and let it be represented by the Hebrew letter ב ("B"), a helpful icon (Fig. 2.10).

In general, and in the most abstract terms, any scheme of recurrence is constituted by the interrelation between two sets of sequences, a *horizontal* and a *vertical* sequence. Bet structure may be formally defined as *the relation between the vertical and the horizontal sequences inherent to schemes of recurrence.* Given a horizontal sequence a, b, c, d . . ., recurrence is expressed by the sequences a_1, a_2, a_3 . . .; b_1, b_2, b_3 . . .; c_1, c_2, c_3 . . .; and so on. So, *to establish a bet structure is to establish the relation between the two sequences.*[33] When having a shower, for instance, the horizontal sequence is the one enumerated above—undressing, turning on hot tap, turning on cold tap, stepping into shower, soaping, rinsing, and so forth. There are several vertical sequences—undressing$_1$, undressing$_2$, undressing$_3$, and so on; turning-on-hot-tap$_1$, turning-on-hot-tap$_2$, turning-on-hot-tap$_3$, and so on. This

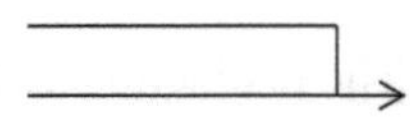

Fig. 2.10 Establishing bet structure or ב

[33] See Mayer, "From Distinction to Anticipation."

vertical relation is so for any of the events, or parts thereof, in the horizontal sequence. The vertical relation obtains for all schemes of recurrence.

Because this relation is somewhat arbitrary, it takes living beings to establish it. And it involves anticipation. Literally, it involves treating a_2 as a repetition of a_1, a_3 as a repetition of a_1 and a_2, and so on. The same is true for b_1, b_2, b_3 . . .; c_1, c_2, c_3 . . .; and so on. Organisms do not make up these repetitions from thin air. They pertain to cycles in their environment. They are designed to respond successfully to these cycles. These designs result from having established bet structure in the past and, so long as the cycles continue, they will serve to do so in the future. Thus, ב is inherently anticipative because the very physical design of organisms treats future events as past ones. It takes a living being to synthesize the relations $a_{1\rightarrow n}$, $b_{1\rightarrow n}$, $c_{1\rightarrow n}$, and so on, or the sets {a}, {b}, {c}, and so on. Concretely, it is by ב that the synthesis of subject and object yields distinctions.

Earth's rotation is experienced by us as the horizontal sequence *8 am, 9 am, 10 am*, and so on. By ב, this yields the vertical relations *Monday 8 am, Tuesday 8 am, Wednesday 8 am, . . .; Monday 9 am, Tuesday 9 am, Wednesday 9 am . . .; and so on.* Because schemes of recurrence are bet-structured they may be expressed in matrix form (Fig. 2.11).

This matrix represents what occurs on the side of the subject. In any such matrix any number of bet structures may be established. As the matrix shows, in any given scheme of recurrence an event can be said to recur if, and only if, it fulfills two conditions: the event follows an event different from itself—for instance, in the horizontal sequence, 9 am follows 8 am—and it iterates a previous instance of the same event in the vertical sequence—Monday 9 am, Tuesday 9 am, and so on. This is only possible because events have duration. Thus, a given event either continues to be itself or not. To recur, an event must first come to an end. But the beginning and ending of events takes place on the side of the subject, not on the side of the object.

The sun's trajectory across the firmament is a continuum that may be parsed by subjects into discrete events. Even the sunrise and the sunset are continuous events whose beginning, and ending are determined by a subject. Of course, the Earth's rotation does not owe its own existence to the living subjects that experiences it. Still, though ontologically independent of living beings, it is adverted to as a cycle or a scheme of recurrence through the synthetic acts of living subjects. The segregation of continuums into discrete sections, such as day-night-day-night is, necessarily arbitrary to some degree. Because it makes a difference to them, both the owl and the mouse know it is dawn. It is subjects who group multiple such events into a single class of event, thus rendering individual occurrences as instances. This is only so because that which takes place on the side of the object is experienced as

Fig. 2.11 Bet structure matrix

Mon	8am	9am	10am
Tue	8am	9am	10am
Wed	8am	9am	10am

Fig. 2.12 Establishing bet structure (ב) spiral

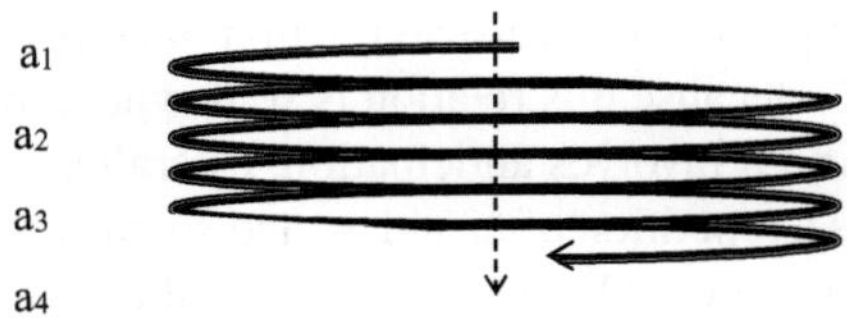

cycling. Indeed, it shall be argued later that this cycling is the fundamental environmental condition that allowed the emergence of living being in the first place.

It is helpful to have names to differentiate between what takes place on the side of the object and how subjects experience it. So, I shall refer to *cycling* on the side of the object and *spirallic* on the side of the subject. Living beings construct their experience of cycling spiralically. A spiral provides an image for ongoing ב, which is the unique character of the living mode of being. The image shows the horizontal/vertical relation between sequences whereby A's in the analogous positions are at once the same and not the same (A/Ā, A=A) (Fig. 2.12).

So, in lieu of *ongoing* ב one might simply say *spirallic*.[34] Lived experience, or aliveness, cannot but be spirallic in a transparent rotating universe. This is how we construct daily experience, such as a daily shower.

A corollary follows from this. If the nature of experience is to establish bet structure, then experience *cannot not* be bet structured. Even if we choose to name it so, as I am now doing, we cannot experience the world's cycling; we can only have our own spirallic experience of this cycling. Obvious as this is, these two are not so easily differentiated. Our spirallic experience points to our nature as living beings while the non-bet-structured cycling of the world points to the nature of the nonliving. For instance, making sense of the world involves some version of duration and extension, no matter how elementary. But these two attributes are spirallic; they pertain to distinguishing and, therefore, to our anticipative nature. Lived experience is always already anticipative. This is not so of the nonliving. Again, living beings *cannot not* experience anticipatively. We cannot not experience the nonliving anticipatively, even as anticipation is radically alien to it. We are spirallic, and the nonliving is not. Yet, for the most part, we elide this and then elide this elision. It turns out, then, that the shoe is on the other foot. Whereas primatologists must guard against the fallacy of anthropomorphism, hard empirical scientists appear to be oblivious to their own biomorphism—the attribution of life-like attributes to the nonliving, a biomorphic fallacy, and a conflation between kinds of things.

We intuitively grasp the spirallic. This is beautifully captured in this fragment by the great Cuban poet José Lezama Lima:

Crea el ser su caracol
y va a la tierra ligera a su canción,
desnuda por la espalda del caracol

[34] A spiral is usually thought of in 2D as a circling with the radius growing at a regular rate. A spiral in 3D is a circling in the x-y plane that moves uniformly in the z direction. The radius of a 3D spiral may or not be constant. The constant radius in Fig. 2.12 produces a helix, whereas a growing radius would produce a cone growing in the z direction.

se muestra ondulante por el agua y sus ramajes.
El ser nace y su nacimiento cumple la mirada,
sus vapores con agua se deshacen,
su dureza se cierra con su aurora.[35]

The bet structure matrix shown above is a calendar, no less, the tool by which we manage and anticipate our experience on Earth as it rotates on its axis and orbits around the Sun. Our reliance on calendars and clocks affirms this spirallic nature—yesterday's 9 o'clock and today's 9 o'clock and tomorrow's 9 o'clock are at once the same and not the same, as are last spring, this spring, and next spring. In passing, it is striking that, of all the cycles we live by—the year, the seasons, the lunar month, the day—the week is not one of them. And yet, in contrast with these natural cycles, it is conceivable that, from outer space, the change of pace at the weekend would be the most visible marker of the living cycles on the planet.

We are spirallic. Recurrences, cycles—these are inherently spirallic. Phenomena, to use Kant's term, is spirallic, because we cannot help but project our living nature onto our experience, even when it pertains to the nonliving. This presents a crucial difficulty. To understand living being is to understand how it differentiates from the nonliving and how this radical differentiation can occur.

So ב makes living being—the living mode of being—inherently anticipative. Schemes of recurrence are constructed by [subject], most often as an oscillation between two states. The segregation of a continuum into events constitutes a distinction accomplished by delimiting a portion of that continuum. As a synthesis of subject and object, distinction itself may be defined as *the concatenation of continuity-discontinuity* or simply *continuity-discontinuity*. This and "scheme of recurrence" highlight different aspects of distinction. It is not a tautology to say that a scheme of recurrence requires continuity-discontinuity, and that continuity-discontinuity requires a scheme of recurrence. Neither is possible without the A/Ā, A=A of living beings. Both explicate an aspect of A/Ā, A=A and thereby of living being.

In conclusion, the condition for the possibility for A=A, and therefore for continuums to be split, the possibility of sameness, recurrence, repetition, twoness, is ב—establishing bet structure. At the most primary level—in both the mathematical and the biological sense—without ב there is no such thing as twoness. Moreover, and for this very reason, twoness is foundational for both the mathematical and the biological. Likewise, though perhaps it is not as obvious, oneness too calls for A/Ā, A=A. Only by ב can a continuum, limitless in principle, or a multiplicity of numerically different elements endless in principle, be taken as one. This is so because oneness itself calls for a distinction, the distinction between all/nothing. That is, the distinction *all there is/ nothing beyond*. In short, ב is at once the foundation of the biological and thereby of the mathematical. The mathematical is born with aliveness. This topic is taken up in detail in Chaps. 5 and 6.

[35] Lima, "Danza de la jerigonza," *Refutación*, 90–98. [Being creates its snail / and the earth goes lightly to its song, naked on the back of the snail / it shows itself undulating by the water and its branches. Being is born and its birth fulfills the look / its vapors of water fall apart, its hardness is closed by its dawn. (Trans. Daniel Mayer)]

Forness is part and parcel of A/Ā, A=A, since it is only for a [subject] that differences make a difference. To define this term, consider one of Peirce's definitions of a sign: "A sign, or *representamen*, is something which stands to somebody *for* something in some respect or capacity."[36] This definition obtains for all [subjects]. There are no signs without distinction, the more so because distinction involves anticipation. Thus, distinction is the foundation of signification. If distinctions are differences that make a difference *for* a [subject], then forness may also be defined as *the relation between a living being and the differences that make a difference to it.*

When examining the nature of living being, the question of life's origin must inevitably come up. To this, which is an ontological question, corresponds an epistemological one. At what point, during the process of life's emergence, must the highest conjugate form come into play? The problem cannot but arise since the early stages of biogenesis pertain to the physical and chemical. But biogenesis is, precisely, the emergence of the spirallic, the *what* that goes beyond the physical and the chemical. In inquiry into biogenesis, it is indispensable to guard against a false attribution of the spirallic to the nonliving since the very point of this inquiry is the emergence of the spirallic. An example of this false attribution at the point where it matters most is provided in the following chapter.

The overall conclusion of this section is as simple as it is radical. The notion of ב is foundational and the intelligibility of virtually all issues pertaining to living being rest on in.

Distinguishing as Scheme of Recurrence

It would seem to be intuitive that grasping the *what* of aliveness cannot be circumscribed to arcane physics or chemistry nor subtle metaphysics. The higher conjugate form of aliveness must be deeply familiar to us since we are ourselves this mode of being. And it is. We are at every moment distinctions that distinguish. We embody layers of distinctions profusely distinguishing on many levels. Thus, though it is so obvious that it goes without saying, we can access intimate knowledge of the distinctional in our inquiry into aliveness.

Take the definition of distinction as perfect continence. Flawed as it has been shown to be, it is a notion not to be lightly dismissed, if only because we take it for granted. It is intuitive and commonsensical. In fact, it points to something fundamental—there are levels of distinction. Automatic systems and bacteria distinguish at the most primary level, as an oscillation between two states. This binary switching takes place along the single dimension of time: on, off, on, off. Perfect continence requires three dimensions, a two-dimensional territory and time. The Great Wall of China is a line segment in time, a two-dimensional distinction that is more than an oscillation but less than a territory. So, there are levels of distinction and the possibilities of its unfolding. This is how the single form of the distinctional works in evolution, beginning at the origin of life and unfolding with the growing

[36] Peirce, "The Logic of Signs," 122 (emphasis added).

complexity of organisms. In short, inherent to the distinctional is its unfolding. And, to understand it, and thereby aliveness, we can examine the deeply familiar to us.

Imagine a dreadful scenario. The unfortunate victim of a kidnapping, blindfolded and with hands tied behind his back, is released by being callously thrown off a car. Injured, afraid, and deeply shocked, but greatly relieved at finding himself alive and free at last, he immediately struggles to his feet, only to realize he is surrounded by the sounds of heavy traffic. No sooner does he step forward there is loud honking on his left side. Swiftly taking a step back, he hears more honking, now on his right. But when he steps forward again, the honking ceases. Collecting himself, he makes a 90° turn and cautiously steps into what he guesses is a space parallel to traffic on either side. From the lack of honking, our subject infers he is walking in the middle of a very busy two-way road. Blindfolded as he is, he walks on, with the occasional honking alerting him that he is veering onto the road on either side. This binary honking/no-honking is the only data available to him.

Compare this to a bacterium. It is unlikely that from its discrete probes into the environment it infers continuums and even less likely that it constructs a map. But this is exactly what our subject does. From the recurrence of honking and no-honking he begins to build continua such as directions. Indeed, begins to anticipate them. This and that direction are dangerous because they take me into the flow of traffic; but this direction and its opposite are safe because they are parallel to the traffic. Assembling these anticipations, he begins to build a map. In Chap. 5, I will show that *LoF* formally demonstrated that such nesting of distinctions leads to the construction of space and time. In this example, the directions forward/backward and right/left are two-dimensional linear distinctions because they are binary options of movement that remain steady through time. To start from honking/no honking and build a sense of directions, and then from these linear distinctions to assemble a map, is the capacity to nest sets of distinctions, anticipations, inferences.

Can distinction as perfect continence be derived from such nesting? Yes, it can. Perfect continence is a circle. Again, *LoF* defines it like this:

> *Distinction is perfect continence.* That is to say, a distinction is drawn by arranging a boundary with separate sides so that a point on one side cannot reach the other side without crossing the boundary. For example, in a plane space a circle draws a distinction.[37]

Having derived directions, our unfortunate victim can now follow paths. He may follow one that leads back to where he began. He discovers himself "going around in circles". It is disturbing because he could walk along this path infinitely in either direction and recurrently arrive back where he began. It so happens that such a path forms the boundary between two territories, an inside and an outside one, A and Ā. Both are perfectly contained because it is not possible to get from one to the other without crossing this path.

Thus, the nesting of linear distinctions yields distinction as closure. Of course, this implies the binary distinction path-whose-ends-meet/path-whose-ends-do-not-meet.

At this point it is worth noting that linear distinctions were left out of the account of the automatic heating system. It, too, nests distinctions. During the summer the

[37] Spencer-Brown, *Laws of Form*, 1.

heating system is unplugged. Regardless, the thermostat still crosses the boundary at 72 °F, opening and closing the circuit that stops and starts the heater. But, being unplugged, nothing happens. Plugging in the heating system is to cross a linear distinction—the flow of electricity is a continuum. The system is now on, which is why it can turn on and off automatically. This plugging in into *on* is the underlying linear distinction that allows the thermostat to operate its binary on/off.

Linear distinctions occur throughout our lives. A traffic light switches from red to green, a basic oscillation. But then, green light is a continuum of time, and so is the red. Amber is too. It is required because driving forward at speed is a continuum, just as standing still is. The transition from moving at speed to standing still requires another continuum—slowing down, what the amber light is for.

These are all linear distinctions. They apply identically to filling a jug of water. Like with the heating system, the tap may be turned on and off regardless of whether there is running water. If there is no water, it is like the heater being unplugged. If there is, one fills a jug and asks oneself the questions: how much longer must I keep the tap open? When do I slow down the pace of the flowing water? Putting a baby to sleep by rocking and singing lullabies brings up identical distinctions: how much longer shall I continue? Shall I sing more quietly for a bit? All instances have one thing in common; there are nested layers of on and off. The same is true of the unfortunate victim as he walks blindfolded listening for honks and sounds of vehicles, continuums that start and stop.

Linear distinctions are the heart of processes, all continua. This fits perfectly with the definition *distinguishing is a scheme of recurrence.* Linear distinctions allow us to grasp the nesting of distinctions, which easily transposes into the nesting of processes. So, *schemes of recurrence* is synonymous with *continuity-discontinuity*. Nested continuity-discontinuity may stand as the nesting of processes. Nestings build up. So, we can dig down to their foundation as we have done, to the conditions for the possibility of schemes of recurrence.

The distinctional is palpable throughout our daily lives. And it is but a small leap from our own lived experience to the nature of experience as such, to aliveness as such.

Nesting of Distinctions

In principle, there is no limit to how many layers of distinctions may be nested. If distinction, as perfect continence, is a circle on a flat plain, further binary distinctions lead to a sphere in space and time.[38] If our horror story is to have a happy ending, the victim must find his way off the road. To do so he adds distinctions that further map his environment. These may include fast-traffic/slow-traffic, pedestrians-present/pedestrians-absent, intersections/no-intersections, heavy-traffic/light-traffic, sound-other-than-traffic/no-sound-other-than-traffic. Attempting to draw the attention of drivers, he may add vehicle-slowing/vehicle-not-slowing,

[38] A sphere that remains in time is a four-dimensional distinction.

different-kind-of-honking/no-different-kind-of-honking, and so forth. So long as these are independent variables, they constitute dimensions of the environment. Being independent from each other, they are orthogonal, like the dimensions of space-time.

The derivation of complex distinctions from elementary is inherent to the synthesis of subject and object. Later we shall see why anticipation begets anticipation, which brings with it differentiations of new independent variables. In simple terms, once differences make a difference, differences within those may also be detected that make a further difference. This latter specific difference is independent from the original one. So, distinctions are made within distinctions. Thus, relating to a more differentiated environment, organisms make themselves ever more complex. This is simultaneously true at various levels of unfolding—evolutionary, developmental, historical, and biographical.

This unfolding raises various questions about the nesting of distinctions. For instance, the number and manner by which nestings of distinction as oscillation results in linear distinctions and onto closure in a plane, closure in a three-dimensional space, and so on. Similarly, it raises questions about memory function. What is the role of memory in the addition and combination of distinctions? As we saw with our unfortunate victim, data points of elementary distinctions nest as they are combined and grouped into sets. Orthogonality is central to such groupings and nestings. Since the value of distinctions is verified by repetition, once established they tend to remain. Distinctions determined by independent variables are, in the long run, more robust and therefore more long-lasting. So, further distinctions may be built upon them. Grouping of distinctions along independent variables of the environment call for embodiments capable of organizing them. Thus, [subject] and [object] play upon each other to construct *Umwelt*—keeping in mind that the environment is primarily made up of other organisms, their products, and byproducts. Underlying the play [subject]/[object] is, therefore, largely the play [subject]/[subject].

Distinguishing, or ב, then, is the root of the evolutionary-ecological unfolding of [subject]/[object]. If distinguishing is the foundation of mind, by placing at its foundation the highest conjugate form, biology becomes the disciplined inquiry into mind and its unfolding at all levels. The play [subject]/[subject] is relationships. The distinctional, then, is the seed that eventually flowers into joinings—a topic we shall take up later.

Repetition and Monkeys

Evidence for ב comes in the form of 'monkey explanations', a term I coined at the Monkey Sanctuary. When the monkeys were startled or frightened by some unfamiliar noise, say a loud clang or an unusually shrill drilling sound, the best way to end their anxious *yuk-yuking* was by repeating the offensive action in full view of them. Of course, they scowled but the mystery was removed, and calm was restored. As it turns out, such explanations satisfy other animals too. I have given many similar explanations to our dogs and cats. Living beings are at home in routines. Repetition is the foundation of sense making.

Chapter 3
ב and the Emergence of Living Being

The First ב

The origin of life is one of the great outstanding questions in biology. If living beings are distinguishing distinctions, how could they have emerged in the first place? This chapter is a test case. Biogenesis is discussed to assess the usefulness of the notion of the distinctional. In terms of this highest conjugate form, the question about life's origins may be stated, How was bet structure first established? What allowed for this possibility? This inquiry, then, is not only biological but also metaphysical. We seek to rigorously differentiate, ontologically and epistemologically, the threshold between the nonliving and aliveness. This is the leap from cause/effect to event/response with the emergence of concrete residue arbitrarily neglected. It is the advent of agency and the Big Bang of anticipation. How did the spirallic come about? Does the distinctional help make current research into biogenesis intelligible? Is there empirical evidence for ב? Are there benefits to conceiving organisms as distinguishing distinctions? It will also be shown why inquiry into biogenesis calls for caution with our own biomorphism, to avoid projecting our own spirallic nature onto the nonliving.

How did spirallic being come about? The Earth spins on its axis and orbits around the Sun. This is expressed in our clocks and calendars. The underlying scheme, our solar system, has recurred about 4.6 billion years. Though over this time the day-night oscillation has gradually slowed, this set of rotations is the dominant feature of our environment.[1] It appears that life owes its origin to this regular cycling and to

[1] Coffey, "How Old is the Universe?" "Hints of inconsistent Earthly timekeeping come through natural calendars preserved in fossils. Corals, for example, go through daily and seasonal growing cycles that form bands akin to growth rings in trees; counting them shows how many days passed in a year. In the early Carboniferous period, some 350 million years ago, an Earth year was around 385 days, ancient corals indicate, meaning not that it took longer for the planet to revolve around the sun, but that a day–night cycle was less than 23 hours long." (Adam Hadhazy, "Fact or Fiction:

© The Author(s), under exclusive license to Springer Nature Switzerland AG 2023

D. C. Mayer-Foulkes, *The Nature of Living Being*, Biosemiotics 26,
https://doi.org/10.1007/978-3-031-24789-7_3

conditions largely traceable to it. I call them cycles with the caveat that it is our own spirallic nature that renders them such.

The Sun is the primary source of energy on Earth. Its impact is palpable in most schemes of recurrence that life depends on. With notable exceptions—such as the ocean tides, which also depend on the Moon—our environment is governed by the Sun. This involves both obvious and subtle variations, such as the shifting durations and angles of sunrays hitting the Earth. These cycles result in the planet's warming and cooling, which in turn cause fluctuations in pressure and temperature that originate ocean and wind currents, rain, humidity, salinity, and so on. All these cycles together, but primarily the day-night cycle, determine the nature of living being. Just as the ocean intermittently bathes the inhabitants of the inter-tidal zone on the shore, so light and dark endlessly wash over all life on earth. A crucial point must be noted, however. Though living being copes with *ever changing* conditions, it is primarily designed around *ever recurring* conditions. The very notion of *environment* may be defined as *a recurring set of conditions*. The overarching and ubiquitous feature of the natural world on a spinning planet of a spinning solar system in a spinning galaxy in a spinning universe, is this underlying condition that, with the emergence of anticipation, is experienced as recurrence.

Thus, given cycling, one may ask, What kind of organization of chemical compounds is required to establish bet structure? ב is to construe events as members of a set, as tokens of a type. This is repetition, and what calls for the arbitrariness inherent to forness. With ב, concrete particulars are treated as abstract generals. So, for instance, *E. coli* bacteria are designed as a whole system—each individual together with the species as a whole—to respond in a standard way during chemotaxis to concentrations of repellants and attractants. This design is the abstract general that detects and responds to all concrete particular instances of concentrations. Thus, individual organisms are inherently anticipative. They have been designed to be so because organic life is founded on ב, itself inherently anticipative. It is inherent to ב, through arbitrary neglect of concrete residue, to treat future events as past ones. The key condition for this possibility is the cycling of the environment.

What Is Life? How Did Life Emerge?

Of course, implicit in any inquiry into biogenesis is a definition of life. In an appendix to his book *The Organic Codes*, Marcello Barbieri presented no less than 63 definitions of life, starting with one by Jean Baptiste Lamarck in1802 and concluding with one by Edward Trifonov in 2002. Appraising their relative merit is not our task. However, the present inquiry itself amounts to stating a definition of life. Like all the others, such a definition has direct bearing on inquiry into life and its origins. Consider Ernst Haeckel's (1866) definition:

The Days and Nights Are Getting Longer," *Scientific American,* posted June 14, 2010, http://www.scientificamerican.com/article/earth-rotation-summer-solstice/).

> Any detailed hypothesis concerning the origin of life must, yet, be considered worthless, because up till now we have no satisfactory information concerning the extremely peculiar conditions which prevailed on the earth at the time when the first organisms developed.[2]

No doubt there is merit to this view. Nonetheless, understanding life's nature is also required to determine the conditions necessary for its emergence. How else would one know what to seek? Haeckel's definition only touches on the *how*; it does not propose *what* life is and therefore what it must be capable of doing. Recalling Kant's remark on definition, both inquiries go together, a definition of life points to its nature and thereby to the conditions for its emergence. A *what* is clearly expressed in the definition by Francisco Varela (1996): "A physical system can be said to be living if it is able to transform external energy/matter into an internal process of self-maintenance and self-generation."[3] A key implication of this definition is the liminal character of life and thereby the centrality of distinguishing. A key question about this definition is whether it acknowledges that, apart from life, there is no such thing as 'internal' or 'external.' The advent of life is the advent of internal/external. As shall be discussed below, for followers of Francisco Varela and Alberto Maturana—known as the Santiago school—inquiry into life's origin involves the paramount issue of outside/inside, here even more relevant because it closely converges with the notion of distinguishing distinctions.[4] The discussion ends up turning on the effect of our own spirallic nature slipping by unnoticed.

A definition of life follows from the notion of ב: *life is that which establishes bet structure*, or simply, *life establishes bet structure*. This definition suggests questions that ought to satisfy Haeckel. What are the peculiar conditions that prevailed when/where ב first came about? An alternative—and equivalent—definition is this: *living beings are distinguishing distinctions*. Assuming these terms are clarified, the same questions follow.

The sheer number of definitions presented in Barbieri's appendix shows that the nature and origin of life remain hotly debated topics that involve many specialized fields. However, the present argument suggests that this inquiry involves a stark choice: (a) either life is accounted for in the proper higher-level conjugate form, or (b) it is implicitly assumed that living being is not qualitatively different from the nonliving and, therefore, living processes can be accounted for in physical and chemical conjugate forms. Any form of vitalism aside, my aim is to show that ב is the key to a higher-level conjugate form. Or, at the very least, that a higher-level conjugate form is indispensable for inquiry into life.

Among the definitions of life in Barbieri's appendix, some do indeed call for a higher-level conjugate form. Among these is the one already mentioned by Jesper Hoffmeyer (1996) from the budding field of biosemiotics—*the basic unit of life is the sign, not the molecule*.[5] Signs pertain exclusively to living beings. As mentioned

---

[2] Barbieri, *The Organic Codes*, 255.

[3] Barbieri, 262.

[4] For a characterization of the Santiago School, see the section that follows, "Autopoiesis".

[5] Barbieri, *The Organic Codes*, 262. Haeckel (1866); Varela (1996).

above, the proposition that life establishes bet structure is consistent and closely allied with biosemiotics. If ב is the condition for the possibility of tokens and types or concrete particulars being treated as abstract generals, then distinction is a condition for the possibility of signs. This is a key point I shall return to. Hoffmeyer's definition is significant because most current models of biogenesis fit into the molecular or chemical evolution frame.[6]

First proposed by Alexander Oparin in 1924 and John Haldane in 1925, in this frame, molecules constituting the earliest cells "were synthesized under natural conditions by a slow process of molecular evolution, and these molecules then organized into the first molecular system with properties with biological order."[7] The Miller–Urey experiments in 1953, and later variations of these, have reproduced in the lab the likely chemical conditions at the time of life's emergence, demonstrating the formation of amino acids, which are the building blocks of proteins. John D. Bernal postulated that life originated in three stages: (1) biological monomers, (2) biological polymers, and (3) the evolution from molecules to cells.[8] Manfred Eigen and Sol Spiegelman demonstrated *nonliving* evolutionary chemical processes that include replication, variation, and natural selection—a point we shall return to and whose outstanding importance cannot be underestimated. It is generally assumed that this chemical evolution was followed by a biological evolution with the emergence of the first cells.[9] Each stage of life's emergence faces specific obstacles articulated by Bernal.[10] Emergence of organic molecules in the first stage is now well understood in thermodynamic terms, the key conditions being a source and sink of energy and a fluid medium. The second stage, the ordering of organic monomers into biologically active polymers, also requires sources and sinks of energy. Recent theories of how this might first have occurred are presented below, along with the problem of the origin of replication that is also being addressed by work in the RNA world. Bernal regarded the third stage as the most difficult—incorporation of biological reactions behind cell walls. Current research focuses on how membranes self-assemble that could be precursors to living cells.[11] Below we discuss one instance of research in the lab and recent findings that suggests how this might have taken place in nature.

The final stage, when aliveness emerges, is the one we are particularly interested in. Can current findings be construed as the emergence of ב? We begin with a model of membrane self-assembly, for its own merit, but also to go on to discuss a paper based on it that asks when such a molecular process may be regarded as being alive. It will be shown that the notion of ב does indeed clarify key points and helps overcome the obstacle of our own biomorphism. Ironically, it is our own spirallic nature that obscures a critical difference between the nonliving and living being.

---

[6] Bahadur, "Photochemical Formation." The rest of this paragraph relies on this entry and its citations.

[7] "Abiogenesis."

[8] Bernal, "The Physical Basis of Life."

[9] Follmann and Brownson, "Darwin's Warm Little Pond Revisited."

[10] Bernal, *The Origin of Life.*

[11] Russell, *Origins*; Kauffman, *At Home in the Universe.*

Autopoiesis

All known life is made up of cells. As Luigi Luisi explains, a cell does something relatively simple: a semi-permeable spherical membrane sets the cell apart from its medium so that certain substances (nutrients and other chemicals) and not others can permeate the interior. A cell is therefore *structurally open but functionally closed*: "the system is open to the input of matter but closed with regard to the dynamics of the relations that generate it."[12] The network of chemical processes in the cell's boundaries maintains its identity and homeostasis. This approach to the study of cells—what Maturana and Varela, founders of the Santiago school, name *autopoiesis*—leads to a very specific framing of the question, What is life?[13]

Semi-permeable membranes that separate cells from their medium by carrying chemical selection or recognition is what makes cells *distinguishing distinctions*, semiautonomous entities in functional closure that are loci of distinguishing activity. This has huge consequences. Because ב involves anticipation and the arbitrary neglect of concrete residue, it makes cells [subjects] with agency whose acts are intelligible in terms of event-response. This makes cells *nodes of causation* since their responses set off causes-effects. Therefore, multicellular organisms, as loci of multiple distinguishing activity, are thereby multiple nodes of causation. In this fundamental regard, all life is isomorphic. The biosphere comprises myriad nodes of causation. In the biosphere, throughout it, and embedded within organisms, dynamic complexity arises from the interrelating of semi-autonomous units, all nodes of causation. This, in a word, is why organisms are organized. The dynamics of nodes of causation is the dynamics of agents, agents in interrelationships. This is the topic of joinings, to be treated at length. The key point here is that joinings result from functional closure. Functional closure makes distinguishing distinctions or autopoietic systems semi-autonomous agents and the encounter between them proper relationships.

Let us return to the basics of functional closure. Figure 3.1 show the basic process-structure of cellular life according to Luisi.

Functional closure is cyclical, and systems in functional closure produce their own rules of existence and generate their own components through a network process in their boundary.[14] This is what allows cells to entrain with the cyclical environment.

---

[12] Maturana, "The Knowledge of Knowledge." Cited in Luisi, *The Emergence of Life*, 158 (slightly modified).

[13] This approach has been the basis for a research program in wet chemistry, most notably by Luisi. The notion of autopoiesis has also been applied to systems that are not alive. Such use is subject to the same considerations raised in the current discussion. These are examined further in Chap. 7 *Four Kinds of Things*, and Chap. 9 *Living Being* in relation to kinds of wholes. The issue is intimately related to the discussion of the distinguishing of systems that are not alive, for instance social systems.

[14] Luisi, *The Emergence of Life*, 159.

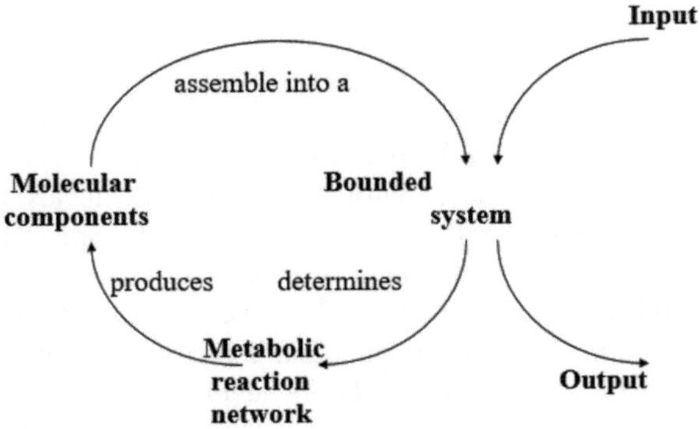

Fig. 3.1 The cyclic logic of cellular life. As an autopoietic unit, the cell is an organized bounded system that determines a network of reactions that in turn produces molecular components that assemble into the organized system that determines the reaction network that… and so on (Luisi, *The Emergence of Life* (slightly modified). For one of many simulations available online see http://www.youtube.com/watch?v=lm-dAvbl330)

The spontaneous self-assembly of protocells, rudimentary autopoietic systems, can be observed in the lab. Luisi presents instances of liposomal structures that self-organize into cell-like compartments. Lipids, chemical compounds that are amphiphilic, are so called because they have "both loves," as the Greek root of the word indicates. They are at once hydrophilic (water loving) and lipophilic (fat-loving). As a result, lipids spontaneously self-assemble into large structures that are more stable than their single components.[15] A few billion of these molecules tend to form spherical bubbles similar in size to biological cells. Moreover, similar structures form within. According to Luisi:

> There is the emergence of compartments, with an inside that is physically different from the outside. The discrimination between inside and outside applicable to compartments, is the first structural prerequisite for the living cell and the living in general. All these notions are beautifully applied to nature. Life is dominated by water as a background medium; however, living beings contain a considerable number of lipophilic compounds that are per se insoluble in water—lipids.[16]

The foregoing statement certainly points towards cells as distinguishing distinctions. However, though the statement may appear uncontroversial, it is not. Anticipating the discussion later, the key terms are problematic. Is "discrimination between inside and outside" *for* the human researchers or *for* a nonliving system? Only the former distinguish. So, *for* what or *for* whom does "discrimination between inside and outside" take place? It occurs for the researchers, certainly. But does it occur *for* the spherical bubbles? We shall return to this crucial point.

Amphiphilic molecules give rise to bilayers or two layered surfaces. An important finding in the lab is that "…the more surface bilayer is formed, the more the

[15] Luisi, 159.

[16] Luisi, 185.

process speeds up, because there is more and more active surface where the next steps of aggregation can take place." Thus, the formation of spherical bubbles is an autocatalytic process because "the product of the reaction … speeds up further self-organization." Self-organization in chemical systems is usually attended by autocatalytic behavior (classic examples of self-organization being DNA self-assembly and t-RNA folding).[17]

These, then, are protocells. Though fully functional protocells are yet to be achieved in a lab, their elementary physicochemical properties approximate cellular behaviors. This includes both nascent Darwinian competition and energy storage. Stabilized membranes are likely to present a selective advantage for the evolution of cross-linked fatty acids that make up phospholipids. Interactions between the membrane and the encapsulated content increases the likelihood of transitioning from replicating molecules to true cells. Encapsulation results in a precursor to metabolism as small molecules cross the membrane while larger ones are prevented from doing so. Thus, encapsulation leads to increased solubility within, and energy storage in the form of a chemical gradient.[18]

Homeostasis, Growth, Decay

With an oleic acid surfactant system, Luisi has implemented a protocell in the lab that is a minimal homeostatic autopoietic system. The system involves two competitive reactions. One uses a nutrient to build the components of the semipermeable boundary; the other one destroys this nutrient. It is a two-component regulatory system with three possible outcomes: (1) homeostasis, a dynamic equilibrium of the two reactions that does not modify the identity of the unit; (2) greater velocity of the reaction that builds the oleate, leading to growth and eventually self-reproduction; and (3) faster velocity of the reaction that destroys the oleate, eventually leading to the system's demise. In all three cases, the primordial property of the two reactions is a spontaneously formed system that includes a boundary (Fig. 3.2).[19]

For Luisi, this autopoietic system is already a rudimentary metabolism. Its pair of ongoing chemical reactions shows the inherent nature and crucial role that timing plays in determining the development and ultimate outcome of the system. Consider the three possible outcomes of the system. With decay, units simply disappear. Homeostatic units remain stable, at least for a time. Then they either decay or split. Those that split, in effect, self-replicate. This results in variation that would hypothetically lend itself to Darwinian selection. We return to this below. This ongoing process of selection itself either decays or tends to optimize a combination

[17] Luisi, 92.

[18] "Protocell."

[19] Capra and Luisi, *The Systems View of Life*, Fig, 7.4, 138 and Luisi, *The Emergence of Life*, 161.

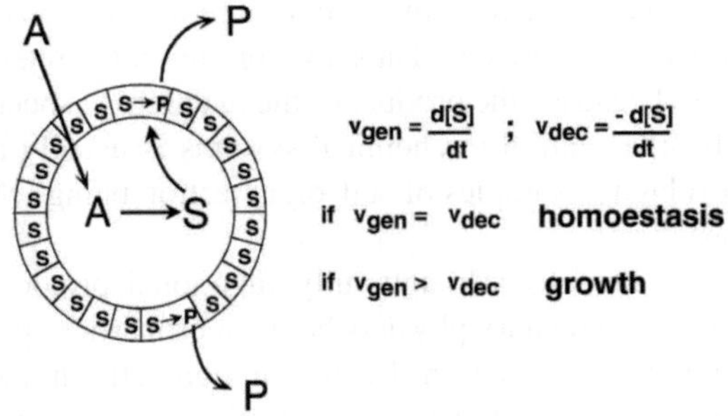

Fig. 3.2 The minimal autopoietic system. This system is characterized by two competitive reactions. One builds the components of the semipermeable boundary S from nutrient A. Another destroys it, turning S into byproduct P. According to the relative value of these two velocity constants, the system can be in homeostasis, grow, or die (Capra and Luisi, *The Systems View of Life*, Fig, 7.4, 138 and Luisi, *The Emergence of Life*, 161)

of homeostasis and replication. This goes well beyond what is currently implemented in the lab. But one may reasonably speculate that, if the spontaneous emergence of these autopoietic structures even roughly resembles how life began, the process of replication and modification opens the possibility for optimization, and thereby the possibility of increasing complexity, from which living autopoietic units may eventually emerge.

Quite clearly, Luisi's approach to the question of life's origin dovetails beautifully with the notion of distinguishing distinctions. It thereby has much to offer it. This is so in two senses. First, is the obvious convergence between the notion of functional closure and that of distinguishing distinction. Both mean, precisely, a system open to the input of matter but closed regarding the dynamics of the relations that generate it. Much can be built upon this convergence. Second, Luisi—and his colleague Bitbol—also offer a striking example of the dangers of our own biomorphism. I shall pursue both points in tandem. On the one hand, how their analysis strengthens the notion of distinguishing distinction and ב, on the other, how our own spirallic nature can easily slip in unnoticed.

Autopoiesis with or without Cognition I

To begin with, for some authors but not for others, autopoiesis and living being are identical. Not so for Luisi and Bitbol, who ask themselves the questions: Are all autopoietic units alive? At what point can autopoietic units be considered to be alive?[20] Much hangs on this question so we must consider it carefully.

[20] Bitbol and Luisi, "Autopoiesis With or Without Cognition."

A semi-permeable spherical membrane forms a boundary that is maintained as the membrane allows certain substances, but not others, to permeate into the cell. Thus, a protocell as the one presented by Luisi, appears to make distinctions. But this would imply that it establishes bet structure. Whether this is so or not is what we must consider. This, then, is a good opportunity to test the notion of ב. The network of chemical processes in the cell's boundaries that maintain homeostasis are, *par excellence*, schemes of recurrence. Strictly speaking it is not the membrane alone but the cell, as a scheme of recurrence, that maintains this membrane and constitutes a functionally closed system.

This recalls the automatic heating system. It was the whole system that was examined, not just the thermostat. Distinguishing requires forness. But, as was shown, there is a crucial difference between the forness of the automatic heating system and that of a living cell. The forness of the heating system, a product, is introduced by a living being. In contrast, the living cell's forness is the means whereby it continues to be itself. Thus, its own forness is what allows a cell to maintain itself as a concrete distinction. As the foundational case of functional closure, a cell is the epitome of a concrete distinction. Its own forness allows the cell to be *itself*, to be an *it*. By its own forness the cell *is*, it *is for itself*. Forness is central to how and *why a cell is*. So, the question arises: Is this the case with the autopoietic unit implemented in the lab?

This is where things get tricky. A cell's forness is fundamental, not only biologically but also metaphysically. This is so because forness pertains to a cell that is *for* itself. It is not merely our attribution. Again, this feature of aliveness is foundational both ontologically and epistemologically. To take for granted that a protocell in the lab is an *it*, and, moreover, an it *for* itself, is to already project aliveness onto this system. But it is crucial not to do so here, when it matters most, because the question is, no more and no less, whether the system is alive or not. That is, the question is, precisely, whether it is an *it for itself*, or not. That living cells are distinctions that distinguish is both an ontological and an epistemological point. Ontologically, distinguishing is the beginning of knowing—epistemology. Likewise, the beginning of distinguishing, the beginning of knowing—epistemology—is the origin of being as a distinction, being as a proper *it*, the living kind of being—ontology. Metaphysically, this is what underlies *distinguishing distinctions.*

Since it is a central issue, we must pause to consider the notion of an *it*, an *itself.*

What Is an *Itself*?

Consider the geyser Old Faithful, an excellent instance of a regularly recurring physical phenomenon. Is Old Faithful an *itself*, a thing *for* itself? This involves two questions: (1) What kind of thing is a geyser? and, (2) Is this the kind of thing that can be considered an *itself*?

A geyser is like a giant pressure cooker. Through fissures in the ground, water migrates miles down into the geyser's plumbing system. This water is under huge pressure from the weight of the water above. This raises the boiling point, so that much more energy (in the form of heat) is required to push up bubbles. A huge amount of energy must be stored before the water starts to boil. As heat is transferred by the magma at the base and pockets of water begin to reach boiling point and become turbulent, water begins to push out the opening of the geyser. This begins to release the pressure from the weight of water, lowering the boiling point. Water that turns into steam expands 1500 times, making water erupt high above the mouth of a geyser. This eruption ends when the water runs out or when it cools. The cycle recurs repeatedly.[21] A geyser is a thermodynamic system. What does this mean? According to Atkins:

> The part of the universe that is at the center of attention in thermodynamics is called the *system*. A system may be a block of iron, a beaker of water, an engine, a human body. It may even be a circumscribed part of each of these entities. The rest of the universe is called the *surroundings* … . The system and the surroundings jointly make up the *universe*.[22]

A thermodynamic *system*, then, amounts to our own distinction. Where is the boundary of a geyser as a system? Does a geyser produce a boundary *for itself*? Clearly, it does not. It is not a functionally closed system. There is filtering of water into the system, but this is not a semi-permeable membrane. The system expends no energy maintaining a boundary that would allow it to remain far from thermodynamic equilibrium, as a living cell does, by selectively controlling inputs and outputs. Lacking self-organization, there is no significant negentropy at work in a geyser. Nothing about a geyser is at odds with the second law of thermodynamics—precisely the opposite of what one expects of self-organizing systems, typically living systems. This is why the functioning of a geyser is intelligible in purely physical terms.

In short, the geyser is not a system *for itself*, it is only a system *for us*. Our distinguishing activity aside, a geyser is but a blip in the earth's hydrosphere. It is we who arbitrarily consider it a system and determine where it starts and where it ends, what it is and what it is not. It is no more for itself than an ephemeral bubble flung up for a fleeting instant when an ocean wave crashes upon the shore. As all thermodynamic systems distinguished by us, geysers and bubbles are alike and utterly different from a living cell. For our own purposes, we regard each as an itself. Only the living cell is not only for us, but also *for itself*.

A shorthand way to state what self-maintaining systems do to remain in functional closure is $\bar{A} \rightarrow A \rightarrow / \rightarrow \bar{A}$. An outside component ($\bar{A}$) crosses a semipermeable boundary to become a component (A) integrated into a system-process in functional closure (/), from which it will eventually again cross the boundary to cease being part of the system ($\bar{A}$). This is precisely what geysers do not do. To be a thing for itself, a system's boundary must be constituted by itself.

For common sense, a bubble is a superb 3D instance of *perfect continence*, a distinction that is real and already out there now, independently of a knowing

[21] Atteberry, "How Geysers Work."

[22] Atkins, *Four Laws*, 3.

subject. However, not doing $\bar{A}\rightarrow A\rightarrow/\rightarrow\bar{A}$, it is not for itself but only for us. It is important to note, however, that this is not an anti-realist solipsistic argument. There is no question that there are real bubbles and real geysers already out there now. What is at issue is not their status as real, but the kind of thing they are—given that their boundary is determined by us and not by them. The question is crucial to our current topic because the origin of life is nothing less than the transition from one kind of thing to another. As we shall soon see, the second law of thermodynamics helps differentiate between kinds of things.

Liposomal Structures and Flow

Having clarified the notion of an *itself*, we can return to inquiry into biogenesis, which takes a special interest in liposomal structures that self-organize into cell-like compartments. In appearance, these are bubbles. So, for instance, Capra and Luisi discuss micelles and liposomes: "There are two remarkable aspects of this process: first, the spontaneous formation of local order, attended by an overall increase of entropy (or disorder), and second, the formation of spherical compartments, which are very important as a model for biological cells."[23] Such processes point to the crucial metaphysical point that we are interested in. Capra and Luisi provide the overall context of the discussion:

> Today there is a broad consensus among biologists that biological evolution was preceded by a molecular, or 'prebiotic,' type of evolution in which the transition from nonliving matter was brought about by a gradual and spontaneous increase of molecular complexity until the first living cells emerged about 3.5 billion years ago.[24]

It was Alexander Oparin who, in 1924, first boldly stated that this increase in complexity "appears to be at odds with the second law of thermodynamics."[24] Nonetheless, as Capra and Luisi explain, there are "quite a few processes that bring about an increase of molecular complexity in perfect agreement with thermodynamics" for which the general term used is "self-organization" or "self-assembly."[24] Among these, "the most interesting self-organizing systems, including many centrally important to the life of the cell, are dynamic; that is, they are non-equilibrium systems that form their characteristic order while dissipating entropy."[24] The condition for this possibility is as simple as it is elegant—a boundary that allows for functional closure. This is why inquiry into biogenesis takes such interest in the protocells formed by self-organized, endogenously ordered, spherical collection of lipids. But, as we can now see, this also brings up the crucial metaphysical question: When is such a molecular structure for itself and not merely for us? That is, when is its activity entelecheia, the process of being-at-work-staying-itself? Capra and Luisi are fully aware of this differentiation. They qualify the term "self" and explain that

---

[23] Capra and Luisi, *The Systemic View of Life*, 146 (slightly altered).

[24] Capra and Luisi, 144.

it denotes endogenous processes dictated by rules internal to the system. This is in contrast to a structure organized by external forces, which is thereby not self-organizing. As instances of the latter, they mention the assembly of a TV set or the pagination of a document. Instances of the former are protein folding of the helix formation of the DNA. Significantly, they say that "the *formation of soap bubbles*, and crystallization, are all examples of self-organization under thermodynamic control."[25]

Let us be clear, no structure can be self-organizing in an absolute sense. All structures, both living and nonliving, are subject to "external imposing forces" such as gravity, pressure, heat, and so forth. They all owe their existence to earlier events all the way back to the Big Bang. They all form in the context of the laws of thermodynamics, specifically the second law, the natural tendency toward entropy. The formation of a soap bubble through an endogenous process dictated by "internal rules" is under thermodynamic control. A bubble is a local and temporary tendency toward order that is not at odds with the second law of thermodynamics. What is more, it is evidence of this law. Again, at issue is not the law itself, but the blurring of meaning that potentially arises from the usage of notions such as "internal" and "self" when thinking about phenomena that result from this law.

A helpful way to think about the second law is in terms of flow. Entropy involves a process in the transition from order to disorder; it does not happen all at once. Flow brings about dissipative structures characterized by symmetry breaking. As Adrian Bejan pointed out, this explains spontaneous designs in nature. They are not the result of chance. They arise naturally, spontaneously, because they enhance access to flow in time."[26] This explains why the second law of thermodynamics, the increase in entropy, is the most creative law of the universe. It is the reason why there is *any* change at all. It is the reason why engines run, and why chemical reactions occur. Moreover, it is the foundation to understand their consequences, as well as the acts of literary, artistic, and musical creativity.[27]

All flow processes are optimized through branching vascular patterns—think of trees and river tributaries. River basins vascularize, producing network designs of evenly distributed large and small streams and islands that are the most efficient pattern to disperse water. Smoke too disperses with optimal flow spontaneously producing extraordinary patterns.

Spontaneous designs with an even distribution of large and small pathways are everywhere. Next time you cook rice, as you remove the lid notice the even distribution of large and small holes on the surface of the rice. These are the outlets of the vertical ducts through which the steam flowed during cooking. Or serve a beer slowly and observe the even distribution of large and small bubble streams that form. Similar patterns form in a shower as rivulets of water trickle down the wall.

---

[25] Capra and Luisi, 145 (emphasis added).

[26] Bejan and Zane, *Design in Nature,* 3.

[27] Atkins, *Four Laws*, 49. Emphasis in original.

Fig. 3.3 Spontaneous designs that result from flow (Daniel C. Mayer-Foulkes)

Watch the formation of droplets as a window fogs up. For a glorious spectacle of emerging designs caused by the spontaneous optimization of flow, take your car to a carwash and watch your windshield. All these spontaneous designs adhere to the *constructal law* enunciated by Bejan: "For a finite-size system to persist in time, its configuration must evolve so it provides easier access to the currents that flow through it"[28] (Fig. 3.3).

Though spontaneous, flow is also predictable. Most critically, if patterns did *not* optimize flow, they would certainly be at odds with the second law of thermodynamics. If this were the case, they would be *holding back* the transition from order to disorder. The second law, then, underlies all designs for efficient flow, and the constructal law applies throughout nature. Optimization through vascular design is at work in all circulatory systems, whether nonliving and living, from river deltas to the distribution of alveoli in lungs. It can be observed in street systems of cities, whether planned or not, and in distribution strategies, for instance by postal systems or courier companies. Designs tend to optimize for efficient distribution or flow. This is mostly spontaneous and the result of recurrence. Optimal flow design dates to the dispersion following the Big Bang. This is how and why galaxies, clusters of galaxies, and super clusters of galaxies formed. All are the result of the optimization

[28] Bejan and Zane, *Design in Nature,* 3 (slightly altered).

of flow. Stars, planets, and the droplets on fogged windows emerge from this flow process. These objects, and the flow process that produces them, are real as already out there now. Nonetheless, they constitute distinctions for us—not for themselves. Therefore, to gratuitously regard any given sector produced by that flow, such as galaxies, stars, pebbles, or bubbles, as autonomous, amounts to what Whitehead calls "the fallacy of misplaced concreteness."[29] Living beings aside, this is the fundamental problem with the notion of "self-formation." Sectors of the nonliving that result from the spontaneous optimization of flow are one kind of thing; semi-autonomous systems that self-maintain are another, very different, kind of thing. How could something be a self if it is not, in a minimal sense, *for* itself? When an endogenous process or system is not autonomous in this sense, it is we who establish its "selfness."

In short, the second law and the process of flow that result from it are real, both descriptively and in an explanatory sense. Sectors of flow that result from vascularization are real. However, it is we who regard these sectors as *bodies* out there now—galaxies, stars, planets, pebbles, bubbles. But this is misplaced concreteness; the boundary of a super cluster of galaxies is ours, our biomorphism, and not the result of a body that self-maintains as an autonomous unit.

One way to acknowledge that this is our distinction is to refer to 'bodies' or a 'body', in quotation marks. A super cluster of galaxies is a 'body,' and so are the Solar System, the Sun, the Earth, and the Moon. The hydrosphere too is a 'body,' along with river deltas, geysers, and water bubbles. We distinguish them and assign as 'bodies' certain sectors of flow. In all these instances, we cannot help our own biomorphism determining what is 'external' and 'internal' to these 'bodies.' Likewise, their 'internal structure' and 'processes' are determined at our own convenience. This matters little for common sense. But it could not be in sharper contrast to a genuine body that does Ā→A→/→Ā. We must keep our biomorphism in check when it interferes with understanding in the explanatory sense.

So, having clarified the key difference between an *it* or an *itself* for us, and one that is not only for us but also *for itself*, now also in a thermodynamic sense, let us return to Capra and Luisi. There are unique instances of flow processes that happen to feedback on themselves. Such is the case of autocatalytic processes of self-assembly of liposomal molecules as they form protocells. But what kind of thing are these sophisticated spherical structures that result from endogenous process under thermodynamic control?[30] Are they more like soap bubbles or like living cells? How close are they to throughput—doing Ā→A→/→Ā? Are they 'bodies' or bodies?

This is to ask the fundamental question in biogenesis but with a twist. The Santiago school framed this classic question like this: When can a molecular process be said to be autopoietic? Instead, we can ask: When can a system be said to establish bet structure? The question—are all autopoietic systems alive?—may be

[29] Whitehead, *Science and the Modern World*, 52.

[30] For further discussion on thermodynamics – directionality, probability, and the nature of lived experience – see Chap. 7, section, "The Nonliving."

asked this way: What is the relation between autopoietic systems that distinguish and those that do not? Or, more precisely, What is the relation between autopoietic systems that establish bet structure and those that do not? Recall Fig. 3.2 above. Luisi, regards this as an autopoietic system. Can ב serve as a criterion to determine whether this autopoietic system is alive or not?

By now it must be clear that we must be keenly aware of our own biomorphism when seeking answers to these questions. Closely related is the further question: How do we differentiate between a system that is a "self" for us and one that is a *self for itself*? When do 'bodies' become bodies? Consider Luisi and Capra's framing:

> One important question is whether autopoiesis is an equivalent to life – namely, whether it is the necessary and sufficient condition for biological life. The primary literature answers this question affirmatively, while more recent studies (Bitbol and Luisi 2004; Bourgine and Stewart 2004) have shown that it is more proper to say that autopoiesis is only the necessary condition and not a sufficient one, as a few artificial systems have been found that are autopoietic but not living. In these recent papers, particularly Bitbol and Luisi (2004) the relations between autopoiesis and metabolism is also discussed. However, for the entire realm of the biological world, the equivalence between autopoiesis and life holds true, and we may safely adopt this generalization. This means that, to determine whether a given system is living or not, it is sufficient to see whether it is autopoietic.[31]

Several issues are at stake here. First, according to the authors, for practical purposes, the issue could be laid to rest. However, the paper mentioned by Bitbol and Luisi and discussed below, fails to differentiate between a system for us and a system for itself. Needless to say, they do not discuss our own biomorphism. Second, it is not trivial that the few systems that have been found to be autopoietic, but not living, are artificial. Strictly, pending our discussion of the four kinds of things in Chap. 7, geysers and bubbles are nonliving, living cells are living beings, and these systems produced in a lab are, at least in some sense, *products*. The automatic heating system has shown us that the forness of a product is due to the living being who produced it. That is the nature of products, they are produced by a living being *for* something. Autopoietic systems implemented in a lab may be uniquely helpful to reflect on the nature of living being, but they are products, nonetheless. This makes them radically different kinds of things, with all the attendant metaphysical considerations this implies. In the discussion that follows, we shall touch on the sharp contrast between scenarios in nature where life may have emerged and implementations in the lab. Hypothetically, the differences between them might be overcome. But for the Santiago school, living beings and their environment co-emerge. Theoretically, if not practically, this places severe constraints on how biogenesis can be tested in the lab.

In conclusion, crucial to inquiry into the final stage of biogenesis, is the point at which a system begins to be for itself and not only for us. A key potential obstacle to this inquiry is our own biomorphism. The problem is not that, distinguishing distinctions ourselves, we are inquiring into the emergence of distinguishing distinctions. Rather, it is our lack of full cognizance of all that follows from this fact.

---

[31] Capra and Luisi, *The Systemic View of Life*, 138 (slightly altered).

Namely, we are not cognizant of our spirallic nature and our proclivity to project it upon the world. But this is not merely an obstacle for inquiry into biogenesis. On the contrary, how it plays out in that inquiry points to a far larger problem. We are not fully cognizant of what it means to be alive, what aliveness involves. So habituated are we to making sense of aliveness in lower physical and chemical conjugate forms, that we do not fathom the limitlessness of the nonliving. In stark contrast to it, our own distinctional nature is a limited mode of being that can advert to the world.

Autopoiesis With or Without Cognition II

We arrive, then, to our discussion of researchers failing to clearly differentiate between their own spirallic nature and that of their object of study, just at the point where it matters most.

For the Santiago school, an autopoietic system is already a rudimentary metabolism. As Capra and Luisi explain, this makes metabolism equivalent to cognition:

> …for the biological cell it is the metabolism itself that is the link between the internal and the outside world. Metabolism is the result of the internal organization of the cellular system, but it is implemented by the interaction with the environment that feeds the cell and accepts its expelled by-products. Thus, metabolism itself is equivalent to cognition in the simple case of unicellular organisms.[32]

Let us return to the key question as stated by the Santiago school. If autopoietic systems are not necessarily alive, the question is, At what point do autopoietic units transition into being alive?[33] Is the autopoietic unit in Fig. 3.2 alive? In my own terms, Does this unit establish bet structure? This implementation in the lab offers important insights into what is required for ב and how this may have first come about. I propose to show that it supports the notion of ב and of the distinctional as the form of aliveness.

The Santiago school's notion of *autopoiesis* is a higher-level conjugate form, not unlike Hoffmeyer's sign: "…the compounds that the living organism extracts from the environment can be seen as something that the organism itself lacks for implementing its life. The appropriation of these missing parts is what gives 'meaning' and links the autopoietic unit with its world. At this point the Santiago school introduces the term 'cognition.'"[34]

That being so, the hazy boundary between the nonliving and living being is the boundary between autopoietic systems that are not cognitional and those that are. Cognition as used here, like distinguishing, refers to the full range of knowing, starting from the most rudimentary. So, for Luisi:

> Cognition … applies only to living entities, and not to the inanimate world … . Chemical *recognition* in the molecule world is not cognition." He adds, "Among all these cognitive

[32] Luisi, *The Emergence of Life*, 167.

[33] Bitbol and Luisi, "Autopoiesis With or Without Cognition."

[34] Luisi, *The Emergence of Life*, 165.

> interactions between autopoietic entities and the environment, some are particularly important because they are recursive, i.e., they happen *repeatedly*.[35]

Let us consider these terms carefully. Doing so requires a few paragraphs of close textual analysis. The exercise is valuable for five reasons: first, the notion of the distinctional dovetails thoroughly with the Santiago School even as, second, the text in question provides an excellent example of the error of biomorphism at the point where it matters most. Third, we shall see classical Darwinism come to the rescue, even as it exhibits its own limitations. Yet, doing so, fourth, it offers the key to how ב could have emerged in the first place. Finally, an additional critique of an altogether different order sets up the discussion for a key differentiation between emergent and developmental processes. Both are central for inquiry into aliveness. They are taken up in subsequent sections.

Our starting point is the fundamental notion of repetition. A first minor point to clear is the error of equating recursion with repetition. Recursion is a hot topic. Its meaning and effects are debated, not least because various authors mean different things by it. Recursion is a very specific kind of repetition, namely self-similar repetition—whether it pertains to processes or to objects.[36] Most significant for our

---

[35] Luisi, 165 (emphasis added).

[36] Recursion plays a crucial role in living being. I shall here restrict myself to Corballis, starting with his discussion in *The Recursive Mind* (pp 2–12), where he cites Steven Pinker and Ray Jackendoff's definition:

> "… a procedure that calls itself, or a constituent that contains a constituent of the same kind." The second part of this definition is important, especially in language, because it allows that recursive constructions need not involve the embedding of the *same* constituents, but may contain constituents of the same *kind*—a process sometimes known as "self-similar embedding."

In a section entitled "What Recursion Is Not", Corballis treats repetition and iteration—of which feedback systems are an instance. He notes that for his purposes iteration does not qualify as true recursion because outputs are discarded, and this does not add complexity. He compares iterative procedures in mathematics to feedback systems such as thermostats and notes that the body is awash with these to "maintain homeostasis of temperature, iron, energy, blood composition, and so on." In iteration, each beginning is initiated by a previous outcome, which is then forgotten, whereas in proper recursion, as defined, is like an ever-deepening whirlpool in which all previous instances remain. So, while the functioning of two-element control systems as such is not recursion, the layering of this kind of feedback systems that allow ever higher levels of anticipation, is an instance of recursion.

Indeed, for Corbalis, the forming of fractals by the nesting of self-similar patterns is a recursive process. So are the Matryoshka dolls on the cover of his book. A system-like reproduction that feedbacks upon itself is a recursive process, and self-similar nested structures, like the images in parallel mirrors, are a recursive structure. Recall that parallel mirrors were used to explain bet structure which, in turn, led to the notion of the spirallic. Spirals are recursive. Therefore, they can be mathematically expressed through the Fibonacci sequence. This sequence also underlies the golden ratio, another recursive structure. Fibonacci arrived at his sequence to answer a simple question about a recursive process: "If a pair of rabbits is placed in an enclosed area, how many rabbits will be born there if we assume that every month a pair of rabbits produces another pair, and that rabbits begin to bear young 2 months after their birth?"

purposes is that ב is recursive because it is repetition of previous repetition. Thus, not only is simple repetition inherent to living being. Recursion must build upon repetition, but it is a specific kind of repetition. For instance, anticipation involves previous anticipations, which in turn results in the unfolding of anticipation. Since anticipation is recursive, features of organisms and organisms themselves, that depend on anticipation, are recursive structures. As ongoing processes, nesting of distinguishing distinctions within distinguishing distinctions results in the ongoing emergence of complexity. In short, recursive structure-processes result in ever more complex nesting of distinctional structure-processes.

Involving self-reference as it does, self-similar repetition subverts classical logic; it generates paradox. This has profound implications that are at the very heart of the nature of aliveness which must be central to its investigation. Living beings are also self-referent in that they act not only in relation to the environment but also in relation to themselves. In this fundamental sense, an organism is the cause of its own effects. In organic life generally, structure follows process. Again, because the fundamental living process is distinguishing—which is anticipative—and the fundamental structure is the nesting of distinctions, organic process and organic structure are inherently recursive. This raises a fundamental epistemological question: if recursion subverts logic, what is it to *understand* (when it comes to) organisms? I am gradually building the case that to understand life is to understand relationships, joinings.[37]

Let us return to the key point: repetition. A differentiation is required here that is analogous to the *cycling* on the side of the object that we experience *spirallically* on the side of the subject. Our spirallic nature is what constructs repetition out of the 'ongoingness' of the nonliving. In the paragraph above, Luisi and Bitbol fail to detect their own biomorphism when they refer to "chemical recognition." Recognition, with its underlying repetition, does not obtain for nonliving molecular structures any more than it does for a dripping faucet. It does not obtain for nonliving structures, autopoietic or not. I therefore opt for bracketing "repetition" and "recognition". We must vigorously guard against our biomorphism if we are to avoid the most fundamental feature of aliveness—repetition, itselfness—to slip into the nonliving. Again: in very simple terms, the threshold between the nonliving and living being is the transition from "repetition" to repetition.

Luisi and Bitbol proposed a "stratified conception of cognition and its relation with life." The endpoint of the transition is this: "When an amoeba or any other living cell chooses the metabolites from the environment and rejects catabolites in it, this corresponds to a dynamic interaction that permits the enacting and the coming to being of both the living organism and the environment."[38] Clearly, "choosing" metabolites and "rejecting" catabolites amounts to distinguishing, as does "discrim-

[37] Gregory Bateson wrote: "In biology, many regularities are part of—contribute to—their own determination. This *recursiveness* is close to the root of the notion of "structure" (*Angel's Fear*, 161, his emphasis). That is, structure follows process. This is embedding, as included in Steven Pinker and Ray Jackendoff's definition.

[38] Bitbol and Luisi, "Autopoiesis With or Without Cognition," 102.

ination". But is this what they really have in mind? Their account for each stage is by analogy to Piaget's notions of "assimilation" and "accommodation."[39]

Let us focus on their three subdivisions, starting from the first:

> (0) The null stage of cognition corresponds to the case when the system of *self*-maintenance is either unaltered or superficially deformed by the irruption of a new environmental factor without assimilating it. This would correspond to the passive *in*corporation of the neutral molecule *into* an autopoietic *unit*, be it a vesicle system or a bacterium.[40]

According to the authors, this autopoietic system is nonliving because it is non-cognitional. However, notice their characterization of the system as a "unit', its "*self*-maintenance," and its "*in*corporation *into*" even as such a system is only for us and not for itself. It follows that its boundary, its inside and outside, are for us, not for it—whether the chemistry of this autocatalytic molecular structure results in a sphere or not. They state that such a "system … is either unaltered or superficially deformed by the irruption of a new environmental factor without assimilating it." This non-assimilation or "passive incorporation" in Piaget's sense, is non-cognitional. There is no ב here, any more than there is in magnets whose poles consistently repel or attract. If there is passive and non-cognitional "repetition", it is established by our own ב.

> (1) Stage one of cognition, which has been compared to Piaget's 'assimilation,' involves integration of an environmental factor (obstacle or molecule) within the *pre-existing processes* of an autopoietic unit that is able to make use of such a factor as part of its defining network. We consider it as the very minimal condition for the concept of cognition to make sense and, accordingly, a basic condition for life. Insistence on the 'cognitive' status of the normal metabolism, which both *maintains the identity of an organized unit* and implies dynamical interaction with the environment, is a specificity of the present paper.[41]

Is this autopoietic "unit" for itself or not? Does it *maintain the identity of an organized unit itself* or is this our own attribution? Again, autopoietic or not, if it is nonliving it does not establish bet structure and it is not a proper metabolism. It if does, it has been optimized to do so and has thereby crossed the threshold into living being. With an identity to maintain by means of an organized process, there is differentiation *for it* between it and the environment. But then, how, and when did it cross the threshold? And what of these "pre-existing processes" that are mentioned? How can they pre-exist? These processes cannot pre-exist in these units unless they established bet structure. But how could they do so if they are not alive? Of course, if they are, then the stages cease to make sense and we still lack an account for the shift from nonliving to living. As this discussion shows, ב can (1) serve to identify

---

[39] Bitbol and Luisi, 105. In Chap. 9 we shall examine the status of such analogies in the method of biology as inquiry into aliveness. The authors say, "One major implication is that the present synthesis drawn from Maturana and Varela and from Piaget can, in principle, be extrapolated toward the higher forms of behavior, thus arriving at a stratified conception of cognition and its relation with life."

[40] Bitbol and Luisi, "Autopoiesis With or Without Cognition," 105.

[41] Bitbol and Luisi, 105 (emphasis added).

biogenesis because (2) it postulates the key feature of life, upon which (3) a new hermeneutic can be built to understand aliveness.

We have arrived at the key question: How does the leap take place from nonliving "autopoiesis"—let us bracket the term—to living autopoiesis? Soon we shall see how Darwinian theory comes to the rescue. Folded into this problem is a larger, subtler one. "Pre-existing" In the lab and "pre-existing" in "the wild," so to speak, are radically different—the more so since, according to the Santiago school, there is "a dynamic interaction that permits the enacting and the coming to being of both the living organism and the environment."[42] In other words, organisms and their environment co-emerge. Does this apply to their initial emergence? At the threshold into living being, might nonliving "autopoietic" molecular structures contribute to the environmental conditions to cross it? This is a key reason why human biomorphism must be strictly guarded against in relation to conditions in the lab.

But let us return to Luisi and Bitbol's notion of "pre-existing structures:"

> (2) Stage two of cognition, which has been compared with Piaget's 'accommodation,' implies enduring modification of the network of processes of an autopoietic unit that then becomes permanently redefined and reaches a new steady state of mutual co-adaptation with its environment. Accommodation is a form of evolution based on stable molecular or dynamic support. It may yield strongly *'anticipative'* behavior such as [motility]. With its memory-like structures and adaptive features, it provides another crucial dimension to cognition.[43]

Clearly, at this stage such a unit can establish bet structure and, therefore, anticipate. Moreover, it can shift from establishing one bet structure to establishing another. It has been optimized by natural selection to optimize itself; from anticipating in one situation, it can switch to anticipating in another. Notice, however, that the reference to "strongly 'anticipative' behavior" carries an unresolved ambiguity—is anticipative behavior specific to this stage or not? Was it or was it not prefigured at stage 1 with its "pre-existing processes"? In short, what is the answer to the fundamental question in biogenesis? At which stage is bet structure first established?

To be sure, once systems do establish bet structure, further "memory-like structures and adaptive features" can be selected for. Selection for anticipation is the critical reason for its unfolding. But how was it selected for in the first place? The example Luisi and Bitbol provide of motility as "strongly 'anticipative' behavior" speaks volumes. For one thing, it implies a direct correlation between anticipation and biological form. It underscores that anticipation is the driving force of adaptation. Popa, for instance, suggested that motility is not merely an ancillary feature of organisms but a direct consequence of the boundaries of primordial vesicles. Indispensable as vesicles are, they are also a curse because living systems must avoid their own waste to avoid poisoning themselves. This, he suggested, is why motion emerged in the first place.[44] We recall *E. coli's* chemotaxis to maintain

---

[42] Bitbol and Luisi, 102.

[43] Bitbol and Luisi, 105 (emphasis added).

[44] Popa, *Between Necessity and Probability*, 57 (Popa credits a 2002 personal communication from Nealson).

essential variables in viable boundaries. Incidentally, Popa, thereby, points to the foundational character of *byproducts*, the fourth kind of thing besides the nonliving, living being, and products. They are expressed by the second Ā of throughput: Ā→A→/→Ā.

If the genesis of motility is anticipation, it is worth digressing briefly to ponder the wide-ranging, indeed the overwhelming importance, of anticipation throughout evolution. Consider teeth. Teeth first appeared on the evolutionary stage with conodonts whose fossils are between 500 million to 250 million years old. They emerged to eat other organisms and signal a new way of making a living and a new way of making organs. Teeth develop from the interaction of two layers of skin, a process that also underlies scales, hair, feathers, sweat glands, and even mammary glands. With teeth, the fish-eat-fish world began in earnest, an arms race that led to jaws, armors, stronger jaws to crack them, harder armors, and so on. The earliest skulls, belonging to ostracoderms, were made entirely of little teeth. Thus, with teeth arose the hard tissues of vertebrates that largely determine who we are, how we eat, how we move, how we breathe, and even how we metabolize certain minerals.[45]

A single word describes an arms race: anticipation. Layer upon layer, anticipation drives evolution, from fundamental features such as motility, teeth, and bones to inter and intra species relationships. Thus, ב and its unfolding through anticipation determines biological form. This alone is substantial corroboration that distinguishing-distinction is the proper conjugate form to understand living being. This form, the condition for the possibility for metabolism-cognition, points to what is unique about life, and explains the dynamic of its unfolding. A point still pending is that inherent to distinguishing distinctions, is their nature as joinings. As we shall soon see, they emerged in interrelationship. For now, the question remains: How did anticipation emerge? How was bet structure first established?

Nonliving Species

There can be no objection to Luisi and Bitbol's account once anticipation is in place. Stage 3 of cognition "relies on highly complex types of accommodative changes;" and stage 4 "may finally involve several social aspects which transform it into genuine knowledge."[46] However, the question remains: How did anticipation emerge in the first place?

Consider again Luisi's' minimal autopoietic system in the lab (see Fig. 3.2), at the null stage of cognition. Though this "autopoietic" system is not anticipative because it does not establish bet structure for itself, it is highly unique. Imagine a population of these systems with some units that disappear and others that grow and subdivide. By subdividing, they replicate what constitutes a population that may

[45] Shubin, *Your Inner Fish.*

[46] Bitbol and Luisi, "Autopoiesis With or Without Cognition," 105.

properly be considered a species. This population presents variations amenable to natural selection leading to a tendency to optimize entrainment to the environment.

Enter Manfred Eigen and Sol Spiegelman who demonstrated just that: nonliving chemical processes that replicate present variations amenable to natural selection. The outstanding importance of this now becomes clear. Though nonliving, the process of transformation of a nonliving species is intelligible *by us* in terms of bet structure. The units cannot be said to self-regulate because they are units only for us, even if they are maintained by endogenous processes. Still, the population constitutes a species that can entrain with the environment.

Four crucial notions are involved in the notion of species that result from replication: species, individual, replication, and replicator. Think of the two latter terms in relation to copyists of biblical texts. Replicators produce replicas. For simplicity, imagine a scenario in which each copyist or replicator produces a single copy or replica of the bible to pass on to the next generation. Assume also that few errors creep into any replica. However, because a skilled copyist replicates faithfully, previous mistakes are replicated even as new mistakes crop into each replica. Noticing these mistakes centuries later, an attentive researcher can infer a line of ancestry. This lineage of replicas is an ontological relation—one that is radically different from other possible relations between them.

Leaving aside the issue of the replicator, this lineage relation is the one that obtains between molecular structures that "self"-replicate. Biblical research refers to *copies of the same text*; biological research refers to *members of the same species*. Needless to say, it is highly unusual for units of nonliving molecular structures to constitute a species. But this is just what one would expect during biogenesis—nonliving linages of molecular structures that antecede living species.

Now, we can ask, What plays the role of replicator in these nonliving molecular structures? Imagine an oil-spill on a beach and that, rolled to and for by small waves, the sticky, oily mud produces sausage-like lumps. By accretion, these lumps grow in width and length until they break up. Once they do, each fragment continues to grow, again by accretion, until further subdivision takes place. In this case, the waves upon the shore are the replicator. This is how lipid vesicles replicate. They grow larger by accretion and twist about until they break up. Though lineage may be difficult to establish, they also collectively constitute a species.

A consequence of lineage is natural selection, regardless of whether units are alive or not. This is how, collectively, units make up a population that goes through modifications. This is inevitable, given the ongoing encounter between a species and—here is the crucial point—*the cyclical nature of the environment*. Random modifications result in various outcomes, and we can define as *more robust* those structures whose variations tend to allow them to propagate and whose descendants are, therefore, more likely to persist through future cycles of the environment. Thus, a sector of the population is more robust because, fortuitously, it is optimized for future circumstances by past modifications. In short, given the cyclic nature of the environment, robustness amounts to a fortuitous "establishing of bet structure" *at the species level*. The cyclical nature of the environment is the replicator that fortuitously modifies units, allowing a species to fortuitously "establish bet structure."

Natural selection results strictly from the concatenation of species and cyclical environment.

I propose that this fortuitous "establishing of bet structure" at the species level constitutes the "pre-existing processes" of Luisi and Bitbol's at stage 1. At this stage we have a nonliving species Eventually, increasing robustness will increase the likelihood of properly self-regulating anticipative units. In short, in addition to homeostasis and replication as modeled in the lab, the indispensable condition for the advent of life is a cyclical environment. This is the conditions that allows Darwinian selection to act upon species, nonliving and living.

In conclusion biogenesis occurs with the doubling up of ב, when natural selection, or the fortuitous "establishing of bet structure" at the species level, brings about proper ב at the individual level. Our textual analysis has shown that the notion of the distinctional dovetails with the Santiago School even as it has illustrated the danger of biomorphism. Classical Darwinism explains how ב might have emerged in the first place. This textual analysis now complete, we briefly discuss the limitations of classical Darwinism. In subsequent sections we shall return to the additional final point, the difference between emergent and developmental processes.

ב and the Limitations of Narrow Darwinism

To this day, natural selection, or the fortuitous "establishing of bet structure" at the species level, still dominates mainstream biology. Despite Darwin's own insistence on sexual selection—which necessarily involves behavior, and therefore the agency of individual organisms—only very recently have evolutionary studies begun to shift away from reduction to natural selection, to add three more dimensions: epigenetic, behavioral, and symbolic inheritance systems.[47]

Purely in terms of species or populations, biology cannot account for the unfolding of autopoietic individuals who establish bet structure for themselves. Natural selection is, no doubt, an extremely powerful explanatory scheme. But it has its limits, as is plainly obvious in the need for other explanatory schemes to account for the events in human history. While a narrow Darwinian paradigm can make sense of the adaptive value of anticipation, it can only very partially account for its emergence and, more significantly, for its unfolding. Reducing life to abstract members of a population renders individuals and their agency alien to inquiry in biology.

Beyond their key biological significance, nonliving species have a metaphysical significance. It calls on us to acknowledge our own biomorphism, and thereby crystalizes the fundamental difference between nonliving units of a species and living individuals. We have seen how critical this unblurring is. Clearly, dis-eliding our biomorphism is indispensable for inquiry into biogenesis. This transition from the physicochemical to the biological calls for a corresponding metaphysical transition,

---

[47] Jablonka and Lamb, *Evolution in Four Dimensions*.

a shift from categories appropriate for the nonliving to those proportionate to living being. Failure to make this shift blurs the founding differentiation of biology. The reduction to narrow Darwinian natural selection is a symptom of this blurring.

Two uses of the term 'selection' epitomize the issue: natural *selection*, as equivalent to "bet structure established" at the species level, versus *selection*, as equivalent to individuals establishing bet structure—or distinguishing. Natural selection alone, then, is insufficient as a higher conjugate form of aliveness. A properly founded biology acknowledges both senses of *selection*, assumes the nesting of the latter in the former; and can thereby account for the unfolding of anticipation.

To regard living individual members of a population and their features as one does the units of a nonliving species, makes little sense. It amounts to treating their aliveness as incidental and removes the heart of biology. While ב is prefigured at the species level, ב proper occurs only in living individuals. Once living individuals have emerged, natural selection, at best, explains why certain bet structures, routinely established, contribute to the robustness of a population. But it cannot explain how and why they were established in the first place, let alone how anticipation unfolds. Aliveness is the doubling of "ב" and ב working at once. Organisms are robust because they are at once fortuitously and non-fortuitously optimized for future circumstances by past modifications. They have individual agency, and their species continues to be molded through natural selection.

Chapter 4
Empirical Evidence for ב

Life Springs

Three bodies of empirical evidence provide support for ב: a scenario for biogenesis, circadian clocks, and Earliest Indication, how organisms anticipate.

Van Kranendonk et al. propose a volcanic pool scenario for the emergence of life.[1] Their hypothesis is summarized in the 2-part illustration based on theirs in Figs. 4.1 and 4.2. This hypothesis pulls together much that has been discussed because it hinges on *cycling* (see the second sentence in Fig. 4.1).

At the foot of the second illustration (see Fig. 4.2), notice the statement, "The cycle repeats again and again." Returning to Fig. 4.1, notice the statement, "The compounds are concentrated in tiny vesicles made of simple molecules called lipids." This is an implementation in nature of Luisi's in the lab. From membranes collected in pools and films "organic building blocks bond together to form polymers . . . . When the pool refills, the films rehydrate and bud off trillions of protocells and membranes that encapsulate collections of random polymers". These survive the changing conditions and then group together.[2] *Cycling* and *membranes* drive the whole scenario, a species of nonliving vesicular structures fortuitously establishing bet structure which, in turn, leads to the emergence of individuals, proper self-regulating distinguishing distinctions that establish bet structure for themselves. The key transitional stage is 4e:

> *Progenotes.* The cycle repeats, again and again. Every time it does, protocells interact, compete for resources, and evolve more complex functional polymers until a 'progenote' community emerges that is able to exchange adaptive molecules, developing ever more sophisticated functions.[3]

---

[1] Van Kranendonk et al., "Life on Earth." Figs. 4.1 and 4.2 by this author follow (and quote) those in the article by Van Kranendonk et al.

[2] See Fig. 4.2b, c.

[3] See Fig. 3.5, e. Cited from diagram in Van Kranendonk et al.

© The Author(s), under exclusive license to Springer Nature Switzerland AG 2023
D. C. Mayer-Foulkes, *The Nature of Living Being*, Biosemiotics 26,
https://doi.org/10.1007/978-3-031-24789-7_4

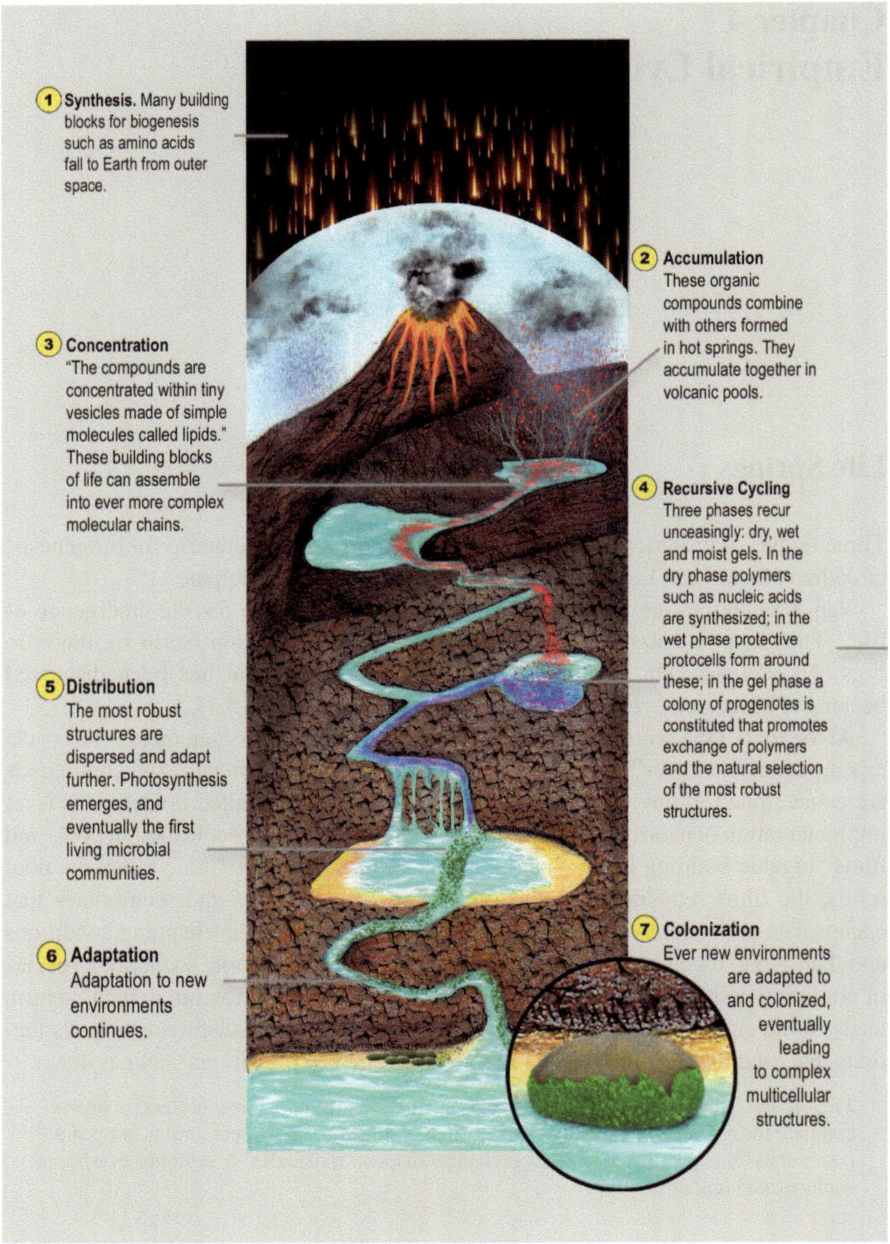

Fig. 4.1 Volcanic pool scenario for the emergence of life. (The author)

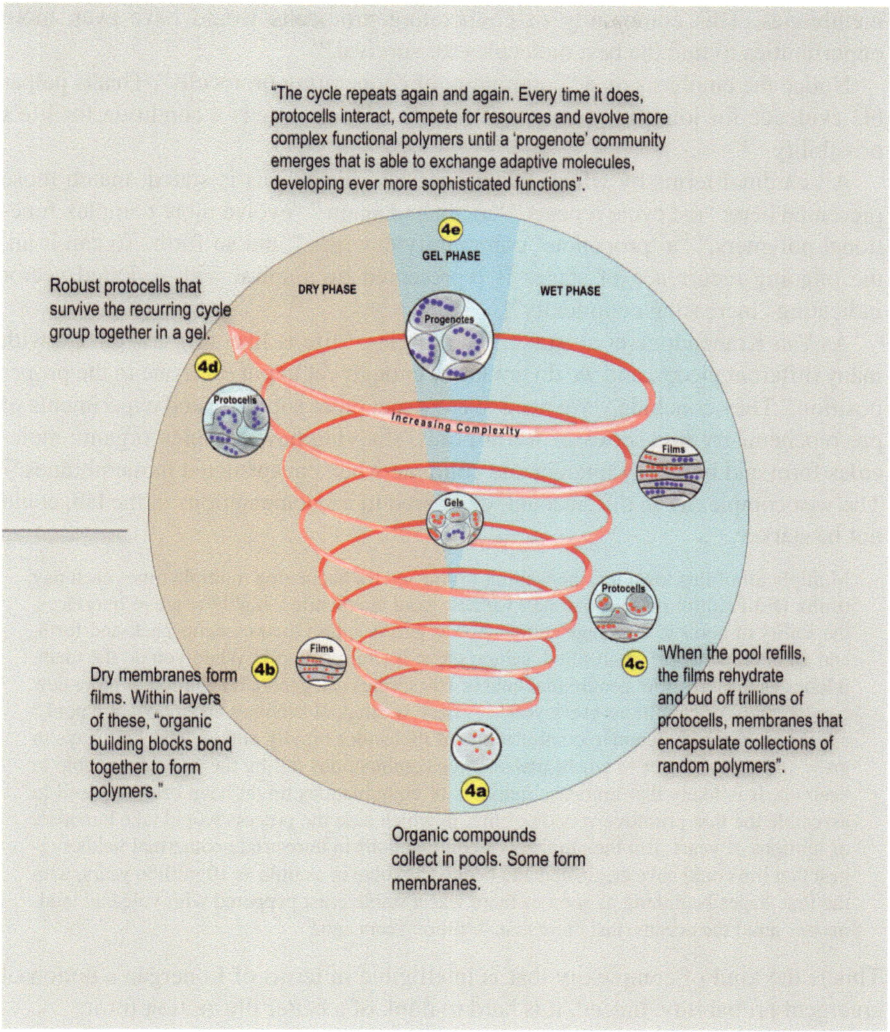

Fig. 4.2 Cycling stage in volcanic pool scenario. (The author)

The outstanding importance of membranes will be discussed in subsequent chapters.

The scenario speaks directly to the co-emergence of environment and living beings. The moist phase could not come closer to bet structure established at the species level because as "... surviving protocells crowd together into a similar gel, polymers and nutrient molecules mix and exchange across the barriers of lipid membranes. This community of cooperating protocells would have even more opportunities to find the best molecules for survival."[4]

Notice the emphasis on a "community of cooperating protocells". This is palpable evidence for joinings at the origin of life. Community is a condition for life's possibility.

All cardinal terms by which this scenario of biogenesis is stated, match those presented here: "the cycle repeats," "again and again," "evolve more complex functional polymers," "a 'progenote' community emerges," and so forth. To cap it all, the ongoing recurrence of stages is represented by a spiral with a dotted vector indicating "increasing complexity"!

As Van Kranendonk et al. state: "The origin of life is like a jigsaw puzzle with many different pieces, and we do not know enough yet to put each one in the proper position." They conclude, "We need to construct more sophisticated experiments of pre-biochemistry in a series of warm pools, studying how complex organic molecules form and how they interact and combine when encapsulated in membranes."[5] The vast complexity in this account, compared to implementations in the lab, could not be starker:

> Multiply all of this together: the waiting-drying cycles happening multiple times each day (think Old Faithful in Yellowstone), variable pool chemistries, highly reactive interfaces, the ability of pools to exchange compounds as gazers splash their contents back-and-forth, and an interconnected, fluid-filled, subterranean fracture network. When you do the math, it looks as if a terrestrial geothermal field of 100 springs can generate 1 million or more new combinations of conditions every year! . . . Each warm pool becomes an "innovation pool," a test bed in which adaptive combinations of molecules rapidly emerge and find ways to grow and reproduce or in which maladaptive combinations fall by the wayside, unable to keep up. It is likely that immense numbers of combinations might have been required to assemble the first primitive version of life, in which case the process would take hundreds of millions of years. But the numbers of combinations in terrestrial geothermal fields suggest that life could have originated and begun to evolve in as little as 10 million years, with the first stages beginning as soon as there was a stable crust peppered with volcanic landmasses amid the oceans, just more than 4 billion years ago.[6]

This is the kind of complexity that is intelligible in terms of Lonergan's notion of emergent probability. Indeed, it is hard to think of a better illustration for it:

> Given sufficient numbers and sufficient time, even slight probabilities become assured . . . Low probabilities are offset by large numbers of occasions, so that what is probable only once on a million occasions is to be expected a million times on a million million occasions. In like manner, the rarity of occasions is offset by long intervals of time.[7]

---

[4] Van Kranendonk et al., "Life on Earth," 32 (slightly altered).

[5] Van Kranendonk et al., 35.

[6] Van Kranendonk et al., "Life on Earth," 34–35.

[7] Lonergan, *Insight*, 136.

Emergent Probability

The volcanic pool scenario brings together all the necessary conditions for the emergence of ב. A superb example of emergent probability, it will also serve to differentiate it from developmental processes. Recall that Luisi and Bitbol's paper is framed in developmental terms proposed by Piaget. This is a fundamental mistake and the likely reason why the failed to incorporate natural selection, which pertains, not to a developmental but to an emergent probabilistic process. Our conclusion above is that biogenesis occurs with the doubling up of ב, when natural selection, or the fortuitous "establishing of bet structure" at the species level, brings about proper ב at the individual level. In short, biogenesis occurs when emergent probability leads to individuals and their development. In organic life the two are always already at work together. So, they must be clearly differentiated. In this section I discuss Lonergan's treatment of the former, and in the following two the latter.

Formulated by Lonergan in the late 1940's, emergent probability is today considered "an early model of complexity."[8] This is the basic idea:

> Schemes of recurrence are conjoined dynamic activities where, in simplest form, each element generates the next action, which in turn generates the next, until the last dynamic regenerates the first one again, locking the whole scheme into long-term stable equilibrium.[9]

This is obviously applicable to the co-emergence of life and its environment and, concretely, to the autocatalytic and spontaneous self-assembly of liposomal molecules into protocells which form large structures that are far more stable than their single components, and from which arise minimal homeostatic "autopoietic" systems.[10] According to Bretz,

> Bernard Lonergan modeled generic growth as the successive appearance of *conditioned recurrent schemes*, each of which comes into function with high probability once all required prior schemes have become functional.[11]

Once liposomal assemblies have emerged, along with the assemblies that form in them, compartments can emerge whose inside is physically different from the outside.[12] This conditioned recurrent scheme is only made possible by the recurrent scheme of liposomal self-assembly into protocells that came before. Once there are compartments, further conditioned schemes of recurrence can emerge: "Conditioned recurrent schemes can be treated as dynamic black cells of activity which

[8] Bretz, "Emergent Probability," 1. For additional information on Bretz's work on emergent probability see https://arxiv.org/pdf/cond-mat/0207241.pdf

[9] Bretz, 1.

[10] Luisi, *Emergence of Life*, 161.

[11] Bretz, "Emergent Probability," 1.

[12] Luisi, *Emergence of Life*, 185.

themselves may contain internal structures and dynamic schemes of arbitrary complexity."[13]

Encapsulation eventually leads to functional closure, which in turn is a precursor to metabolism. Now, small molecules can cross membranes while larger ones cannot. As solubility increases within, storage of energy occurs in the form of a chemical gradient. Complex recurrent schemes result in further conditioned recurrent schemes.

In general, ". . . emergent probability is a generic heuristic model. Application to specific physical problems requires detailed knowledge of the recurrent schemes' makeup and interrelationships."[14] Again, this is why the origin of life is a "like a jigsaw puzzle with many different pieces, and we do not know enough yet to put each in the proper position."[15] "The universe of possibilities available to any open-ended development process is vast, and the interrelation among the elements can become arbitrarily convoluted as the process proceeds."[16] This applies to the volcanic pool scenario: "All growth in that universe, no matter how complex, depends upon a suitable underlying environmental 'situation,' and an ecology, or niche, to thrive."[17] This is why the co-emergence of life and its cycling environment is so difficult to reproduce in the lab:

> Any full simulation of emergent probability must allow for adaptation to changing situations and ecology, so that recovery can occur when events disrupt underlying schemes, or when separate growths vie among themselves for dominance in a stressed environment.[17]

Clearly, a multitude of hot volcanic pools is an ideal scenario for the relatively rapid process of emergent probability that led to life.

In this light, consider briefly an alternative recent hypothesis that life originated in deep-sea hydrothermal vents. Discovery of these vents along the East Pacific Rise in 1977 was transformative for the fields of geological, biological, and chemical oceanography.[18] Life near these vents is sustained through chemosynthesis. The use of energy released by inorganic chemical reactions to produce food is primarily carried out by bacteria—prokaryotes—at the bottom of the food chain.[19]

> Hydrothermal vent environments experience unique extremes in temperature variation, levels of acidity (pH as low as 2.8), and flow of toxic metals and sulfide. Venting fluids expelled from the seafloor can be as high as 752° F (400 °C) but do not boil because of the exceptional pressure of the deep sea.[20]

Assuming the centrality of ב, oscillation is indispensable. According to Van Dover:

> In all situations, temperate fluctuates or "flickers" over short time scales, as warm vent fluids and cold seawater undergo turbulent mixing ... This kind of short-term variability is

[13] Bretz, "Emergent Probability," 1.

[14] Bretz, 1.

[15] Van Kranendonk et al., "Life on Earth," 35.

[16] Bretz, 2. (Strictly, "development" does not obtain in this process of emergent probability.)

[17] Bretz, "Emergent Probability," 2.

[18] Joye Research Group, "Hydrothermal Vents."

[19] U.S. Department of Commerce, "Chemosynthesis and Hydrothermal Vent Life."

[20] Joye Research Group, "Hydrothermal Vents."

extremely important for organisms requiring simultaneous delivery of o-rich seawater for aerobic respiration and sulfide-rich fluids for support of microbial primary production (Johnson et al.1988a). On longer timescales, many records show periodic influences of tidal motions on temperature, due presumably to some combination of tidally induced pumping and tidal oscillations in near-bottom currents (e.g., Johnson and Tunnicliffe 1985; Little et al. 1988; Chevaldonné et al. 1991).[21]

However, the vent theory has problems. The big one is the vastness of the ocean. "The needed molecules might spread out too quickly to interact and form cell membranes and primitive metabolisms."[22] In other words, this scenario makes emergent probability far less likely. These combinations are many orders smaller compared to the volcanic pool scenario, which in turn lowers the probability of ongoing assembly of further sets of conditions. This would make it less likely that a community of progenotes would form, diminishing the overall probability of "fortuitous ב" at the species level, and unlikelier still at the individual level.

Circadian Rhythms

A second body of evidence for ב is circadian clocks. If ב is at the origin of life, evidence of it must ubiquitous. Is it? Is such a feature shared by all organisms, all cells? According to Wolpert, cells can tell time. We directly experience circadian rhythms with jet lag. This rhythm affects our sleep-like cycle but also hormone secretion, liver and kidney function.

Light intensities set this circadian 24-hour clock. When light reaches the retina in our eyes, signals are sent to a special region of our brain that is the major timekeeper in the body. But it is not just in your brain that there is a clock; it now seems that every cell in the body may have a 24-hour clock ticking away. In tissues as different as liver and lung there are genes that oscillate by being turned on and off with a daily cycle. Cells that have been in culture for twenty years can be made to oscillate again. It is not inconceivable that the clock-like activity in our cells may be more relevant to our health than previously thought. One need

[21] Van Dover, *Deep-sea Hydrothermal Vents*, 86. Van Dover cites, in order: Kenneth S. Johnson, James J. Childress, and Carl L Beehler,"Short-term Temperature Variability in the Rose Garden Hydrothermal Vents Field: An Unstable Deep-see Environment," *Deep-Sea Research Part A: Oceanographic Research Papers* 35 (October–November, 1988): 1711–1722, doi:10.1016/0198-0149(88)90045-3; H. Paul Johnson and Verena Tunnicliffe, "Time-series Measurement of Hypothermal Vent Activity on Northern Juan de Fuca Ridge," *Geophysical Research Letters* 12 (October 1985)::685–688, doi:10.1029/GL012i010p00685; Sarah A. Little, Keith D. Stolzenbach, and Frederick J. Grassle, "Tidal Current Effects of Temperature in Diffuse Hydrothermal Flow: Guaymas Basin," *Geophysical Research Letters* 15 (December 1988):1491–1494, doi:10.1029/GL015i013p01491; Pierre Chevaldonné, Daniel Desbruyères, and Michael LeHaitre, "Time-series of Temperature from Deep-sea Hydrothermal Sites," Deep Sea Research Part A: *Oceanographic Research Papers* 38 (November 1991): 1417–1430, doi:10.1016/0198-0149(91)90014-7.

[22] Van Kranendonk et al., "Life on Earth," 30.

> only recall the sleep problems of depressed patients and how awful jetlag can be. Perhaps in the future there will be clever ways to reset all the clocks in all the cells.[23]

Predating circadian rhythms is the ruling feature Wolpert mentions of genes that oscillate by being turned on and off. Even before genes, this is the foundation of self-regulation. Still, since light is central to circadian rhythms, one may ask how far back it goes. As Schwab observed:

> Our story begins when life-forms recognized light. Although we cannot know exactly when that occurred, it had to be early in life's history—probably in the first few million years, if not earlier.[24]

Cyanobacteria, which emerged between 3.75 and 2.6 billion years ago, make sophisticated use of sensitivity to light:

> Cyanobacteria are photoreceptive through a compound they possess—chlorophyll, a photopigment. Chlorophyll may well have been the first photoreceptive witness, although there are other candidates. Using chlorophyll, cyanobacteria employ light to produce energy in the form of a carbohydrate. Cyanobacteria are our first *known* proof, or witness, of photoreception with a response to the light, shaped by the early sun's energy and the climate of the day. Cyanobacteria exhibit one key feature of early life on earth that would be essential in later evolution. That is, they are capable of photosynthesis in which organisms make nutritional substances internally with the use of sunlight.[25]

Cyanobacteria are still around—aquatic, photosynthetic, usually unicellular—and they comprise one of the largest and most important groups of bacteria on earth. These bacteria have profoundly shaped the course of evolution and ecology throughout Earth's history. They produced our oxygen atmosphere during the Archaean and Proterozoic Eras. Before that, the chemistry of the atmosphere was unsuitable for life as we know it. They also originated plants; the chloroplasts by which plants make food are cyanobacteria living in their cells. (Though known as "blue-green algae," cyanobacteria are not algae; only the chloroplast in eukaryotic algae is related to them.) Sometime in the late Proterozoic or early Cambrian, cyanobacteria took up residence in some eukaryote cells. Known as endosymbiosis, this is how light-sensitive eukaryotic mitochondria arose.[26] Being critically important to eukaryotes, sensitivity to light is critical to any organism whose cells consists of a membrane containing a nucleus and other structures (organelles)—that is, all animals, plants, and fungi.

Is sensitivity to light associated to ב? According to Loza-Correa, et al.:

> Circadian clock genes are vital features of eukaryotes that have evolved such that organisms *can adapt to our planet's rotation to anticipate* the coming day or night as well as unfavorable seasons... Many living organisms from bacteria to eukaryotes have evolved internal mechanisms to synchronize their physiological functions with *the Earth's rotation cycle*.[27]

[23] Wolpert, *How We Live and Why We Die,* 65–66.

[24] Ian Schwab, *Evolution's Witness,* 5

[25] Schwab, *Evolution's Witness,* 8 (slightly modified).

[26] "Introduction to Cyanobacteria."

[27] Loza-Correa et al., "Circadian Clock Proteins," abstract (emphasis added).

Synchronizing functions with the Earth's rotation is not restricted to eukaryotes. Prokaryotes, of which bacteria are the most familiar to us, do so too. Unlike eukaryotes, prokaryotes are single-celled organisms that lack a membrane-bound nucleus, mitochondria, or any other membrane-bound organelles. Prokaryotic cells were the earliest life on Earth. They are the smallest form of life that can live independently. Most are tiny single cells though some form larger, multi-celled structures. They gave rise to the domains of Bacteria and Archaea. Some of these organisms, mainly from the latter domain, can live in extreme environments such as the hot volcanic pools and the hydrothermal vents proposed as sites for biogenesis. Eukaryotic cells emerged later. Many biologists believe that, if life exists elsewhere in our solar system, or beyond, it must resemble prokaryotes. It is the lipid-protein structure membrane of prokaryotes that Luisi has approximated in the lab, along with hydrophobic substances that move through the lipid layers by diffusion and hydrophilic substances that approximate the role of membrane proteins that move molecules against a concentration gradient.[28]

So, just how far back does light-sensitivity go? Since when are prokaryotes sensitive to light? It seems to go back to shortly after life originated. This is stated in what follows in very technical terms: "*[T]he bias in favor of 12C* [Carbon isotope 12] *during photosynthesis*" was first carried out by *microfossils of prokaryotic affinity*. Notice the emphasized terms suggesting this might be one of the earliest instances of distinguishing:

> With the currently available geological record at hand, the existence of life on this planet as from at least 3.8 Gyr ago seems so firmly established as to be virtually unassailable. Specifically, various disparate lines of evidence have merged to indicate (1) that the surface of the Archaean Earth had hosted prolific microbial ecosystems as is testified by a quasi-continuous record of microbialites ('stromatolites') and associated *microfossils of prokaryotic affinity* over 3.5, if not 3.8 Gyr of geological history, and (2) that the sedimentary carbon record has preserved the isotopic signature of autotrophic (notably *photosynthetic*) carbon fixation over the same time span. With the observed enrichment of isotopically light carbon in sedimentary organic matter largely consonant with *the bias in favor of 12C during photosynthesis,* the mainstream of the carbon isotope record can be best explained as geochemical manifestation of the *isotope discriminating properties of the ribulose-1,5-bisphosphate (RuBP) carboxylase reaction of the Calvin cycle* suggesting an extreme degree of evolutionary conservatism in the biochemistry of autotrophic carbon fixation. As a consequence, partial biological control of the geochemical carbon cycle was established already during Early Archaean times and fully operative by the time of formation of the Earth's earliest sediments.[29]

Notice again the terms used: "bias in favor of," "discriminating properties," "biological control." Do these point to the researcher's biomorphism or to properties of the earliest forms of life? We now know this is a fundamental question. If these properties are attributed to nonliving molecular structures, it would imply that reactivity

[28] "Prokaryote."

[29] Schidlowski, "The Initiation of Biological Processes on Earth," 143–156. In the early Archean the Earth's crust had cooled and continents had formed. The earliest fossils of organism go back to this time.

to light precedes and could be a condition for the possibility of life. If, on the other hand, it pertains to living beings, this would point to an intimate and extremely early relation between life and light. This is the hazy boundary between nonliving "autopoiesis" and living autopoiesis. Of course, the earliest ב may date back further and be unrelated to light, a result of some other bias or discriminating property.

In any case, the evidence conclusively shows that ב is ubiquitous. Also, its relation to light goes back to the earliest fossils known. Seeking evidence for ב seems to be akin to searching for actual triangles to verify that the sum of their inner angles adds up to two right angles. The second law of thermodynamics requires a boundary to separate inner and outer processes and functional closure determines what is allowed in and what is not. This is a cell. So, a cell is a distinguishing distinction, and this implies ב.

In conclusion, the empirical evidence here presented is clearly amenable to intelligibility in terms of the fundamental notions of distinguishing distinction and ב, thereby reinforcing these notions as foundational for a higher conjugate form to understand aliveness.

What Constitutes a Circadian Clock?

Circadian biological clocks, it has just been shown, are an essential link between an organism and its environment. We are interested in them, not only in their own right as in the previous section, but also because their underlying working principle is an archetype of self-regulation. Regarding the simplest clocks known that belong to prokaryotic cyanobacteria, recent research demonstrates *Synechococcus elongatus'* clock can be reconstituted in vitro with three proteins (KaiA, KaiB, KaiC). Upon the addition of ATP, an oscillator starts that sustains a 22-hour rhythm over several days.[30] Such clocks

> are self-sustained biochemical oscillations that are temperature compensated and persist in absence of external environmental cues. [They can] respond to changes in the environment like intensities of light and dark and to entrain its phase to the local environmental cues.[31]

That the oscillator links the organism to its environment is a critical point: circadian rhythms are *both* internally generated *and* continually synchronized to the daily cycle. Light is the primary synchronizer.[32] In human cells, such an auto-regulatory cycle is a feedback loop in which one set of genes turns on another set of genes, and then these turn off the original set of genes.[33] In general, ". . . both prokaryotic and eukaryotic circadian systems involve a *negative feedback loop* with genes encoding

30 Hut and Beersma, "Evolution of Time-keeping Mechanisms."

31 Loza-Correa et al., "Circadian Clock Proteins," 30.

32 Takahashi, "The Clock in our Genes."

33 Takahashi, "The Clock in our Genes," see minute 9:00 to 10:00.

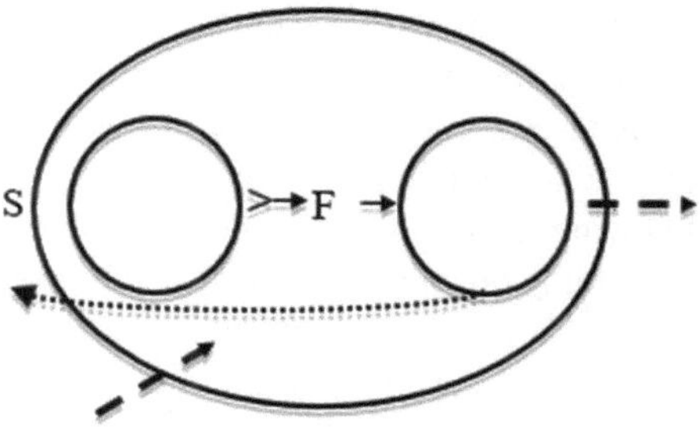

Fig. 4.3 A negative feedback loop that interacts. The system self-regulates by interacting with the world beyond its boundaries

proteins that repress their own transcription."[34] In other words, these two-element regulatory systems, which are a primary form of recursive self-reference, are foundational to living being. This is the *what* that they accomplish.

The fundamental diagram for a circadian clock is identical to the negative feedback loop of an automatic heating system. While in the former the boundary is the room whose temperature is regulated, in a biological clock the components are in a bounded system. Both result in oscillation.

As basic components, call them C1, the two components are independent. Together, call it C2, the two components mutually self-regulate. Altogether, call it C3, C1 and C2 serve a function for the larger system. C3, or forness, is inherent to the functioning of C1 and C2 (Fig. 4.3).

In congruence to what has been maintained throughout this study, C3, the highest level, is indispensable to the intelligibility of the whole system. The system is both self-referent and refers systems beyond it. This kind of nested and unfolding structure-processes may be referred to as a *C123*. The dynamic interaction between C1, C2 and C3 is mostly stable, yet small variations allow systems to make small adjustments and therefore unfold:

> Circadian rhythms allow organisms to anticipate and prepare for precise and regular environmental changes.... [R]hythmicity appears to be as important in regulating and coordinating *internal* metabolic processes, as in coordinating with the *environment*."[35]

That a self-regulating cell interacts with itself via its environment, and particularly via other cells in is surroundings, is a fundamental metaphysical point, given that it blurs the neat distinction between 'outer' and 'inner.' This is the topic of joinings that will be discussed at length in chapters that follow. As we saw, the condition for the emergence of functional closure was multiplicity in, nonliving species. It follows that in living distinguishing distinctions, the boundary between inner and outer is always already hazy and fluid. It is therefore a fundamental error to overdetermine their autonomy by abstractly distinguishing between them.

---

[34] Loza-Correa et al., 5 (emphasis added).

[35] *Wikipedia*, "Circadian Rhythm" (emphasis in original), cites Sharma, "Adaptive Significance." There is a vast literature and ongoing research to support these claims as reported in peer-reviewed specialized journals such as *Chronobiology International*.

This brings us back to the radical ontological and epistemological difference between abstract distinctions and functional closure that goes to the very foundations of biology. It is simply not possible to understand aliveness merely through abstract distinction. Obviously, clear and distinct ideas have a key role to play in knowing aliveness—it is the very exercise we are engaging in here and now. But something beyond them must also be involved.

In will be shown that the self-referential and oscillating character of circadian clocks is at the outer limit of conventional logic. *Laws of Form* makes sense of such self-referential oscillation and suggest how one might determine the depth of nested distinctions required to build a minimal clock. Since negative feedback loops are concretely carried out by two-component regulatory systems, this points to the deep relation between aliveness and logic. The root is the simple fact that both are founded on distinction. The spirallic nature of circadian clocks, like all two-component control systems, points to ב as the proper conjugate for inquiry into aliveness. We now turn to its inherently anticipative character.

Earliest Indication (EI)

We conclude this chapter returning to the difference between emergent probability and development. Their fundamental and overriding difference is that development is founded on anticipation. With the advent of life, emergent probability gave birth to development. Our biosphere, then, is made up of myriad nodes of causation in process of development.

We begin with a puzzle.

It was a complete mystery how Alpha, our female Labrador retriever, anticipated the children's imminent arrival home from school. It took about a minute from the time their school bus pulled up in the street to when they walked in the front door. However, at least half an hour earlier, Alpha began to get agitated. How could she know? Did she have a 'sixth sense'? Or should we deliberately ignore what we observed? This went on day after day, month after month. For years we wondered how she did it until, at long last, we understood. The answer was so obvious it was almost embarrassing.

When the kids were due home from school the whole neighborhood filled with school buses, all like theirs, following similar routes, with similar stopping and starting patterns and similar smells. No doubt Alpha would have done the same with any set of events that systematically preceded the children's arrival—a factory whistle, the unique sound of a train going by, and so forth. She had identified the *earliest indication*—henceforth EI—that announced their arrival.

Alpha is subtly attuned to her environment and the progression of the day—the lights, sounds, and temperature of the yard, the household routines, and innumerable other events known only to her. Since she was a puppy, the children have let themselves out into the garden to play with her. Looking forward to this, she learns that before doing so they approach the glass door, through which she can see them.

So, she learns to observe their movements before they open it. The day comes when she associates the sound of the front door opening and closing with their arrival, whether they come out to play or not. She is delighted when they arrive home. Later, she notices the sound of the outer gate. When she associates this with the front door buzzer, she can next notice the preceding sounds and smells of their school bus arriving. Eventually, she notices this is preceded by the sounds and smells of multiple school buses all around the neighborhood. During school days—when the children leave early and are absent most of the day—she verifies this pattern regularly. So, over time, a clearly identifiable progression of smells and sounds indicates their imminent return.

It could be said that Alpha was "conditioned," Pavlov's name for this innate capacity to associate events, though, more accurately, she "conditioned herself" because she loves being with the children. Underlying her capacity to associate is an innate drive to do so, and, therefore, to do so recurrently. Alpha associates new associations to previous ones *recursively*, in a process that allows her to build a chain of ever-earlier events until she identifies the EI that the children's arrival is imminent. EI, then, refers both to the set of earliest indications arrived at and to the recursive heuristic structure or vector by which they are arrived at. This is Alpha's heuristic structure—the method by which she arrives at EI. By the way, this is also how she communicates with us. If she wants to go for a walk she stands by the door. Later she learns to sniff at the drawer where her leash is kept. Eventually, she may even rush excitedly back and forth from door to drawer to help us understand. EI therefore accounts, not only for how the environment is made intelligible, but also how we use this intelligibility to communicate. So, EI is also the heuristic structure by which we ourselves understand Alpha's heuristic structure—her understanding and her communicating. Our own search for explanations, or antecedent events, follows a similar vector. Thus, EI is the heuristic structure by which living beings recursively associate events, and nest these associations, and how we do it ourselves in all aspects of our lives, including when we seek to understand.

In Lonergan's terms, and in relation to animals, EI is a specific heuristic structure to understand psychic or intelligent operations as *flexible series of schemes of recurrence*.[36] Children going and coming home from school is such a flexible scheme—as are the sounds and smells of traffic in a neighborhood. So is the interrelation between them—as established both by Alpha and by us. Noticing ongoing regularities in her continually changing, yet cycling, environment, Alpha entrains what is of interest to her. Anticipating, she displays her capacity to establish bet structure, just as we do when we grasp that this is what she is doing.

Founded on ב, then, EI comprises the recursiveness of anticipation. In all organisms, ב unfolds through EI, whether it is the cells of an immune system, an individual monkey, or the monkey group as a whole. Recalling the Santiago School, this is the concrete equivalence between metabolism and cognition. Ever more complex metabolism is made possible through deeper and ever more complex nesting of

[36] Lonergan, *Insight*, see section "Genetic Method," 484.

anticipation and, conversely, ever more complex nesting of anticipation allows ever more complex metabolism. More generally, complexity in organization calls for, and results in, complexity in anticipation. Earliest indication—both the provisional construct and the heuristic—emerged with ב. It is how the unfolding of ב is operationalized in practice. This unfolding began with the advent of life 4.28 billion years ago and it is at work at this very moment, as we probe that beginning.

EI as Integrator-Operator

Lonergan proposed that the unfolding of biological processes (i.e., growth, development, learning) must be accounted for through a genetic or developmental heuristic structure. Concretely, this is done through successive integrations in terms of an operator. The operator accounts for the sequence of prior integrations being replaced by more developed successor integrations:

> The relevant heuristic structure is 'Specify the operator.' In general, development is higher system on the move. The operator is the higher system known by grasping the interrelated set, not as the integrator of a given stage, but as the source of the differences that appear in the next stage . . . The matter may be clarified by an illustration. There seems to be a general principle of development that is named the law of effect. It contends that development takes place along lines of successful functioning. Thus, a tree in a forest puts forth branches and leaves not to its sides but at its top. Now, such a principle offers a specification of the operator. For the operator is the higher system on the move. The higher system is the ground of the flexible circle of schemes of recurrence in which the organism functions. The law of effect states that the ground of functioning advances to a new ground of functioning where functioning occurs successfully.[37]

The law of effect cited by Lonergan was advanced by Edward Thorndike in 1905:

> Responses that produce a satisfying effect in a particular situation become more likely to occur again in that situation, and responses that produce a discomforting effect become less likely to occur again in that situation.[38]

Like natural selection, character traits that provide an advantage persist, not those that do not.[39] (In operant conditioning theory, the law of effect was transposed in terms of "reinforcing" and "punishing".) Satisfying behavior is more likely to recur, dissatisfying behavior less so.[40] Because this is determined behaviorally, it cannot be accurately predicted. Thorndike refuted the idea that anecdotal evidence of behavior is weak because he hypothesized that animals understand their physical environment by trial and error.[41] Lonergan illustrates the law of effect with the growth of trees and generalizes it into the notions of integrator and operator, suited

[37] Lonergan, 491–92.

[38] Gray, *Psychology*, 108–109.

[39] Schacter et al., *Psychology*.

[40] Mazur, *Learning and Behavior*.

[41] Mazur.

to the full range of biological inquiry. He says: "one's understanding of the operator begins to be an instance of higher system on the move in the development of scientific knowledge of development."[42]

This too is the case when understanding EI. Given [subjects] that establish bet structure, EI as operator may be specified as follows. Consider the events K, L and M:

- First, an organism is affected by a recurring event M.
- Second, by bet structure, the organism establishes M as recurring.
- Third, once M is identified, a recurring event L that regularly precedes it may be associated with M's recurrence. L may be within or without the organism.
- Fourth, an organism may take L as indicator of the imminent occurrence of M.
- Fifth, the previous sequence may recur in relation to L. Thus, a recurring event K that precedes L may itself be identified by the organism as indicator of M. Each in turn is an EI of M.
- Sixth, this recurring identification of an EI, or EI vector, is flexible and tends to produce ever-earlier indicators.[43]

EI as operator, then, similarly to the law of effect, refers to the earliest marker as well as to the vector, the recurring process by which markers are arrived at. Together these two aspects of EI define it as a higher system potentially always on the move. Markers such as K, L and M, can be any difference that makes a difference, either within or without an organism—a vibration, the presence of certain particles, shifts in temperature, pressure, pH, and so forth. While an EI must be a correlation, it need not be causal. Moreover, EI is recursive in the strict sense defined above because a new EI is derived from a previous EI. The set of EIs nest within each other precisely like matryoshka dolls. After adverting to her most recent EI, Alpha verifies all her previous EIs. This is inherent to EI, the vector.

What makes EI an operator is the ongoing possibility of ב. Because she is a dog, we refer to bet structures established by Alpha as 'associations.' Given her innate propensity to associate—that is, her innate propensity to identify these markers and link them, forming vectors—any EI as a marker that turns out to systematically indicate a difference that makes a difference to Alpha, such as the imminent return of the children, is reinforced in two ways. It both secures previous anticipations (integrator) and sets up the possibility for new, more anticipative ones (operator). Together, these two recurrently set up the possibility for ever earlier markers, the vector for further anticipations. Each instance of EI *verifies* Alpha's construction of her experience, allowing her to further *interrogate* this experience. This dovetails with what we advert to in our own cognitional activity: a question is an operator that brings further questions and perhaps further insights. Thus, for instance, the insight here and now into EI as operator, like the law of effect, brings up further questions

[42] Lonergan, *Insight,* 492.

[43] Note that "may" is used throughout to underscore the inherent unpredictability of behavior. Most significantly, EI applies to both inner and outer sense data. This points to the fundamental liminality of organisms.

about the nature of living being. It accounts for Alpha's distinguishing activity and shows why it serves as a general genetic heuristic to investigate that activity. Notice, first, the value of this anecdote; and second, that it refutes how operant conditioning theory transposes the law of effect, simply because Alpha does not "reinforce" or "punish" herself. Though, quite clearly, she is gratified by the outcome of her inquiries.

A key question that comes up is the following. Given an organism that adverts to M, by what process does L enter its horizon, and subsequently K, and so on? So, for instance, how does Alpha jump from the EI (children open glass door to the garden) to EI (children enter front door to the house)?

Anticipating discussion in later chapters, and generalizing, this can be accounted for in terms of Hegel-Lonergan's nest of terms, the progression in our understanding that moves implicit➔explicit➔abstract➔alien➔mediation➔reconciliation. In this case it pertains to the account of how something comes to be part of the wherewithal by which a subject confronts the environment. This is a highly differentiated version of how, in the sequence Ā → A→/→Ā, Ā becomes / (and then Ā). This is the process by which a new EI is arrived at, in this case, the transition from M to L:

- Implicit—L, though there all along, has not been adverted to.
- Explicit—L is adverted to.
- Abstract—L is noticed and interrogated.
- Alien—L is resisted.
- Mediation—elements begin to arise by which resistance to L is overcome.
- Reconciliation—answers, verification that L recurrently antecedes M.[44]

For living beings, the world becomes progressively intelligible as new EIs are acknowledged, and old ones are subsumed within them. EI provides an account of progressive sense making, not only in the lives of individual organisms, but also in evolutionary terms, starting from the most elementary level in single primitive living cells, through ever deeper and nested layers of anticipation, right up to our own questions and insights.

And this brings us to EI, such as Alpha's heuristic structure as a whole, as *integrator*.

EI functions as integrator because it unfolds on a second level, also self-referentially and therefore recursively. This is a third level, above the two previously explained in terms of the two uses of the *select*. In *natural selection*, at the species level, character traits that provide an advantage persist. Through EI, individual organisms select features of the environment that make a difference to them. Though the law of effect refers to the latter, it applies just as well to behaviors of populations naturally selected for. This was Thorndike's point, as quoted above and it is also what determines a specie's *Umwelt*. A third level emerges when organisms are complex enough to evaluate their previous selections. This typically occurs with social interaction, which leads to group selection—in both senses of the term.

[44] This very process occurs with the reader as she or he confront a new idea.

The name for this is culture. A famous instance of this is Imoko, a dominant female macaque monkey in the Japanese island of Koshima, who started dipping her potatoes in river water to wash them. The idea soon spread to her family, then to her group, and eventually, over a decade, to the entire population of macaques on the island. The same happened with another insight of Imoko's, dipping her potatoes in saltwater for added taste.[45]

As a higher system on the move, the two higher level selections act together instrumentalize anticipation and its process of unfolding. Recall that ב is to treat events as recurring, as members of a set, as tokens of a type. But there is in A=A, A≠A an inherent process of drift. A species that feeds on one insect may, for some reason, generalize to feeding on a variety of insects. This may occur at both the natural selection and the individual selection level. Natural selection is about the increase or decrease of selective fitness, whereas in the case of individuals selections, individuals recall a taste and select to again eat a new insect or not. On the third level, this new diet becomes the norm. Throughout, given food = food, generalization may be expressed as [A=A]➔[A = A] or simply []➔[ ].

[]>[ ] expresses the drift inherent to A=A that allows for the unfolding of anticipation, and therefore the higher-level vector of anticipation itself as a higher system on the move. This is the vector from a current integrator-operator [] to another []. Therefore, EI as operator in individual organisms turns, by natural selection or through social interaction, into an integrator that in a population opens a door to higher levels of EI.

In short, as layers build one upon another, EI becomes an ever more sophisticated integrator-operator. In Chap. 10 it is shown how anticipation folds back upon itself, resulting in higher levels of integrator-operator.

Individuation itself unfolds by expanding the range of possible modes of interaction with the environment. It too is thereby an integrator-operator. With the unfolding of self-determination and interrelation, organisms unfold as nodes of causation. As organisms and environment are increasingly monitored and mediated through EI, the environment becomes an ever more integrated network of EIs at ever higher levels. Layers of inner and outer monitoring build upon each layer. As further layers of arbitrariness are introduced, it becomes palpably obvious that organisms could never be said to react in the mechanical sense. Rather, organisms co-create as they inter-act.

This has been inherent to life from the outset. Recall how the colony of progenotes, exposed to the environment's cycles and fluctuations, transitions from emergent probability—operating at the species level—to the genetic development of living autopoietic cells. From the moment life emerges each individual cell is an operator that carries the potential to be an integrator. This is precisely the bifurcation when, to the "fortuitous bet structure established" at the species level through the natural selection of nonliving molecular structures, is added the bet structure established at the individual level by organisms capable of EI. Cells that do EI, no

[45] Kawai, Masao, "Newly-acquired Pre-cultural Behavior…"

matter how minimally, are likely to be favored by natural selection. The effect of this ripples through the whole population, gradually setting up networks of EI. Every slight additional capacity for EI at once secures previous capacity for anticipation and sets up the possibility for even greater anticipation *in its own lifetime*. Like molecular autocatalytic processes, but now on a higher level, EI speeds up EI. Echoing Hoffmeyer one might say:

> The basic unit of life is ב, not the molecule.

Or, slightly differently,

> The basic unit of life is EI, not the molecule.

EI sets up new and ever greater possibilities to entrain with the environment, and such entrainment tends to be ever more varied and open-ended. Thus, the long-term formation of the species is ever more determined by the EI of cells, groups of cells, groups of groups of cells, and so forth. With the recurrence of EI sensitivity grows and sense organs emerge and develop. Structure follows process.

The foregoing shows why EI is a heuristic structure. Like naming x as the unknown in algebra, it aids by stating the How to be sought, given the What. Inquiring into biogenesis, it aids in filling the blanks of the elements at work in its emergent probability. In inquiry into the stages of development of organisms and their evolution, it helps to fill in the blanks about vectors of anticipation and their potential interactions.

Recall the Santiago school's notion of metabolism as the correlate of cognition. Rather than a *descriptive* account by analogy to Piaget, EI can provide an *explanatory* account of the "stratified conception of cognition and its relation with life":[46]

- At the null stage of cognition, no bet structure is established. Molecular structures appear as 'autopoietic' to us, but they are not so *for themselves*.
- (0.5) An intermediary level is introduced between the null state and the first state. Molecular structures appear as units to us but, being "autopoietic," they are not *for themselves*. However, their replication allows "ב" fortuitously at the species level. This, as has been suggested, is a protean anticipativeness that predates proper autopoiesis and is a necessary condition for its emergence.
- After crossing a hazy boundary, bet structure is established and EI begins. This is anticipation at the individual level. Now anticipative, the units themselves can be said to be *for themselves,* properly autopoietic. These are the first cells. The two levels of selection at work together make EI an integrator-operator []➔[ ].
- In stage 2, the survival of cells depends on the unfolding of individual EI. This results in progressive individuation which in turn sets up the possibility for the unfolding of anticipation itself, a higher level []➔[ ].

EI as an explanatory account fully justifies the Santiago schools' "insistence on the 'cognitive' status of metabolism, which both maintains the identity of an organized

[46] Bitbol and Luisi, "Autopoiesis With or Without Cognition," 105.

unit and implies dynamical interaction with the environment . . . ."[47] If before it was said that a system is alive if it establishes bet structure, or anticipates, it must now be clear that anticipation unfolds by EI. EI points to the depth of aliveness of organisms because their very form has resulted from sets of flexible schemes of recurrence that can make sense of flexible sets of schemes of recurrence.

We can summarize by taking the whole sweep of this survey so far. EI shows, beyond the functional closure that is the thermodynamic condition for the possibility of organisms, that aliveness, as founded on the arbitrariness inherent to distinguishing, not only makes every living cell a [subject] with agency, and therefore a node of causation, but also a [subject] that is in ongoing process of development. Again, so radically does EI underscore the networked co-emergence of life and environment, that the environments as such—indeed, the whole biosphere—is *qua* cognitive, *qua distinctional*. Living beings do not inhabit a physicochemical world but an ecology of mind. We inhabit the living world.

Monkey EIs and Social Life

The monkeys were deeply superstitious of our hand hidden under our jumper gently thumping on our chest. It was apparently an uncanny menace to them and even to us. If something moves the monkeys must immediately find out what it is. Occasionally we resorted to this gesture to shoo them along a runway that needed to be shut for cleaning. If they didn't budge, one of us would resolutely set off towards the monkey kitchen to fetch a squirt bottle, which they detested. In only took an "OK, that does it!" and two gritty steps *away* from them to send them off scrambling. Monkey-human relations were a profuse and unending stream of earliest indications.

The Santiago School's theme of the co-creation of organism and environment, and the volcanic pool scenario's dense multiplicity of molecular structures poised to slip into life, brings to mind a question often asked by visitors. How can the Sanctuary's be a natural colony of monkeys in this habitat so different from their rainforest? We were usually standing somewhere by the monkey territory, a set of large outdoor enclosures and heated indoor areas built over years with interconnecting runways and corridors, furnished with beams and ropes, as well as several outdoor wooden huts and two enormous beech trees. True, also their nourishment of mostly fruits and vegetables was provided by us. But wooly monkeys are, just as humans, first and foremost social animals. This is why this species, so notoriously difficult to keep and which had never bred outside the wild, thrives here. By then, some of the monkeys were of the third and fourth generation. Started with rescues from zoos and pets, a proper social group was born with the arrival of the first baby, as males and females learnt to take their place in a nascent hierarchy. With their

[47] Bitbol and Luisi, 105 (emphasis added).

physical needs met, just like humans, it is the monkeys themselves who provide the social environment for each other. And the deep bonds of friendship between monkeys and humans enrich the lives of both, the heart of what Williams called "a study in interliving."

After months of listening to stories about the monkeys, my wife concluded that for a decade I had lived attached to a high-voltage cable. It is true. Entering the spacious monkey room, its floor dressed in a thick layer of fresh wood shavings, their fragrance interlaced with the monkeys' pungent odor and traces of fruit, one found oneself in a quiet world of gymnastic glides on squeaky ropes and the patter of tiny feet racing along wooden beams, accompanied by the almost inaudible puffs, and the snuffs and peeps of females and their offspring interacting. A state that recalls the stillness in the heart of nature best describes the blissful liveliness that instantly overtook you. There, facing you from the ground as a baby monkey, nature reached out requesting to be lifted, and then, squeezing your fingers as a handlebar to swing on, chuckled with a mirthful grin as your nose was tweaked by tiny feet challenging you cheekily to a game of tickle.

Chapter 5
Formal Definition of Distinction

Introduction

Can the mathematical ideal of knowledge apply to biology? Galileo's aphorism, "Mathematics is the language in which God has written the universe," points to the inauguration of a new mathematical ideal of knowledge in Western science. This principle was the motor of the scientific revolution of the seventeenth century and the key animating idea that underwrote the enterprise to comprehend the world in mechanical terms. Thus, reaping the fruits of Galileo and Descartes, in 1687 Newton stated his first three laws of motion in *Philosophiæ Naturalis Principia Mathematica* (Mathematical Principles of Natural Philosophy).[1] Thereby giving birth to the modern science of physics, its success impels other disciplines to embrace its ideal and emulate its methods. Over 300 years later, we can ask whether this ideal can obtain in biology. This chapter argues that the mathematical ideal is at odds with biology if it is to be inquiry into aliveness. It does so by examining Spencer-Brown's *Laws of Form* (*LoF*). Concretely, we do so by examining how ב is at work in *LoF*.

To say that living beings are distinguishing distinctions is to say that aliveness is the modality of being that knows. Aliveness is being-by-knowing. ב is the foundation of being-by-knowing, the shared foundation of living knowing and being. As has been shown, ב is the root of twoness, repetition, recurrence.

Wittingly and unwittingly, *LoF* is a treatise on repetition. How so? We shall analyze *LoF* and find that the unwitting part is what says most. It was mentioned that *LoF* could just as well be named *Laws of Distinction*. So, its definitions of distinction are of inherent interest, as well as what the formalism accomplishes with them. But there are two additional reasons why *LoF* is of great interest. First, because it touches on the foundations of mathematics, and second, because it allows us to examine the nature of mathematical formalisms.

---

[1] See Gorham et al., *The Language of Nature.*

© The Author(s), under exclusive license to Springer Nature Switzerland AG 2023

D. C. Mayer-Foulkes, *The Nature of Living Being*, Biosemiotics 26,
https://doi.org/10.1007/978-3-031-24789-7_5

That being so, *LoF* affords the opportunity to cut to the chase in relation to our key question in relation to the mathematical ideal of knowledge, What at the heart of mathematics is the obstacle for biology? Once this has been examined, we can ask the questions that follow, What must be the ideal of knowledge for biology? What can provide the foundations for biology?

Mathematics, as we shall see, like aliveness, is founded on distinction. In a nutshell, what must be examined is the relation between distinguishing, the verb, and distinction, the noun. Aliveness unfolds distinguishing, mathematics unfolds distinction. This clearly points to the profound relation between living being and mathematics.

Concretely, we shall be examining the relation between ב and logic, specifically, the foundations of logic. As *LoF* shows, distinction is a significant, inner layer, of mathematics. *LoF* presents an arithmetic that may stand as the foundation of logic, and thereby affords us the opportunity to examine this foundation. Doing so will allow us to reflect on whether the mathematical ideal of knowledge is appropriate for inquiry into aliveness.

We begin at the beginning, then, with the formal definition of distinction in terms of two axioms, upon which a powerful mathematical formalism is built. Our aim is to examine how ב is at work in these axioms, therefore in the formalism of *LoF*, and therefore in mathematics as a whole. Since ב pertains to aliveness, this will tell us much about the part played by aliveness in the formalism, in logic, and therefore in mathematics in general. *Mutatis mutandis*, this will also tell us much about distinguishing distinctions and their biomorphism as they parse the world in terms of distinction. These two strands together point us towards method, therefore also the proper heuristic structure for inquiry into aliveness. We start, then, with a brief gloss of *LoF* as it states its axioms.

The Axioms of *Laws of Form*

In Chap. 1 it was mentioned that the definition, *distinction is perfect continence*, is illustrated in *LoF* by a circle drawn on a flat plane. This image helps to grasp the formal definition of distinction. Stated by means of two axioms, these are also the only two operations or laws of transformation of the mathematical formalism. They are *the law of calling* and *the law of crossing*:

> Axiom 1. *The value of a call made again is the value of the call.* That is to say, if a name is called and then is called again, the value indicated by the two calls taken together is the value indicated by one of them. That is to say, for any name, to recall is to call.

> Axiom 2. *The value of a crossing made again is not the value of the crossing.* That is to say, if it is intended to cross a boundary and then it is intended to cross it again, the value indicated by the two intentions taken together is the value indicated by none of them. That is to say, for any boundary, to recross is not to cross.[2]

[2] Spencer-Brown, *Laws of Form*, 1–2.

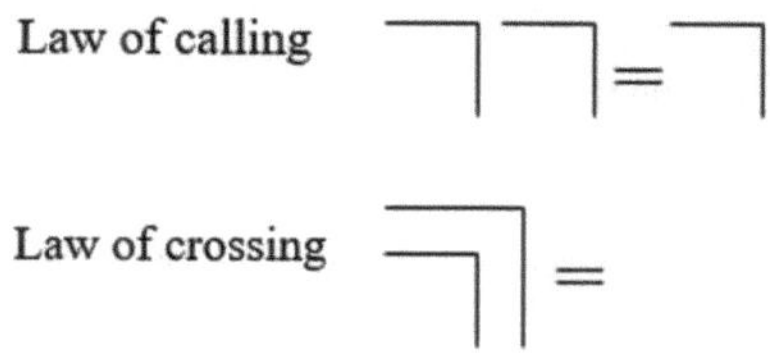

Fig. 5.1 The axioms of *Laws of Form*

The first axiom states that if you call something "A" and then call it "A" again, the value of those two calls is the same as the value of one of them—an A is an A is an A… The second axiom states that if you cross a boundary from A to Ā, and then cross it again, you go from A to Ā and back again, which is the same as not having crossed at all (Fig. 5.1).

Taken together, and in formal notation, these two equations define distinction.[3]

Stating these two axioms in terms of two radicals, the construction of the formalism begins. These radicals are the single mark Γ ('cross') and , its absence.[4] With these, the formalism builds a primary arithmetic by means of operations that follow from the two radicals. Various interpretations of this arithmetic are possible. *LoF* discusses two of them—as the foundation of electronics and as the foundation of logic. It is demonstrated that the primary arithmetic can stand as the "Boolean arithmetic" that underlies Boolean algebra, hitherto never explicitly stated. Indeed, Boole's postulates are proven as theorems in the arithmetic.

Now, it is immediately obvious that repetition is built into both axioms since both are stated it terms of 'again'. Both formally stated axioms involve two crosses. Moreover, not only is 'again' explicitly used in both axioms, but the explanation for each axiom involves repetition. Thus, recurrence is built into the formalism at its very foundation.

In short, the ground of the axioms is ב, the seminal feature of aliveness.

Recall the discussion on *linear distinctions* in Chap. 1. Axiom 1, *the law of calling*, like A=A, may be thought of as the principle of continuity; and Axiom 2, *the law of crossing*, A/Ā, may be thought of as the principle of discontinuity. This converges with the definition of *distinction as continuity-discontinuity*, and by extension, *scheme of recurrence*.[5] So, both distinction as perfect continence, and linear distinctions, are congruent with the axioms. Applied to electronic circuits, for instance, the arithmetic built on the axioms is about On and Off, and the switching

[3] Notice the striking convergence of the etymology of distinction as marking and separating with the two axioms. (See Chap. 2, section "Definitions of distinction".)

[4] For typographical reasons the symbol Γ is used in the text. In *LoF*, the symbol used is the one shown in Fig. 5.1.

[5] Continuity and discontinuity constitute the two underlying currents of mathematics. Together they produce classic problems, amongst them Zeno's paradoxes from the fifth century B.C.E. Little new was said about these for 22 centuries until the advent of calculus, the mathematical tool developed independently by Newton and Leibniz to make sense of *process*. Thus, the title of Robert M. Young's introductory book to the subject is appropriately: *Excursions in Calculus: An Interplay of the Continuous and the Discrete*.

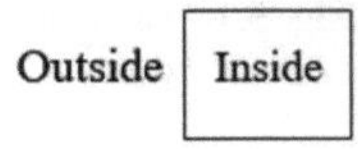

Fig. 5.2 Inside and outside regions

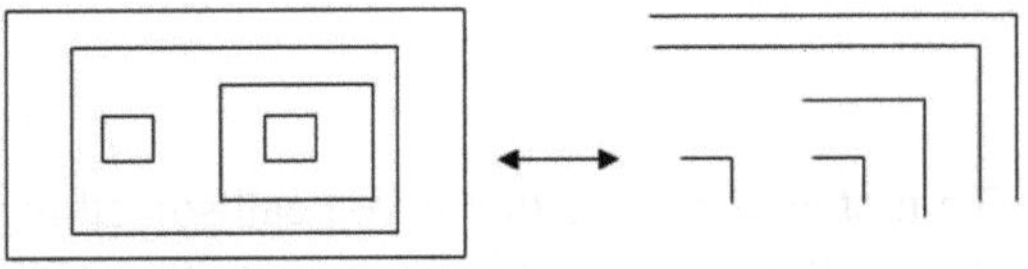

Fig. 5.3 Expressions in Boolean arithmetic

between them (A, Ā). This is the linear *distinction as continuity-discontinuity* that is more primitive than the spatial *distinction as perfect continence*.

Distinction as perfect continence helps understand the radicals. A cross ⌐ may be taken to represent a rectangle on a plane that divides an inside and an outside region (see Fig. 5.2).[6]

This also helps understand how patterns of non-intersecting rectangles yield expressions (see Fig. 5.3). Crosses 'pervade' spaces that they cover. In this example, the two largest crosses pervade all spaces under them. The middle-size cross pervades one of the small crosses but not the other. Consistent with the axioms, marks are unambiguously inside or outside each other.

Implicit Definition

Before proceeding let us pause to consider the generality of the axioms. This is a great strength of the formalism, and key to our purpose since they define distinction. Clearly, for the purpose of this analysis, the more general its definition of distinction, the better. As we have just seen, the axioms are congruent with linear and spatial distinctions. Even more crucial for our purposes, this helps us on our way to the heart of mathematics, the larger context to consider the significance of axioms founded on ב. What *kind* of axioms are these?

Lonergan pointed out that definitions may be of three kinds: descriptive, nominal, and explanatory/implicit. Descriptive definitions, like descriptive accounts in general, regard things as related to our senses and pertain to our needs, interests, and concerns in everyday life. Therefore, they are "expressed, at least in the last analysis, by appealing to the content of some human experience."[7] On the other hand, a nominal definition "merely tells us about the correct uses of names."[8] The foregoing characterizes shortcomings of these kinds of definitions. Thus, distinction as *perfect continence* is clearly a descriptive definition. An explanatory definition, on the other

[6]The use of rectangles to introduce the notation of *LoF* is due to Kaufman's helpful *Laws of Form: An Exploration*, 4–5.

[7]Lonergan, *Insight*, 102.

[8]Lonergan, 35.

hand, regards things, not as related to us, not as related to our senses, not as represented in our imagination, but as understood in their relations to one another. The three kinds of definitions suppose insights, but, whereas the second is only into the proper use of language, the third "supposes a further insight into the objects to which language refers."[9]

Lonergan uses explanatory definition to clarify what he means by implicit definition. To do so, he borrows from David Hilbert's work on the foundations of geometry:

> [T]he meaning of both point and straight line is fixed by the relation that two and only two points determine a straight line. . . . [O]ne may say that implicit definition consists in explanatory definition without nominal definition. It consists in explanatory definition, for the relation that two points determine a straight line is a postulational element such as the equality of all radii in a circle. It omits nominal definition, for one cannot restrict Hilbert's point to the Euclidean meaning of position without magnitude. An ordered pair of numbers satisfies Hilbert's implicit definition of a point, for two such pairs determine a straight line. Similarly, a first-degree equation satisfies Hilbert's implicit definition of a straight line, for such an equation is determined by two ordered pairs of numbers.

This brings us to the crucial point:

> The significance of implicit definition is its complete generality. The omission of nominal definitions is the omission of a restriction to the objects which, in the first instance, one happens to be thinking about. The exclusive use of explanatory or postulational elements concentrates attention upon the use of relationships in which the whole scientific significance is contained.[10]

The foregoing can stand as an account for the pure inner mathematical coherence of the formalism of *LoF* and, at the same time, for why it may be modeled by both propositional calculus and electric circuits. Clearly, in an inquiry into the nature of distinction, the generality of implicit definition is highly desirable, even indispensable. But even more so for our inquiry into the nature of the mathematical, and beyond that, into why the mathematical is an obstacle to explore the biological. Note, by the way, the weight of who Lonergan refers to. Geometry was the first discipline to be fully systematized, that is, fully transposed from a descriptive to an explanatory system. This was Euclid's astonishing accomplishment. He stated the axioms of geometry and then, building theorems upon them, systematized all the geometry known at the time. David Hilbert re-axiomizes geometry through implicit definitions to remove all residual descriptive elements. And now Spencer-Brown in *LoF* re-axiomizes logic, making it possible to prove Boole's postulates—no mean accomplishment.

An implicit definition of distinction is particularly appropriate, indeed necessary for our purposes, given the inevitable performative circularity since, as we saw in Chap. 1, defining always already calls for distinctions. Thus, definitions cannot but reflect the conditions for their possibility, which is the distinguishing activity of

[9] Lonergan, 34.

[10] *Insight*, 37. Reference is to David Hilbert, *The Foundations of Geometry.*

living subjects. That being so, we could expect ב to make an appearance in the axioms of *LoF*. The key question, though, is how this appearance takes place.

This touches on the foundations of logic. As Bertrand Russell pointed out: "Since all terms that are defined are defined by means of other terms, it is clear that human knowledge must always be content to accept some terms as intelligible without definition, in order to have a starting point for its definitions."[11] This is the problem that an implicit definition attempt to overcome formally by co-defining terms. More than curious, it is significant that implicit definition, which resolves the identical formal problem in geometry, could go unmentioned in textbooks on logic—the more so given Russell's point, which is true even of an implicit definition.[12] Even in this case we must be content to accept as intelligible what is not defined, or taken as given. This brings us to what is taken as given in the axioms of *LoF*: ב, *again*, repetition. Repetition is taken as given in the axioms. This is explicitly stated with the use of the term *again*. This points, not only to common sense. Beyond that, it points to the subject doing the defining, the distinguishing, to the verb that brings about the noun. So doing, it points to subjects and their nature, who are the condition for the possibility of all formalisms.

As discussed in the Appendix, the claim could be made that *again* is not taken as given but implicitly defined by the axioms. Though this is a valid claim, it does not hold up to scrutiny. A proper discussion of this claim calls for a close reading of the text of *LoF*, which is why it has been consigned to the Appendix. It is shown there that *again* is not codefined by the axioms and its meaning is taken as given. This argument is also of interest for what it says about mathematical formalisms generally. The Appendix also shows that, consistent with what Spencer-Brown himself states, becoming conscious of the subject's ב as condition for the possibility of mathematics, *is an advancement of mathematics*. That ב is implied tells us that *LoF* is about the compact and *important* foundational level of mathematics at which *indication*, *distinction*, and *repetition* are undifferentiated. Indeed, what could be closer to *LoF*'s imaginary Boolean values than 'again?'

The fundamental conclusion, however, is undisputable. The axioms of *LoF* take as given ב, *again*, repetition. This convergence of the foundations of mathematics and of aliveness is as remarkable as it is significant. It at once suggests just how pertinent *LoF* is for inquiry into the foundations of biology, and why biology—the nature of the subject, what I shall later refer to as *phenomenology*—is the foundation of *LoF.*

Latent in living distinctions is their distinguishing; it takes distinguishing distinctions to distinguish. In what it takes for granted, the implicit definition of distinction proposed in *LoF* is performatively consistent with the distinguishing activity of [subjects]. *Laws of Form* is about any possible distinction, no matter how

---

[11] Cited in Copi and Cohen, *Introduction to Logic*, 92. Interestingly, there is no reference to David Hilbert's notion of implicit definition.

[12] This is the case of Copi and Cohen. "For more than six decades, and for thousands of students, *Introduction to Logic* has been the gold standard in introductory logic texts" (see https://www.amazon.com/Introduction-Logic-Irving-M-Copi/dp/1292024828)

concrete or abstract. This pertains to the distinguishing activity of [subjects] and, even more fundamentally, it pertains to [subjects] *as distinctions*. This is the most crucial link between *LoF* and biology. To fully discuss this link is a central purpose of our examination of *LoF*. However, further preliminary clarification is required about *LoF* and the nature of mathematical formalisms on the one hand, and the nature of aliveness on the other.

The mathematical formalism of *LoF* can stand alone. Still, its power is verified—one is tempted to say 'empirically'—by its models. One of these models is logic. This is the context in which *LoF* does for Boole's Propositional Calculus what Hilbert's *Foundations of Geometry* does for Euclidian geometry—provide it with new foundations by re-axiomatization. It also demonstrates, among other things, that a "(Boolean) arithmetic of distinction" grounds Boolean algebra, the Propositional Calculus we loosely call 'logic.'[13]

---

[13]That being so, it is striking that *LoF* is still not mentioned in many mainstream books on logic, the more so since the equivalence of the formalism of *LoF* with Propositional Calculus has been independently verified. John D. Collier wrote: "The Laws of Form are equivalent to propositional calculus. Spencer-Brown showed LoF ➔ PC in his book. B. Banaschewski showed the opposite entailment in *Notre Dame Journal of Formal Logic*, 3, (1977): 507–509. I always suspected that the Laws of Form and the Propositional Calculus (or Boolean algebra) were equivalent and have treated them as such in my papers on distinction dynamics. It is nice to see that somebody actually proved this result, though. As I guess people will start to wonder by now why we need Spencer-Brown then, if his algebra is merely equivalent to something that is very well known, I [think] the axiom system of *LoF* is much simpler than the usual axiom systems for PC. Moreover, these axioms are derived from two extremely simple and intuitively understandable properties of the act of making a distinction" (John D. Collier, Department of Philosophy, University of KwaZulu-Natal, Durban, South Africa. Retrieved from www.enolagaia.com/GSB.html).

It is beyond the scope of this review to discuss scholarly analysis of *LoF*. A search on the internet soon provides background for those unfamiliar with this text. Randall Whitaker, "George Spencer-Brown and his Laws of Form" (http://www.enolagaia.com/GSB.html) provides a brief introduction. The following are quotations from this site. Sir Stafford Beer's review in *Nature* at the time of *LoF*'s publication stated: "Sheffer reduced Boole's whole logic to five unproven postulates. Spencer-Brown proves all five in a casual couple of pages. As for the famous 'Sheffer stroke,' itself a triumph of mathematical simplification, Spencer-Brown calmly said that 'it may be omitted.' He is right—as I now (but could not earlier) see. Nor would I previously have regarded Gödel's theorem as logically trivial. These revised views are the result of spending two delightful days in St. Mark's Square, immersed in this condensed and exquisitely poised argument."

The key difference between *LoF* and PC is not only that the axioms in *LoF* are simpler and fewer, thereby making it a more elegant construct. More precisely, as has been shown, these axioms are implicitly defined, whereas the axioms of PC require additional nominal definitions. Given its significant achievements, one may wonder about the relative obscurity of *LoF*. In relation to this, two further remarks from this web page are of interest. According to Frances Heylighen, *LoF* is "referenced very frequently in cybernetics (especially the constructivist epistemologists of the 'new' or 'second order' cybernetics school..." He refers to Frances Heylighen, E. Rosseel, & F. Demeyere, eds., *Self-Steering and Cognition in Complex Systems: Toward a New Cybernetics*, New York, NY: Gordon and Breach Science Publishers, 1990). He states that he is himself "very impressed by Spencer-Brown's concept of 'distinction'" which he has elaborated in what he calls a "'distinction dynamics:' a theory of how new distinctions arise out of combinations of existing ones." Further, Eric Oberle wrote: "Brown has had an impact reaching far beyond cybernetics. Not only has he been invoked in recent debates about biological and psychological autopoiesis (as in the works of Varela and Maturana: *Autopoiesis and Cognition: The Realization of the Living*, Dordrecht 1980; Heinz von Forester,

But a crucial question must be asked about the formalism of *LoF*. Stated schematically, this is how the formalism proceeds, starting from what it takes for granted:

ב➔Repetition, *Again*➔Axioms of *LoF*➔Calculus of distinction or "Boolean Arithmetic"➔various models➔Boolean Algebra or Logic.

So, a fundamental question arises about the relation between aliveness and logic, given that logic is one possible model of the calculus of distinctions, which is a primary form of mathematics. Since ב is the condition for the possibility of mathematics and logic, does it make sense to expect an account of ב to be provided in mathematical or logical terms? Could ב be accounted for in mathematical terms, given that ב is the condition for the possibility of mathematics? If aliveness is the foundation of mathematics, can this foundation be accounted for mathematically?

Herein is the enormous value of *LoF*'s with its implicit definition of distinction. Showing what is taken for granted in its axioms, *LoF* makes explicit the role of the living subject in the formalism. Since the formalism is about the foundations of mathematics, it thereby makes explicit the role of the subject in logic, and more generally in mathematics, insofar as it is founded on distinction. And it thereby allows questions to be framed about the foundations of biology vis-à-vis mathematics, specifically whether the mathematical ideal of knowledge can obtain in inquiry into aliveness.

The Formalism of *Laws of Form*

To continue with our gloss of *LoF*, it follows from their implicit definition that the two radicals—which co-define each other and distinction—also co-define the two and only two operations of the formalism. The value of the single written symbol Γ and the non-written symbol , remain indeterminate unless they form part of an expression. On its own, a single written symbol means neither *call* nor *cross*. Only in the context of an expression is it defined either way. Similarly, only in the context of such expressions can the non-written symbol , be a value. This is all consistent with implicit definition—the axioms and the terms by which they are stated all mutually co-define each other. Spencer-Brown mentions this:

> The universe is simply what would appear if it could. Its laws are the laws of the possible, called by Shakyamuni the links of conditioned coproduction, called by me the calculus of indications.[14]

For Spencer-Brown this refers to distinction as the relation between inner, outer, and the boundary between them (A, Ā, /). For us, it says: If a universe appears it is because there is ב. It is highly noteworthy that the notion of conditioned coproduction states

---

Observing Systems, Seaside CA 1981). Brown has also been incorporated into Parsonian sociology in some interesting (and sometimes perverse) ways by Niklas Luhmann..."

The work of Spencer-Brown is often discussed in the journal *Cybernetics of Human Knowing.*

[14] Spencer-Brown, *Laws of Form*, viii.

in two words the nature of joinings that we shall examine at length later. Joinings, or conditioned coproduction, is the nature of [subjects]; it is what allows them to be what they are and do what they do. This is what allows the universe *to appear*—no less. Think colony of progenotes: a multiplicity of progenotes mutually coproduced, allowing aliveness to emerge, with it forness, and thereby appearance. We have now seen that the notion of implicit definition helps us grasp the conditioned coproduction of the axioms in *LoF*. We can now see that their condition coproduction leads us beyond, to the [subjects] that are the condition for their possibility. We must not only accept some terms as intelligible without definition, as Russell suggests. We must thereby also take up the task to arrive at their intelligibility.

To continue this gloss of *LoF*, as in all axiomatic systems, from a highly intuitive beginning, theorems are rigorously demonstrated as the formalism is built. Like all mathematical formalisms, this one too is sufficient unto itself, regardless of applications or interpretations. It has one stated aim—"to clarify the laws serving as a basis for the formation of forms."[15] "Form, in Spencer-Brown's calculus, is foremost about the marking of a difference . . . ."[16] This is why one possible interpretation, or *model*, is the arithmetic or 'reckoning' of distinctions.[17] As variables are incorporated to the arithmetic, a primary algebra is developed that is equivalent to Boolean algebra. The analogy in numerical arithmetic is the transition from $3 + 2 = 5$ to $3a + 2a = 5a$. The *models* of this primary algebra include two-element Boolean algebra and Boolean logic, which is the classical Propositional Calculus. One consequence already noted is that the postulates of Boole's algebra are demonstrated as theorems in the primary arithmetic, and this is what amounts to the re-axiomatization of Boole's Propositional Calculus. But the models for this formalism, which Spencer-Brown also refers to as the *calculus of indications*, are not limited to logic or electronic circuits. We are back to Lonergan's point about the complete generality of implicit definitions. Because the formalism may be taken to model the unfolding of any set of distinctions, and, moreover of organisms, conceived as distinguishing distinctions, the unfolding of biological form is itself a model of the unfolding of distinction. Indeed, the non-numerical mathematics presented in *LoF* is how biology could be at least partially expressed mathematically. But the fundamental point here is not the distinctions made, but the elision of the process of making them. It takes a distinguishing distinction to distinguish. Both the bodies of organisms and the distinguishing activity by which they parse the world are models of the formalism. In short, the laws of form elucidated in *LoF* are at work in biomorphism, which obtains in two senses: the form of living beings, as distinctions themselves, which is the condition for the possibility of aliveness as such; and the forms whereby they parse the world through their distinguishing.[18]

---

[15] Bakken, "George Spencer-Brown," 482.

[16] Bakken, 482.

[17] *Arithmetike* (*tekhne*), from Greek '(the) counting (art),' feminine of *arithmetikos* 'of or for reckoning, arithmetical. Available at https://www.etymonline.com/search?q=arithmetic.

[18] The first, the *form* of aliveness, is referred to below as biomorphism-a or Ba; the second, how living beings parse or form the world, is referred to as biomorphism-w or Bw. The two are inherently interlinked and mutually co-produce, Ba-Bw.

Fig. 5.4 The six logical conjunctions

In words	In the sentential calculus	In the primary algebra
not *a*	~a	$\overline{a}|$
a or *b*	*a* ∨ *b*	Ab
a and *b*	*A* . *b*	$\overline{\overline{a}|\ \overline{b}|}|$
A implies *b*	*a* ⊃ *b*	$\overline{a}|b$

Fig. 5.5 The six logical conjunctions transposed

a.b
b.a
~(~a ∨ ~b)
~(~b ∨ ~a) →
~(a ⊃ ~b)
~(a ⊃ ~b)

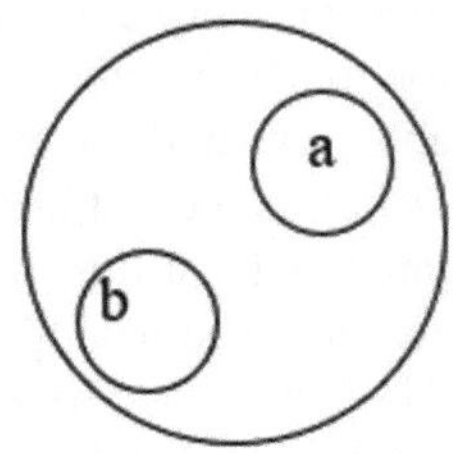

Expressions are arrived at in *LoF* that may be taken to stand for the operators in logic or electronic circuits: "and," "not," "is," "or," "implies," "some," and so forth. Since these expressions are built on the axioms, they can always be simplified to one of the two primary values, either Γ or . One consequence of this is that, as a model for propositional calculus, the algebra renders trivial the calculation of the truth-value of even the most convoluted sets of propositions. As soon as are they stated in algebraic notation, they simplify to either Γ or , and can be read as 'true' or 'false,' or 'on' or 'off.' For a single example, consider again Fig. 5.2. Following axiom 2, the law of calling, the two largest crosses cancel each other. Likewise, the middle cross and the small cross it pervades. This leaves a single cross. If a single cross stands for *on*, then this is the state of the electronic circuit; if it stands for *true*, then this is the truth value of the set of propositions.

Spencer-Brown illustrated the power of the formalism by showing the equivalence between the six logical conjunctions—the two-value logical operations that follow from distinction—in propositional calculus and in the primary algebra (see Fig. 5.4).[19]

He shows that the six expressions may be transposed (see Fig. 5.5).

Spencer-Brown noted, "This is a proper simplification, since the object of making such sentences correspond with these symbols is not representation but

[19] See Spencer-Brown, *Laws of Form*, 115.

calculation."[20] Keeping in mind the statement of the axioms in the formal notation, and Kaufmann's transposition of them in terms of boxes, it is clear that this expression is equivalent to "*a and b*." The other expressions in the figure follow from this.

The significance of this transposition cannot be underestimated. Kauffman writes in relation to *LoF*: "logic is nothing more than the properties of the act of distinction."[21] The diagram unfolds what a distinction does. Once a distinction is drawn, whether in concrete or in conceptual space, logic follows—Boolean Algebra or Logic.

In later chapters of *LoF*, equations of the second-degree are generated that are self-referential. These are modelled by feedback systems, either positive or negative. In positive feedback systems, the positive or negative value remains constant; in negative feedback systems the value oscillates. The interpretation of these includes finite automata and recursive arithmetic. And, of course, the negative feedback systems we have delved into: two-component regulatory systems such as automatic heating systems and circadian clocks. These oscillating systems are precise models of these equations. *LoF* shows, not only why feedback loops are possible, but why they make counting possible and, further, why counting calls for memory–in a nutshell, because memory is that a current state points to a previous one. Most significantly, *LoF* goes on to show how time, space, and number may be derived from these loops, and thereby how these all, in the end, derive from distinction.

Once again, the significance of this cannot be overestimated. Think of it like this: distinction is the foundation of phenomena in Kant's sense, the complexity of which is built layer by layer. No doubt, he would have been astonished at this mathematical demonstration. Even more significant, then, that underlying distinction as implicitly defined by the two axioms, is ב, the distinguishing of [subjects]. This leads to an inevitable conclusion. Strictly, if the presence of ב is not overlooked, *LoF* may be read as a phenomenology, a treatise on appearance, on the appearing.

The second-degree equations are arrived at the end of the construction. However, they were the reason for *LoF* in the first place. As Spencer-Brown explains, when he and his brother worked as civil engineers with British Railways, they were asked to develop safety devices to detect if train cars had accidentally been left behind in tunnels. These special-purpose circuits to count wagon wheels moving backwards and forwards were patented and remain in use to this day.[22] Effective as they are, these switching systems make use of self-referential expressions. Such expressions are banned from logic.[23] Instances of these are:

The statement below is false.
The statement above is true.

---

[20] Spencer-Brown, 115 (emphasis added).

[21] Kauffman, "Peirce's Existential Graphs," 61.

[22] Notice that these are excellent examples of what we have referred to as *linear distinctions*.

[23] Spencer-Brown, *Laws of Form*, 81, and Transcript, 4–5.

Or simply:

This statement is false.

Consider the latter. If the statement is false, as it says it is, then it is true; but if it is true, then, as it says, it must be false. The truth-value of the statement oscillates between true and false. There are many other examples of such self-referring or oscillating statements, the most famous being *Russell's Paradox.*[24] Now, an exactly analogous statement occurs in numerical algebra. Take the simple equation:

$$x^2 + 1 = 0$$

Though innocent looking, this is a troublesome equation. But it is a very useful one: all of phase theory in electricity depends on it. Spencer-Brown pointed out that, though Russell and Whitehead knew of it, they did not notice that it is a direct analogue for what is disallowed by the 'Theory of Types' in *Principia Mathematica.* Transposing, the equation becomes:

$$x^2 = -1$$

Dividing both sides by x gives:

$$x = \frac{-1}{x}$$

[24] This paradox pertains to the class of classes that do not contain themselves as members. Russell came up with this paradox when working on the foundations of mathematics in *Principia Mathematica.* Today it would be stated in terms of sets: *the set of sets that do not contain themselves as members.* It goes like this. Sets can either be members of themselves or not. For instance, the set Words is a member of itself because the word 'word' is itself a word. Similarly, the set Thoughts is itself a thought. So, this too is a set that is a member of itself. Most sets, however, do not contain themselves as members. The set Automobiles, for instance, is not itself an automobile, nor is the set Coins itself a coin. This may appear nonsensical, but most sets are of this kind. We now have two kinds—or sets—of sets: the set of sets that contain themselves as members (Thoughts, Words, and so on) and the set of sets that do not (Automobiles, Coins, and so on). Since both of these are sets, we can ask the same question about them as we asked before: Do either of these two sets contain themselves as members? Consider the set of sets that do not contain themselves as a member—the set (Automobiles, Coins, and so on). Is this set a member of itself or not? Let us say that, like all its members, it does not contain itself as a member. If it does not, then the set *is* a member of the set that do not contain themselves as members. This is paradoxical. On the other hand, let us say that it belongs with the sets that do contain themselves as members—the set (Thoughts, Words, and so on). If this is so, then it is a member of itself, since it is a member of the set of sets that do contain themselves as a member *because* it does not contain itself as a member, which again is paradoxical. In short, if it does not belong to itself as a member, then it does, and if it does belong to itself as a member, then it does not.

The Zermelo–Fraenkel axioms plus the so-called Axiom of Choice, or ZFC, is today the standard form of axiomatic set theory and, thereby, most commonly taken as the foundation of mathematics. Its explicit aim is to avoid Russell's paradox. (See http://plato.stanford.edu/entries/set-theory/index.html.) Though ZFC is not the only option, it is the one for which the notion of distinction is essential.

Now, take $x = 1$. Then:

$$1 = \frac{-1}{1} = -1$$

This is a paradoxical result. On the other hand, take $x = -1$. Then:

$$-1 = \frac{-1}{-1} = 1$$

This is also paradoxical. Spencer-Brown explains:

> Of course, as everybody knows, the paradox in this case is resolved by introducing a fourth class of number: called *imaginary*, so that we can say the roots of the equation above are $\pm\, i$, where i is a new kind of unity that consists of a square root of minus one . . . . [In Chapter 11 of *LoF* this concept is extended] to Boolean algebras, which means that a valid argument may contain not just three classes of statement, but four: true, false, meaningless, and imaginary. The implications of this, in the fields of logic, philosophy, mathematics, and even physics, are profound.[25]

This is the mathematics of the modest electric bell and other oscillating systems. A circuit turns itself on, off, on, off, and so on, by operating a magnet that makes a small hammer hit a metal bell intermittently. The speed of the oscillation is heard as a continuous *rrrrrrriiiiing*. The circuit turns on *because* it turns off and it turns off *because* it turns on. This is critically important. The electric bell functions *because it is* self-referential. It is a two-element regulatory system like those at work in chemotaxis and the automatic heating system. The algebraic equation presented above is itself such a system; the first component assigns a value to x and the second component plugs it into the equation, a new value is generated and a new one is plugged in, and so on. It is a two-component *mathematical* system.

The 'Theory of Types' by which self-referential statements were disallowed in logic was not a theory at all, as Russell himself acknowledged in 1967, but merely an arbitrary rule to disallow this kind of statement.[26] The purpose, and another merit of *LoF*, is its rigorous justification of self-referential values, starting from two highly intuitive axioms that define distinction and from which logic is built.[27] From the highly intuitive definition of distinction in terms of the two axioms inexorably follows Boolean algebra, and from it follow second-order imaginary Boolean values—the mathematics of oscillation.

Like many formal systems, *LoF* was conceived in the inverse order to its presentation. Following the successful application for British Rail, the need to justify these oscillating values eventually led to the formalism, and in turn to the axioms. As for the 'Theory of Types,' its fundamental strategy remains a key method to circumvent

---

[25] Spencer-Brown, *Laws of Form*, xv (slightly altered). We, of course, would add the field of biology.

[26] Spencer-Brown, *Laws of Form*, xiv.

[27] Spencer-Brown, *Transcript, AUM Conference, Session Three*, 7.

paradoxes, which is to build hierarchies. Hierarchies provide a framework to understand structure. Imaginary Boolean values, on the other hand, provide a framework to understand process, along with the logic of nesting as found in chemotaxis and in the mapping of an environment, as the example of our unfortunate victim showed in Chap. 2.

To sum up this gloss, *LoF* accomplishes four things: (a) from distinction as foundation, as defined implicitly through two axioms, (b) a calculus of indications is developed that begins from the hitherto unknown arithmetic that underlies Boolean algebra (or Propositional Calculus), (c) upon which is build an algebra that renders trivial the calculation of highly convoluted sets of propositions, and, built upon this is (d) a second level algebra that provides the mathematics of oscillation. Patently, the preeminence accorded in this essay to *LoF*'s definition of distinction is amply justified. More so if, as proposed here, distinction is to be the foundation of inquiry into aliveness. Beyond that, given its treatment as a formalism, *LoF* offers an opportunity to reflect on the nature of formalisms and mathematics in general, and further to reflect on the question of whether the mathematical ideal of knowledge is appropriate for biology. In his notes to the chapters, Spencer-Brown himself offers reflections about some of these topics.

Notice that, except for (c) above, the other three points are directly relevant to our current topic. And more to the point, all involve ב. (a) involves repetition, which requires ב; in (b), the arithmetic is set up in terms of two radicals, and of these—the cross—acquires two possible meanings depending on whether it is used twice in one of two ways (notice the saturation of *two's* in this statement). (d) concerns oscillation, the alternation between two states (again notice the recurrence of *two* here). Can any doubt remain about the intimate link between distinction and twoness? Twoness is inherent to crossing-separating and calling-marking, and continuity/discontinuity. It must now be clear why it is reasonable to claim that *LoF* opens the door to a reflection on the foundations of logic. As we have just seen, it takes us to the two outer limits of logic: on one hand, its foundations, on the other, the disallowed self-referential values. Moreover, given its topic, the nature of its axioms, and its construction of a formal system, it also opens the door to reflection on the foundations of mathematics. Here too it takes us to the edge of the mathematical and its transition to the phenomenological.

Lower and Upper Schemes: Living Being and Mathematics

Laws of Form starts at such a fundamental level of mathematics that it allows us to shift our attention to mathematics in general as well as to its relation to living being. The reason is as simple as it is profound. If distinction is the foundation of form, it must also be the foundation of the form of organisms, given that they are distinctions. But there are other things that *LoF* makes possible. It provides a point of leverage to discuss the mathematical ideal of knowledge that is the foundation of what we understand as science. So, we can raise the question of just how appropriate this ideal is for inquiry into living being. And also, by clarifying the interrelation between the foundations of mathematics and the foundation of inquiry into living being, issues of method can also be clarified. These possibilities all follow from the key finding that living beings, conceived as distinguishing distinctions, are characterized by ב. Yet, this is what the axioms of *LoF* leave out by taking repetition as given.

Just as with the axioms, using *LoF* as a point of departure to understand living being, both affirms the formalism, and calls for its criticism. Consider the framing of the mathematical argument of *LoF*:

> ***A note on the mathematical approach***
> The theme of this book is that a universe comes into being when a space is severed or taken apart. The skin of a living organism cuts off an outside from an inside. So does the circumference of a circle in a plane. By tracing the way we represent such a severance, we can begin to reconstruct, with an accuracy and coverage that appear almost uncanny, the basic forms underlying linguistic, mathematical, physical, and biological science, and can begin to see how the familiar laws of our own experience follow inexorably from the original act of severance.[28]

Given ב, there is something fundamentally right and something fundamentally wrong with this statement. What is wrong is soon pointed out through an alternate statement:

> ***A note on this essay***
> The theme of the present essay is that a universe comes into being when there is ב, *distinguishing*. The skin of a living organism cuts off an outside from an inside. *Not* so the circumference of a circle in a plane. By tracing the way *living distinguishing takes place*, we can begin to reconstruct, with an accuracy and coverage that appear almost uncanny, the basic forms underlying linguistic, mathematical, physical, and biological science, and can begin to see how the familiar laws of our own experience follow inexorably from our – any organisms' – original act of *distinguishing*.

That a circumference of a circle in a plane does not cut off an outside from an inside was the upshot of the examination of distinction in Chap. 2 and summarized in Fig. 2.4. We can say in a koan-like summary:

The skin of a living organism cuts off an outside from an inside.
Not so the circumference of a circle in a plane.

The radical contrast between circumference and skin hinges on ב.

The fundamental reason why *LoF* has so much to offer our topic is what it is about. Spencer-Brown situates as follows the calculus of indications presented in *LoF*:

> Logic is itself not mathematics; it is an interpretation of a particular branch of mathematics, which is the most important non-numerical branch of mathematics. There are other non-numerical branches of mathematics. Mathematics is not exclusively about number.

[28] Spencer-Brown, *Laws of Form*, xxix. In the original hardcover edition this note opens the book. In subsequent editions, it follows the Introduction and various prefaces. Either way, its placements clearly shows it is at the threshold.

> Mathematics is, in fact, about space and relationships. A number comes into mathematics only as a measure of space and/or relationship. The most fundamental relationships in mathematics, the most fundamental laws of mathematics, are not numerical. Boolean mathematics is prior to numerical mathematics. Numerical mathematics can be constructed out of Boolean mathematics as a special discipline. Boolean mathematics is more important, using the word in its original sense: what is important is what is imported. The most important is, therefore, the inner, what is most inside. Because that is imported further. Boolean mathematics is more important than numerical mathematics simply in the technical sense of the word "important." It is inner, prior to, numerical mathematics – it is deeper.[29]

The foundation of mathematics most widely accepted today is in terms of set theory: the Zermelo–Fraenkel axioms plus the Axiom of Choice, or ZFC. Our discussion of distinction in Chap. 1 led to ב as the condition for the possibility of tokens and types, which is another way of referring to a set and its members. In constructing higher-level distinctions from lower-level ones, our unfortunate victim in Chap. 1 made lower-level distinctions belong to higher-level sets. Thus, whether expressed in terms of sets or as the reckoning of distinctions (or Boolean arithmetic), this basic principle underlies both mathematics and aliveness.

But the quote above from Spencer-Brown brings us straight back to the issue of circularity. How is it at work in *LoF*? The formalism is constructed by demonstrating theorems. These demonstrations are logical, which is to say their validity is demonstrated in terms of Boolean algebra, which, according to *LoF*, is but one of many possible models of the formalism. The problem is depicted in our scheme above:

> ב➔Repetition, *Again*➔Axioms of *LoF*➔Calculus of distinction or "Boolean Arithmetic"➔various models➔Boolean Algebra or Logic.

It turns out that Boolean Algebra or Logic, the last link in this chain and one possible interpretation of the algebra, is the criterion by which its construction is justified, starting from the axioms, through the construction of the arithmetic and onto the algebra that it interprets. What could be more circular? Spencer-Brown said as much himself:

> The arguments used to validate the theorems in *Laws of Form*, as we now begin to see, are themselves validated by the calculus dependent upon those theorems. And yet in no way is the argument a begging the question. . . . *Principia-principii*, begging the question, is not a valid argument; it is a form of fallacy. In no way is the question begged, but in producing a system, in making its latter parts come true, we use them to validate the earlier parts; and so *the system actually comes from nothing and pulls itself up from its own bootstraps, and there it is* . . . . Nowhere does this become more evident than in this first and most primitive system of non-numerical mathematics. . . . [T]hat is why it is a branch of mathematics so neglected hitherto. . . . [It] is a bit too real. It is a bit too evident what game one is playing when one plays the game of mathematics.[30]

The key term of this argument is *validation*. The claim is that later parts of the formalism of *LoF* validate earlier ones. And the validity of this is nowhere more evident than in this first and most primitive system of non-numerical mathematics.

---

[29] Spencer-Brown, *Transcript*, 1.

[30] Spencer-Brown, *Transcript*, 8–9 (emphasis added).

According to Spencer-Brown, this is the reason why it has been neglected. What does Spencer-Brown mean by the "game one is playing when one plays the game of mathematics"? And why is it "a bit too real?"

Like any other formalism, *LoF* cannot avoid the fundamental problem of circularity. But, can a formal system "come from nothing?" Can it "pull itself up from its own bootstraps," appear from nothing—"there it is"? Only in a Platonic reality. That is, a reality where there is pure abstract identity that does not require ב. Which, of course, is precisely what the axioms take for granted. Repetition is taken as given. Perhaps, like Plato, Spencer-Brown considers this reality to be "a bit too real", more real than living reality. In lived reality, *LoF* is there because Spencer-Brown understood something and wrote it and because we read it and understand it.

Why can we understand? Because *LoF* is a formal statement of the reckoning of distinctions inherent to our own cognitional, distinctional activity. Long before that, and much more radically, it is inherent to the embodiment of living beings, distinguishing distinctions. What *LoF*'s system demonstrates formally is performatively demonstrated in every single living cell's ongoing managing of inner and outer. This reckoning is the systematic establishing of bet structure that brings about our spirallic nature. This is the radical and explicit condition for the possibility of the axioms of *LoF*. In non-Platonic living reality, the ב of living beings is the condition for the possibility of any mathematical formalism, for the mathematical as such, in the deepest possible sense. I shall return to this point.

In short, distinguishing distinctions are at once inherently mathematical and the condition for the possibility for the mathematical. Clearly, the problem of circularity is inherently linked to the relation between the mathematical and living being.

In the scheme above that depicts the problem of circularity in *LoF*, let us replace "various models" with "cells as distinctions". This amounts to replacing 'a circle in a plane' with 'the skin of a living organism'. The previous scheme depicted the construction of the formalism and what is taken for granted in the axioms—an epistemological point. With this replacement, the second scheme depicts the ground that makes the first one possible. It does so by depicting what follows from the advent of ב—an ontological point. This then, in the distinctional terms of *LoF*, shows what is involved in the emergence of organisms that are embodied as a distinction:

> ב➔Repetition, *Again*➔Axioms of *LoF*➔Calculus of distinction or "Boolean Arithmetic"➔**Cells as distinctions**➔Boolean Algebra or Logic.

This brings us back to the full significance of Fig. 5.5 on the six logical conjunctions. The diagram unfolds what a distinction does, but this is so whether the set of distinctions depicted is abstract, like the diagram of the figure, or a building, or a living cell. From an embodied living distinction, logic too follows. As I wrote elsewhere about Fig. 5.5.

> . . . an architectural structure *calculates*: in a house one may be in the kitchen or in the bathroom or in neither but not in both. In the diagram above, *a* and *b* state that this is so regardless of further nested structures (the larder, a shower.) This parsing of the world, calculation or classification, either by formal logic or by living cells, is accomplished by

surfaces within surfaces. This is the full significance of "the very architectonic structure of the complete and fully functioning cell." Sign processes (semiosis) = distinctional calculi.[31]

Since from distinction unfolds logic, to be alive, which is to be a distinction for itself, is to reckon distinctions, which is to implement logic. The laws of form underlie the unfolding of living forms. These laws are at work, both in how living beings are—how they are structured—and in how they process the world to self-maintain. These laws are at work in biomorphism in two senses, as was mentioned already—how living beings are materially formed and how they metabolically-distinctionally form the world for themselves.

In short, the second scheme explains why we are content to accept as intelligible repetition, without its definition. Simply, because it is inherent to distinction, both mathematically and biologically. With this the circularity in *LoF* disappears. The two schemes together, and in light of Fig. 5.5, tell us that abstract distinction results from the very form of an organism as a distinction that distinguishes. Here are the two schemes, the second one grounded and the first, or lower scheme:

2. ב➔Repetition, *Again*➔Axioms of *LoF*➔Calculus of distinction or "Boolean Arithmetic"➔ "various models"➔Boolean Algebra or Logic.
1. ב➔Repetition, *Again*➔Axioms of *LoF*➔Calculus of distinction or "Boolean Arithmetic"➔**Cells as distinctions**➔Boolean Algebra or Logic.

LoF itself, as a treatment of distinction, is a specialized expression that belongs to the upper scheme. The ground scheme is *biomorphism aliveness* (hereafter biomorphism-a or simply Ba), the form of living beings as distinctions. This is distinction as embodied. From it follow distinguishing activity (verb), and distinctions (noun). Embodied distinctions are distinguishing distinctions, a structure-process that is at once a noun and a verb. This is what *entelecheia* is all about: *being-at-work* (verb)-*staying-itself* (noun) or *holding together actively* (verb) *as a whole* (noun).

The upper scheme is *biomorphism world* (hereafter biomorphism-w or simply Bw), the construction of the world by organisms in terms of the form of distinction. To both the lower and the upper scheme applies the statement: "logic is nothing more than the properties of the act of distinction." The condition for the possibility for entelecheia of the ground scheme is ב. Ultimately, this is why we take for granted *again* in the upper scheme. *Again* is our very nature as living beings. It is also why logic is the criterion for constructing the formalism from the ground up. It is the form of distinguishing distinctions. As subjects, we are that form, we embody it. We could not *not* take it for granted.

So, the ground scheme validates the upper scheme presented in *LoF*. We are content to take the *again* in the axioms as given. I shall refer to the upper scheme, along with the diagram in Fig. 5.5 as *the form of distinction* or @@L (L for logic). To the ground scheme, partially because of the diagram in Fig. 5.5—we shall soon

[31] Mayer, Daniel, "Significance." The quote at the end is from Hoffmeyer, *Biosemiotics*, from a dazzling chapter of his significantly entitled "Surfaces within Surfaces." This is why the last sentence quoted here links semiosis to the distinctional.

see why only partially—I shall refer to as *the embodied form of distinguishing distinctions* or @@A (A for aliveness). The form of organisms, embodied distinctions that distinguish, always already acquaints them with logic, the/our properties of/as distinction.

Much will be said about @@A and @@L (this is how they shall be written to underscore this sense, they can be read as "aliveness" and "logic"). This differentiation lays open many key issues. It is therefore worth briefly stepping back to ask, why is this so? The ground and the upper scheme explain what we are content to take for granted in rigorous formalisms. What are we content to take for granted, and why?

A crucial link between @@L and @@A is the fourth logical conjunction in Fig. 5.5: *a implies b*. This conjunction tells us that calculating and thereby process is built into (the very form of) form.[32] Thus, *a implies b* is embodied in organisms. Recall the volcanic pool scenario. In a community of progenotes, at some point, protocells will inevitably be encapsulated in other protocells. The architecture depicted in Fig. 5.5 will emerge, a form that can *calculate*. This is the architecture of two-component regulating system—no less. Very likely at first not alive, this architecture sets the stage for the emergence of ב.

But ב points to subjects. *LoF*, like any other mathematical formalism, elides the subject—this too we are content to take for granted. In general, mathematical formalisms treat @@L and elide @@A. This makes sense. Typically, the very delimited purpose of a formalism need not include a treatment of the subject that produces its. But, how, precisely is the elision accomplished? Chap. 6 examines at length how it is done in *LoF*. There is much to learn from the manner of the elision about its effects and about the nature of mathematics.

Recall Spencer-Brown's reference above to the "game one is playing when one plays the game of mathematics." Why is it "a bit too real?" Well, most radically, the game one is playing when one plays the game of mathematics is to be alive—in several senses. First, as the discussion of the two schemes suggests, biomorphism-a (Ba) and biomorphism-w (Bw) are the condition for the possibility of the world appearing. This is as real as it gets. He also refers to his own book, published under the pseudonym James Keys, *Only Two Can Play This Game*. This title may refer both to a relationship between two individuals and to the fundamental nature of distinction. As was shown, these two readings converge. ב, condition for the possibility of twoness and root of the radical contrast between a circumference and skin, points to plurality, joinings, the topic of the following section. Spencer-Brown's point is about the real and the mathematical. How is being alive part of this? We will also take up this point.

A key difference between the ground and the upper scheme is that the ground scheme refers to an analogical process whereas the upper one refers to digital binarity. The ground scheme, embodied or distinguishing distinctions, involves the verb

---

[32] The conjunction *a implies b*, known as *modus ponens*, a postulate and traditionally one of the laws of logic, is demonstrated as a theorem in *LoF*. The reader may wish to read in the Appendix how Lewis Carroll used it to illustrate the limits of axiomatic systems.

and the noun. The upper scheme centers on distinction, the noun. The upper scheme is *LoF*'s axioms as they stand, uncritically, whereas the ground scheme is the source of 'again'. The analogical process of distinguishing of the ground scheme underlies, is expressed by, *requires,* the digital upper scheme. For one thing, this is empirically so. Embodied distinctions that are for themselves must distinguish—the verb—to remain so; and distinguishing—again, the verb—is only possible by embodied distinctions. Beyond that, however, it is also true analytically. An early indication (EI) is a distinction. Therefore, the same analysis obtains for the relation between distinguishing and a distinction as it did for EI as an integrator-operator. The ground scheme is an activity, a vector, a continuum, a verb. During its process, distinctions are arrived at, just as discrete EIs are arrived at. Each of these is a moment, a freeze, a discontinuity, a noun. Thus, the distinctions arrived at in the upper scheme are the means-whereby for the ground scheme. In short, distinguishing, like EI, is an integrator-operator. In distinguishing as integrator-operator the verb arrives at a noun, then at another noun, and so forth. These are transitions from analog to digital. Distinguishing (verb) is analog, distinction (noun) is digital—A/Ā. This, by the way, links directly to the definition of distinction as A=A, A/Ā, and to the metaprocess []➔[ ], which also may be taken to be a general representation of the function integrator-operator. Moreover, if distinguishing, which is analog, is all about the [subject], distinction, which is digital, involves the [subject's] elision.

Early in Chap. 1 it was asked whether the distinctional is the interplay between fuzziness and clarity, and therefore whether distinguishing might be the leap from analog to digital. We can now see that the relation between analog and digital involves the elision of the subject. We shall soon delve into this by examining how this elision occurs in *LoF*. Before doing so, however, we must delve further into the nature of @@A.

Living Beings: Joinings

Consider again Spencer-Brown's "A note on the mathematical approach". The first reading led to our koan:

> The skin of a living organism cuts off an outside from an inside. Not so the circumference of a circle in a plane.

This differentiated between abstract and concrete distinctions, the root of which is ב. A second paraphrasis helps differentiate further between @@L and @@A:

> A key theme of this essay is the kind of being that emerges when two or more living organisms *join together*. The skin of a living organism *both cuts off and joins* outside and inside. By tracing the way we represent such a *cutting/joining*, we can begin to reconstruct, with an accuracy and coverage that appear almost uncanny, the basic forms underlying linguistic, mathematical, physical, and biological, science, and can begin to see how the familiar laws of our own experience follow inexorably from *ongoing acts of cutting/joining*.

In other words, we must examine a fundamental difference between the upper and the ground schemes: the difference between sets of abstract distinctions that originate from severance (@@L) and sets of concrete distinctions that originate from joinings (@@A). We are back to biogenesis as it occurred in a colony of progenotes. Distinction by severance is @@L and belongs to Bw, our parsing of the world, whereas joinings @@A and Ba, are the embodied process-structure of organisms as distinguishing distinctions.

The key point, then, is that embodiment did not emerge in isolation. Therefore, inquiry into living being must be in terms of joinings, that is, in terms of *organization*. And, as it will gradually become clear, organization is not intelligible merely logically because it involves paradox. In turn, it will eventually become clear why this leads to ethics.

Just what joinings are, why they are necessarily organized, and how they constitute living being is therefore a central topic to understand aliveness.

LoF is at once radically correct and radically incorrect. It correctly states:

> ... we can begin to reconstruct, with an accuracy and coverage that appear almost uncanny, the basic forms underlying linguistic, mathematical, physical, and biological science and can begin to see how the familiar laws of our own experience follow inexorably.

But it incorrectly concludes: "... from the original act of *severance*." No. What follows inexorably does so from original and ongoing acts of joining. True, the form of our experience (Bw) follows inexorably from acts of severance. But our form as distinguishing distinctions, [body-subjects] as Ba-Bw's capable of acting, anticipating, experiencing, and indeed, the very possibility of formulating laws—this all results from our being joinings. In short, the upper scheme concerns distinction as severance, the ground scheme the distinction-distinguishing of joinings.

But this brings up the issue of understanding @@A in terms of @@L. How can one do this? Doing this would be to understand the ground scheme in terms of the upper scheme, which is the condition for its possibility. This is the mistake in *LoF* that we have been discussing at length. And, given what *LoF* is about, this applies to mathematics, since any formalisms begins with distinctions. So, this brings up the key question, What is it to understand when it comes to living beings?[33]

Organisms are built in layers. But these layers, which might be described as surfaces within surfaces, far from being abstract geometric circles within circles or boxes within boxes, are layers of skin within skin, flesh within flesh. This flesh is membranes within membranes. While @@L is a necessary condition for surfaces within surfaces calculating, it is insufficient to account for the process of calculation.

The similarity between @@L and @@A has already been mentioned. This brings us to another similarity/difference between them. Clearly, @@A is partially

[33] To paraphrase Lonergan: "What am I doing when I am knowing aliveness? Why is doing that to know aliveness? What do I know of aliveness when I do it?" (Lonergan, *Method*, 25). According to Lonergan, adverting to our own cognitional activity allows us to advert to what are, respectively, our immanent cognitional theory, epistemology, and metaphysics. Yet this very act of self-appropriation leads to a transcendental metaphysics that is an integration of heuristic structures (Lonergan, *Method*).

intelligible in terms of @@L. In both diagrams, the two smaller circles (semiautonomous subsystems) are within the larger circle (a semiautonomous system). Inherent to @@L, and, therefore, necessarily to @@A, are the logical conjunctions: not *a*, *a* or *b*, *a* and *b*, and *a* implies *b*. None of these conjunctions alone, however, is sufficient to account for what @@A conveys—namely a process and, moreover, a process of self-regulation of a two-element regulatory system in functional closure. @@A is never in a single one of the logical states. Any of these single states leaves out what is most significant about @@A, namely that, if it is possible at all to account for it logically, it is the sequence A, Ā, A, Ā . . . . That is, if one is constrained to the six logical conjunctions to account logically for @@A, one must say that @@A is *both* A and Ā, *and* A or Ā; *both* Γ and , *and* Γ or . This is oscillation, *LoF*'s imaginary Boolean values.

This is soon verified with the list of conjunctions in Fig. 5.5, because they apply at once to @@A. Most obviously, it is true of the third conjunction, *a* and *b*, as stated by the figure @@ or *two within one*. The second conjunction accounts for how two-component self-regulatory systems operate overall, with *either a or b* at work. At a lower level, the mode of this operation is stated by the fourth conjunction (which, by the way, includes the first within it), *a implies b*. Transposed as *not a, b*, it points to the concrete stages of the system's operation: *a* on *implies b* off; *b* on *implies a* off. All four conjunctions *together* are oscillation. In this strict sense, @@L and @@A are conditions for the possibility of each other, while, at the same time, @@A is well beyond @@L. @@L is the basic set of relations and @@A the structure-process by which these are concretely actualized. If this is so epistemologically, ontologically, @@A, immanent to which is @@L, is the origin of @@L, how it emerged. From this follows a significant corollary. Oscillation is the condition for the possibility of the pure abstract states A or Ā, which are derivative. Skin, @@A and oscillation are more primitive biologically and more primary logically than @@L, which is the six conjunctions taken individually. Concretely, joinings—@@A—is what originated with biogenesis. And from that could follow logic once embodied distinctions had emerged.

Given that @@L refers to an abstraction and @@A to functional closure, I shall use the terms *relation* when referring to @@L and *relationship* when referring to @@A. The term 'concrete' is woefully inadequate with regards to joinings. I shall therefore, at least on occasion, refer to its etymology, which is particularly appropriate—*con-crēscere,* from the Latin *grow together*.[34]

The derivation of @@L from @@A, which is expressed in the differentiation of oscillation into the pure logical states, is the transition from the analogical to the digital. This derivation, then, is intimately related to the elision of the subject, soon to be discussed.

[34] Partridge, *Origins*, 129.

Foundations, and Proper and Borrowed Content of Distinction

No formalism picks itself up by its own bootstraps. Still, this speaks directly to the boundary between the mathematical and the meta-mathematical. The foundation of mathematics necessarily remains beyond it. Russell implies this when he says we must be content to accept some terms as intelligible without definition. The problem goes beyond mathematics to all fields of knowing. It is the general problem of foundations. We can now say precisely what this means about any body of knowledge insofar as it begins with distinction. A body of knowing built logically falls within Kauffman's statement "logic is nothing more than the properties of the act of distinction".[35] Start with a distinction, and that body of knowledge is grounded on the upper scheme (repeated here for convenience):

> ב➔Repetition, *Again*➔Axioms of *LoF*➔Calculus of distinction or "Boolean Arithmetic"➔ "various models"➔Boolean Algebra or Logic.

Falling outside any body of knowledge—for math this is the metamathematical—the ground scheme provides the foundation:

> ב➔Repetition, *Again*➔Axioms of *LoF*➔Calculus of distinction or "Boolean Arithmetic"➔**Cells as distinctions**➔Boolean Algebra or Logic.

This is the form of our embodiment, as distinguishing distinctions. We have now advanced our inquiry into the ground scheme with the finding that embodiment is all about joinings.

The differentiation between ground and upper scheme converges directly with Lonergan's project, which is to establish foundations upon the process and structure of our own cognitional activity—his version of the ground scheme. Lonergan's is a massive project. Glossing it is well beyond the scope of this study. However, a bridge to it can be established through his own notions of *proper* and *borrowed* content.

These notions pertain to his account of judgment. Through self-appropriation, Lonergan invites us to verify for ourselves that our cognitional activity involves three components inherently interlinked in a self-assembling sequence: experiencing, understanding, and judging. According to Lonergan, human knowing may be defined implicitly because it is a formally dynamic structure:

> It is self-assembling, self-constituting. It puts itself together, one part summoning forth the next, till the whole is reached. And this occurs, not with the blindness of natural process, but consciously, intelligently, rationally. Experience stimulates inquiry, and inquiry is intelligence bringing itself to act; it leads from experience through imagination to insight, and from insight to the concepts that combine in single objects both what has been grasped by insight and what in experience or imagination is relevant to the insight. In turn, concepts

[35] Kauffman, "Peirce's Existential Graphs," 61.

> stimulate reflection, and reflection is the conscious exigence of rationality; it marshals the evidence and weighs it either to judge or else to doubt and so renew inquiry.[36]

This, by the way, is the reason why Lonergan invoked Hilbert's notion of implicit definition. Recall why this is crucial: complete generality. Proposed in what follows is the isomorphism between human cognitional structure-process, as we can advert to it in ourselves, and the living distinctional structure-process, as we can advert to it in any living being, including ourselves. Specifically, the proposition is that the cognitional is a highly differentiated version of the distinctional.

Observe how one can describe the self-assembling, self-constituting process of solving a puzzle using Lonergan's words. In taking up the challenge of a puzzle, our cognitional activity puts itself together, one part summoning forth the next, till a solution is reached. This does not occur with the blindness of natural process but as we consciously examine the data of the puzzle, intelligently seek out patterns, and rationally evaluate whether these and the rest of the data of the puzzle fit together or not. These data are what stimulates inquiry, and the tentative reaching for solutions is our intelligence bringing itself to act. The experience of the data, through imagination, leads to insights that pattern the data, suggesting further possible relevant insights. These stimulate reflection, which is the conscious exigency of rationality that over and again marshals the data to weigh tentative solutions to judge whether the puzzle has been solved, or else to doubt that it has and renew inquiry.[37]

The phase of judgment involves the question whether a proposed solution does indeed solve the puzzle, to which the answer is either yes or no. For Lonergan, *proper content* is this binary character inherent to judgment. Depending on the borrowed content, this binary character is expressed as yes/no, true/false, guilty/not guilty, on/off, and so forth. This binary character is inherent to the function of judgment in the process and structure of knowing. Borrowed content, for its part, refers to whatever is specific to a particular judgment.[38] There is no judgment unless there is a judgment *of something*; that something is necessarily contingent, else judgment would be purely mechanical and arrived at "with the blindness of natural process." Without the contingent borrowed content, there is no such thing as judgment.

Palpably, judging and distinguishing are isomorphic. Indeed, I propose that judgment, in Lonergan's use of the term, is a species of distinction. To demonstrate this rigorously, it need only be shown that the proper content of a judgment and the

---

[36] Lonergan, "Cognitional Structure," 207. These mutually interdependent terms are implicitly defined.

[37] Lonergan names 'insight' both the ah-ha! moment, and the full process of cognitional activity, as just described. The reason for this is not hard to appreciate. Lonergan convincingly shows that previous accounts of the process of understanding have a decisive flaw: oversight of insight. As meticulously shown over 770 pages of *Insight*, the weight of this oversight, and thereby of its reversal through insight into insight, brings a massive shift in our understanding of understanding. Therefore, by synecdoche, the moment of 'insight' stands for the whole process of understanding. So, Lonergan refers to Mendeleev's Periodic Table and Einstein's Theory of Relativity as insights, or sets of insights, even if, strictly in his own terms, they are properly speaking sets of judgments.

[38] See Lonergan, *Insight*, 300.

proper content of distinction are the same, because the borrowed content, whether of distinction or of judgment, is contingent. If this is so, then *distinction* or *distinguishing* is the general category, and *judgment* or *judging* is a subset of this general category. As with any other distinction, the proper content of judgment refers to differences that make a difference. Like distinctions, judgments are arbitrary in principle—which is not to say that they are biased in principle. On the contrary, a judgment in court, for instance, can go either way, which is why legal cases must be adjudicated. The proper content of a judgment in court is guilty or not guilty. Whether and how, according to the evidence—which is the borrowed content—the judgment arrived at is either just or unjust, pertains to this borrowed content. Lonergan referred to the transition from intelligence, the analogical patterning that is borrowed content, to reasoning, the digital either-or that is proper content.

For Lonergan, a judgment is a "virtually unconditioned" arrived at through self-reflective insight by a human subject who asks the key question: Have all pertinent questions been asked? So long as there are no further questions to be asked—including those of others—and all pertinent questions have been asked and answered, a judgment may be arrived at, even if it is necessarily only provisional. This is what he means when he said that a judgment is a virtually unconditioned. Yet all such questions, including the key self-reflective question, belong to the borrowed content. The questions asked and not asked sway a judgment either way, but this does not affect the fundamental binary structure of judgment, its proper content.

Flaws in the borrowed content lead to poor judgments, whether it is in a court of law, in the court of public opinion, or in peer reviews by fellow scientists. Yet, flawed or not, their binary form is the proper content of judgment. Again, the urge to ask question, is what drives the self-assembling, self-constituting, formally dynamic structure of human knowing. Flaws in the process result from bias. Without going into details, Lonergan defined bias as extraneous reasons that block the drive to inquiry, blocking further questions from being asked. Yet, once again, all questions and their answers—that is, all data and all questions about this data, including self-reflective questions to guard against bias—constitute the borrowed content of human judgment.

Thus, while the borrowed content of human judging can be immensely complex, its proper content is identical to that of distinguishing—either A or Ā. In terms of the formalism of *LoF*, proper content is expressed through the binary option Γ or . So, distinction is modeled in different ways—in court as guilty or not guilty, in science and mathematics as true or false, in day-to-day common sense as yes or no, in chemotaxis as running or tumbling, and so on. What concretely makes a difference or not refers to borrowed content or value. Given value—any value—proper content follows. The notion of value is taken up below.

Very significantly, for our purposes, is that Boolean imaginary values are modeled by the opening/closing of cellular membranes, the primary binary function that underlies all acts of living beings. In their distinguishing lies the inherent mathematical and *qua* cognitional content of all organic structures and processes. This, of course, points to the ground scheme that is the foundation of distinction. While there may be myriad actual instances of embodiment—borrowed content—this opening/

closing is the proper content of embodiment, as such. Not only is judgment a species of distinction, but it also results from the actual organic process of distinguishing. In strictly organic terms, given value or differences that make a difference, a human judgment is arrived at by a systematic and vastly complex layered array of membranes opening and closing, neurons touching or not touching, firing, or not firing, that in the end simplify to Γ or . Our building of networks of judgments in our knowing is isomorphic to how any organism builds networks of distinctions. Moreover, it is isomorphic to embodiment, as such, nestings of distinctions within distinctions. That is, there is an isomorphism between the ground scheme (@@A, aliveness) and the upper scheme (@@L, logic) that is foundational. In the upper scheme, this applies to all distinctions drawn by any organism, even as we reserve the name 'judgment' to a process of drawing distinctions that includes the key question, Have all pertinent questions been asked? As to the ground scheme, much more will be said about the nature of joinings that obtains in all embodiments, even as it subverts @@L.

In conclusion, judgment is a highly differentiated version of distinction, and distinction a deeply compact version of judging. This pertains to the upper scheme. Likewise, in the ground scheme, we can examine the self-assembling, self-constituting structure-process of our cognitional activity as a highly differentiated version of our distinguishing, and our distinguishing as a deeply compact version of our cognitional activity. This isomorphism is the foundation of inquiry, as such. It is thereby the foundation of inquiry (by living beings) into living being. This is the overarching importance of our conclusion that a judgment is a kind of distinction, and judging a kind of distinguishing.[39]

This brings us back to phenomenology and its foundation. The full significance of Boolean algebra is that any [subject] may introduce any borrowed content. Recall the parallel to numeric algebra. Borrowed content is the a in 3a + 2a = 5a. The full significance of Boolean arithmetic is that, no matter what borrowed content is introduced, all [subjects] share the underlying proper content. Recall that this arithmetic is intrinsically binary, all expressions simplify to either Γ or . Because borrowed and proper content together are the form of distinction, considering them separately leads to error. To treat borrowed content, as if it alone constitutes distinction, is relativism. Thus, for instance, one may wonder if we all see colors differently and whether we could even find out. The actual colors we see are borrowed content; that we differentiate between colors is proper content. By contrast, to ignore borrowed content is objectivism or naive realism. This is to take for granted that we all see colors the same and that it could not be otherwise. To take contingent borrowed content and necessary proper content together I would venture to call *critical realism*, following Lonergan—but justifying this would alone require a book. In short, we possibly see colors differently, or at least slightly differently, and, further, we might never find out whether we do or what the differences are. Still, we all

[39] The foregoing serves as the metaphysical foundation for *analogy as a source of knowledge*. This was the title of Konrad Lorenz's Nobel Lecture on December 12, 1973, at the Austrian Academy of Sciences Institute of Comparative Behaviour Research in Altenberg, Austria, available at https://www.nobelprize.org/uploads/2018/06/lorenz-lecture.pdf.

distinguish different colors. It could not be otherwise; differences either make a difference or they do not. We cannot know each other's experience of color, but we can know and do know, without a doubt, that we all organize our experience of color identically. We can know and do know that this is how all [subjects] organize experience. Moreover, all [subjects] are part of a plurality that has unfolded in a similar context, environmental, historical, etc. This makes it highly improbable that, as a norm, there are radical differences between them. Thus, all [subjects], *qua subjects*, share a single underlying logical structure-process that may vary somewhat in relation to borrowed content, but is identical in relation to proper content. A *critical realist* inquiry into the experience of color is framed along these lines.[40]

While cognitional structure-process pertains to mental activity, distinctional structure-process pertains to the *entelecheia* of all living beings and at all levels. *LoF* is a mathematical phenomenology—mathematical because its foundation is the proper content of distinction and phenomenological because proper content frames the indispensable borrowed content that [subjects], at all levels, provide. Thus, the implicit definition of distinction takes us deep into the nature of living being.

Louie and the Rope

Consider this problem, I once said to Phil McShane:

A rope-climbing device about the size and weight of a human fist, starting at ground level, must reach a point 3 ft high and 3 ft away. It must climb a rope that hangs from 20 ft by setting it on an elliptical swing. In what direction must the rope be swung? From what point on the ground? At what rate must the rope be climbed? Well, he averred, this is about combined vector forces in elliptical conical movement.

An individual who easily solves for various heights, distances, lengths of rope, rates of displacement and elliptical swings, masters the math and physics involved, right? He agreed.

But the device is a baby woolly monkey. Six-month-old Louie consistently verifies his skill at these puzzles. And he is just getting started. His performance will improve dramatically and with increasingly complex problems. Given his rate of practice (~8 h a day) he will be a world-class expert when he is 4, having by then put in his 10K h.[41] He is not unique. Such mastery is the norm in his community. This incident, a mere instance in monkey locomotion, is one of dozens of events we witnessed every day over years. Louie was idly half-hanging from a rope by hand and tail when he glanced up at a nearby shelf. After some morsel, he pulled the rope taking a few steps back to then push off the ground in a trajectory parallel to his object. Climbing the rope as it swung through two-and-a-half loops, he alighted on

[40] For a thorough recent treatment, see Claude Romano, *De la Couleur*. To succeed, discussion of this topic must include the highest conjugate form, our own lived experience of color.

[41] Herbert Simon and William Chase, cited in Daniel Coyle, *The Talent Code*, 53.

the shelf with a graceful downward skip. An impeccable performance without a moment's hesitation while chewing on a bit of grass.

Knowing I was addressing one, if not *the* disciple of Bernard Lonergan, I staked my claim: Louie *knows* what he is doing. I myself had accumulated 10K h relating to monkeys, body to body, face to face. Plus, another 10K h observing them at very close range. *Knowing* is the highest conjugate form for inquiry into primatology, and, therefore, properly defined, zoology generally. Yet it is systematically left out. Phil agreed.[42]

[42] Daniel Mayer, "McShane on the Transformation of Botany: Outline for a Dissertation," 95. For another account of the monkeys' mathematical mastery of their territory, see the account of Lulu in Chap. 10, section "Inner conflict, transparency/opaqueness and the unfolding of anticipation, transparency/opaqueness".

Chapter 6
The Mathematical Ideal and the Elision of the Subject

Proper and Borrowed Content of @@L and @@A

Let us return to the similarity, difference and interrelation between @@L and @@A, the upper scheme and the ground scheme that provides its foundation. Let *borrowed content* refer to actual concrete values that make some differences make a difference, and let *proper content* refer to the structure of distinction, including the irreducible fact that distinction involves some, any, given value. The six logical conjunctions (Figs. 5.4 and 5.5) are the proper content of distinction. Kauffman refers to this proper content when he says, "logic is nothing more than the properties of the act of distinction."[1] This, then, is the proper content of @@L.

As for @@A, the ground scheme, it is the form of aliveness. @@A too must have a proper and a borrowed content. Indeed, its proper content must go back to multiplicity and to homogeneity—though of a different order. Aliveness, at its inception, is plural. In the colony of progenotes, protocells group together in moist gel. Thus, the condition for the possibility of biogenesis, and therefore the ground scheme, is plurality. Therefore joinings, the fact itself a plurality of individuals, along with their interrelationships, is the *proper content* of @@A. I reserve the term *plurality* to refer to this specific instance of multiplicity: multiple individuals, actual distinguishing distinctions for themselves, in *entelecheia*.

In @@A *conditioned coproduction* obtains. Spencer-Brown's refers to this notion. But, in fact, it could only obtain for pluralities, joinings of distinctions that are for themselves, and leaving our own biomorphism aside. If the proper content of @@L includes the irreducible fact that some, any, borrowed content must be involved in establishing diff-diff, so in @@A the proper content is the irreducible fact of joinings of semiautonomous units that must offer some borrowed content to

[1] Kauffman, "Peirce's Existential Graphs," 61.

© The Author(s), under exclusive license to Springer Nature Switzerland AG 2023
D. C. Mayer-Foulkes, *The Nature of Living Being*, Biosemiotics 26,
https://doi.org/10.1007/978-3-031-24789-7_6

each other. *Con-crēscere,* growing together, expresses the proper content of @@A; it's borrowed content refers to concrete modes of joinings.

But @@L and @@A do not only have similarities and contrasts. They also mutually co-produce. Ontologically, indeed, existentially, the ground and the upper scheme, embodied distinctions (Bw) and their distinguishing (Ba) mutually co-produce. So do structure and process. This is why I refer to living beings as distinguishing distinctions. Joinings as such are Ba-Bw. This too is part of the proper content of @@A. Given the need to differentiate between the being (Ba) and the knowing (Bw) of living beings—an abstract differentiation—Bw may in turn be understood in terms of @@L and @@A. Their mutual interrelation points to a key difficulty that underlies the question, What is it to know when it comes to living beings? There is a palpable and fundamental difference between abstract distinctions by which we reason, on the one hand, and joinings and interdependence of a plurality of living beings, on the other. There are abstract relations and living relationships; the stasis of logic and the dynamics of two-element regulatory systems. I shall return to this contrast in Chap. 10, which will lead us into reason/emotion, right brain/left brain, mythos/logos, and so on. The full explicit differentiation @@A/@@L, which is the development of the upper scheme, is a human differentiation that dates back merely about 2500 years, a mere eyeblink in evolutionary terms. It's occurrence was represented in myth by the Greeks as the birth Athena, daughter of Zeus, with no mother and emerging fully grown from his forehead. This differentiation was the condition for the possibility of our asking ourselves all kinds of questions, among them about the nature of living being. Answers to our questions must therefore be grounded and work through this differentiation.

Value and Name

Paradoxically, *LoF* offers a superb door, at once to the heart of aliveness and to the heart of mathematics, because, starting from its axioms, it points *both* to the need for a [subject] *and* for the [subject's] elision. This points to the fundamental (mathematical) nature of living being and, at the same time, to a key problem with the mathematical ideal of knowledge. Living beings are [subjects] and it is into [subjects] that inquiry about aliveness must be about. The mathematical ideal of knowledge, however, idealizes formal systems—@@L. These aim at universality and generality, accomplished through abstraction. To abstract is, precisely, to elide the knowing subject. Thus, the mathematical ideal of knowledge aims to remove the very [subject] that inquiry into aliveness must be about.

With the use of *again* in its axioms, *LoF* performatively shows, not only that a knowing subject is necessary for its construction, but also that, intrinsic to inquiry as such, is the play between the [subject] and its elision; [subject]/[no-subject].

How, then, is the subject elided in *LoF*? It involves values and names. This is a key point in the link and divergence between @@L and @@A.

According to Keith Devlin, most mathematicians now agree that mathematics is the science of pattern.[2] Inherent to pattern is distinction and repetition, that is, ב. So, distinction and repetition, and therefore ב, are also inherent to the formalisms that account for patterns. Paradoxically, the fewer the symbols a formalism uses, the more these must be repeated. So fundamental is this that it is taken for granted. We have seen that *LoF* makes use of only three graphic symbolic components: cross, its absence, and its placement. Notice that all of the foregoing refers to @@L, the upper scheme. And yet, inherent to it is ב, the ground scheme.

Unusually, despite its covert use of repetition, *LoF* does overtly acknowledge the knowing subject, both at the outset and at the end of the formalism. This acknowledgement marks the link and divergence between @@L and @@A. The final chapter, entitled *Re-entry into the Form,* concludes with this statement:

> An observer, since he distinguishes the space he copies, is also a mark. . . . We see now that the first distinction, the mark, and the observer are not only interchangeable, but, in the form, identical.[3]

"Observer" is synonymous to [subject]. However, the ground and the upper schemes described above are equated, or conflated, or collapsed into one. In any case, the [subject] embodies distinction. This is prefigured from the very outset when distinction is defined. Here it is again, it is the final line that we are interested in:

> **Definition**
> *Distinction is perfect continence.*
> That is to say, a distinction is drawn by arranging a boundary with separate sides so that a point on one side cannot reach the other side without crossing the boundary. For example, in a plane space a circle draws a distinction. Once a distinction is drawn, the spaces, states, or contents on each side of the boundary, being distinct, can be indicated.
>
> *There can be no distinction without motive, and there can be no motive unless contents are seen to differ in value.*[4]

For contents to be of value, differences must make a difference. And differences only make a difference to [subjects]. Though referred to obliquely, somewhat like in the *again* of the axioms, the subject is implicit here. Though mathematically this acknowledgement makes little difference, meta-mathematically it does, a profound difference. It is particularly significant because distinction and repetition apply to all formalisms and therefore to mathematics as a whole. It says that mathematics relies on ב, whether this is explicitly acknowledged or not.

So, this paradox is at the very heart of mathematics. To get off the ground, mathematical constructs require repetition, which can only be provided by living subjects, and yet, to be formal, these subjects must be elided. How is this paradox resolved?

Let us see how it is done in *LoF*. The knowing subject is acknowledged at the outset and then immediately elided; this is done again at the end, though equivocally

---

[2] Devlin, *The Language of Mathematics*, 3.

[3] Spencer-Brown, *Laws of Form*, 76.

[4] Spencer-Brown, 1.

in a way that blurs the boundary between the mathematics and the meta-mathematics. As we have just seen, the subject is implicit in the notion of value yet immediately elided in the next sentence: "If a content is of value, a name can be taken to indicate this value. *Thus, the calling of the name can be identified with the value of the content.*"[5]

So, value, which is a difference that makes a difference for a [subject], becomes identified with a name. The transposition from [subject] to name allows the formalism to proceed. At this key point, *acts of distinguishing* (verbs) become *distinctions* (nouns). Thus, in general, naming, as in "Let x be q . . . ", is a how the knowing subject is elided and a formalism set up. With distinguishing activity transposed into abstract distinctions, the boundary is crossed from meta-mathematics to mathematics. Thus, the paradox of both requiring and eliding the knowing subject is resolved. This is the strength, beauty, and wonder of mathematics—what lends it an air of transcendence. Naming elides the subject and objectivity is claimed. This launches the mathematical ideal of knowledge.

Uniquely in *LoF*, the matching reverse transposition occurs at the end, leading from the mathematical back to the meta-mathematical: ". . . the first distinction, the mark, and the observer are not only interchangeable, but, in the form, identical."[6] If this "first distinction" in this statement is taken to be a noun, then it is true. However, there is a radical difference between the *act* of a [subject] (the verb) and a *mark* (the noun). Because the first distinction must be an act, this statement blurs the difference between [subjects] and their distinctions. It blurs the difference between the ground and the upper scheme. So doing, it also blurs the boundary between mathematics and meta-mathematics.

Since names are produced by living beings, they are products. As a kind of thing, products are ontologically radically different from the living beings that produce them. Formalisms are products too, and therefore so too is the whole of mathematics—a topic we return to in the next chapter. This brings up important questions: Is this transposition from distinguishing to distinctions, and thence to names, something all organisms—[subjects]—do? Is this at the heart of the nature of living being? Is the transposition from verb to noun typical of aliveness, such that there are differentiated and compact versions of it? The answer to all these questions is in the affirmative. We have now established that EI, the process, can stand for distinguishing, the verb, while single EIs may stand for distinctions, the name. Thus, any given EI is a [name]. Likewise, in both distinguishing, and in EI as integrator-operators, []➔[ ] obtains. Thus, analogous to *analog/ digital* is the relation *living being/ [products]*. The process by which EIs are arrived at is a verb, and the ongoing EIs produced, which are signs or [names], are protean products. In distinctional activity, then, the play *verb/noun* is how the play *analog/digital* takes place.

---

[5] Spencer-Brown, 1.

[6] Spencer-Brown, 76.

Liminalism and Substantialism

Something of fundamental importance that happens with naming is the transposition of distinctions (noun) from continence to content. Recall that distinction begins as a boundary:

> a distinction is drawn by arranging a boundary with separate sides so that a point on one side cannot reach the other side without crossing the boundary. For example, in a plane space a circle draws a distinction. Once a distinction is drawn, the spaces, states, or contents on each side of the boundary, being distinct, can be indicated.

But boundary is then transposed as content:

> There can be no distinction without motive, and there can be no motive unless *contents* are seen to differ in value.[7]

This content is associated to a value and this value leads to naming:

> If a content is of value, a name can be taken to indicate this value. *Thus, the calling of the name can be identified with the value of the content.*[8]

This transposition can be illuminated with figures from Chap. 2 reproduced here. A distinction appears as shown on the right in Fig. 6.1 and the arrows run in the opposite direction. From there it goes to Fig. 6.2. There, the content is identified with a name, and this name is a product.

The [subject], however, moves in the opposite direction, beginning as something substantial in Fig. 6.2 and moving in the direction of the arrows in Fig. 6.1 until it is elided all together. So, not only does a name serve to elide the [subject]. It also brings about a transposition from the liminal to the substantial. For substantialism, things are 'bodies', already out there now, concrete, and bounded clearly and distinctly. For liminalism things are intelligently grasped and rationally affirmed in relational terms that may be fuzzy, ambiguous, paradoxical. This transition from liminalism to substantialism is a profound metaphysical shift, with radical consequences for biology. We will see examples of this in the following chapters. Perhaps this is most easily understood in terms of systems, as we shall see. The systemic approach to understanding organization is an invitation to reverse the transposition and reframe from substantialism to liminalism.

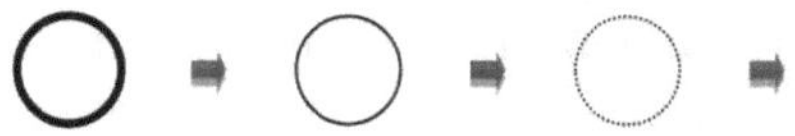

Fig. 6.1 Distinction as circumference

Fig. 6.2 Territory on a plane

[7] Spencer-Brown, 1.

[8] Spencer-Brown, 1.

Most radically, understanding aliveness in terms of joinings is a reframing as liminalism. This is the reason why @@A is about *relationships*. It is why @@A is only partially intelligible in terms of @@L. Relations are abstract and pertain to @@L, they are Bw. The proper content of @@L is logic, the properties that follow from distinction, including the irreducible fact that some borrowed content is required for differences to make a difference. Relations, and therefore the logical conjunctions, express the proper content of @@L. Relationships, for their part, are recurring two-way and reciprocally anticipative acts between organisms. These pertain to the proper content of @@A, Ba. In @@L, the two smaller circles within the larger one, refer to substances, whereas in @@A, the two semiautonomous systems within a larger one refers to structure-processes in entelecheia and in *con-crēscere* relationships. This is why logical conjunctions alone cannot account for @@A. As was shown, it takes all the conjunctions working simultaneously to account for oscillation, which is intrinsically liminal. Again, skin, @@A and oscillation, belonging to the ground scheme, are more primitive biologically and more primary logically than @@L. They are all liminal. The liminal-substantial precedes the substantial. Biogenesis emerged, not from single protocells, but from their relationships as a plurality in a colony of progenotes. This all pertains to the proper content of @@A.

The liminal nature of @@A as relationships exceeds @@L and leads to paradox. Perhaps the most radical obstacle to accepting the paradoxical nature of relationships, is that substantial entities are identified by a name and thereby conceived as products; and products inevitably invite, not only substantial, but mechanical thinking.

Naming as Integrator-Operator

The centrality of naming has been understood for millennia. Spencer-Brown was roundly criticized, even mocked, for the epigraph heading *LoF* in Chinese characters: 無名天地之始, "Heaven and Earth's nameless origin." Today, half a century later, it is worth dwelling on this statement, which is the opening of the *Dao De Jing*. As we have seen, naming plays a key role in *LoF*. Indeed, beyond what Spencer-Brown intended, it suggests why *LoF* may be read as a phenomenology. As it turns out, it is a highly appropriate epigraph because Daoism treats naming as a process. Here are some translations of the epigraph:

- The unnamable is the origin of heaven and earth;

  named, it is the mother of the thousand things.[9]

- [Tao is both Named and Nameless]

  As Nameless, it is the origin of all things
As Named, it is the mother of all things[10]

[9] Lao Tzu, *Tao Te Ching: A New Translation,* 3.

[10] Lao Tzu, *Tao Te Ching, The Definitive Edition*, 14.

- The nameless is the origin of the myriad creatures;

  The named is mother of the myriad creatures.[11]

- There were no names in the beginning of heaven and earth;

  Attribution of names is the matrix of myriad things.[12]

- The nameless (*wuming*) is the fetal beginning of everything that is happening (*wanwu*), While that which is named is their mother.[13]

The last translation shows that the meaning of the verse hangs on the term *wuming*. Because it speaks eloquently, indeed uncannily, to our topic generally, and to the topic of naming specifically, the following explanation is quoted at length:

> *Wu*-forms are pervasive in the *Daodejing*. One additional *wu*-form, for example, is *wuming* 無名: translated as "the nameless," but actually suggesting a kind of naming the does not assign fixed reference to things. To function effectively in negotiating our environment, we need to rely upon our ability to make distinctions. These distinctions in themselves are certainly functional and enabling, but can distort the way in which we understand our world. We can easily fall into the fallacy of what Whitehead describes as "misplaced concreteness," reifying what is abstract and treating these hypostatized "things" as more real than the changing events of our experience. We can easily and at real expense overdetermine the continuity within the life process as some underlying and unchanging foundation. Such linguistic habits can institutionalize and enforce an overly static vision of the world, and in so doing, deprive both language and life of their creative possibilities. The referential use of language as someone's technical morality—expressing the way the world *ought* to be—can too easily lay claim to the power and control that would make it so. Naming as power undermines the importantly creative aspect in the effective use of names. In a processual world—a world ever under construction—to be able to name something is to be able to trace out its concrete relation to you and the world, and on that basis, respond to it productively. While naming can be understood as an abstractive and isolating gesture, Daoist naming personalizes a relationship and, abjuring any temptation to fix what is referenced, instead understands the name as a shared ground of growing intimacy. Such naming is presentational rather than just representational, normative rather than just descriptive, perlocutionary rather than just locutionary, a doing and a knowing rather than just a saying. Naming as knowing must have the provisionality to accommodate engaged relationships as in their "doing and undergoing" they deepen and become increasingly robust. Such knowing is dependent upon an awareness of the indeterminate aspect of things. The ongoing shaping of experience requires a degree of imagination and creative projection that does not reference the world as it is, but anticipates what it might become.[14]

[11] Lao Tzu, *Tao Te Ching, The Classic Book of Integrity and The Way*, 59.

[12] Soho, *Tao Te Ching, Zen Teachings on the Taoist Classic,* 1.

[13] Lao Tzu, *Dao De Jing, A Philosophical Translation*, 77.

[14] Lao Tzu, 45–46. Ames and Hall define *wuming* as "naming without fixed reference" 67. Converging remarkably with the notion of 'bodies' and the discussion of flow in the previous chapter, they write, "The key to understanding of *wuyu*—indeed of all these *wu* forms that comprise the Daoist disposition—lies in the contrast between 'objects' and 'objectivity.' Using Western epistemological terms, the thoughts about the world expressed in both the *Zhuangzi* and the *Daodejing* represent what we might call a realistic perspective. Beyond the mediating confusions introduced by language, and by layers of our own distorted perceptions and tendentious categorizations, there is nevertheless, with properly Daoist qualifications, an 'objectively' real world. Our

Written in terms almost shockingly alike to those in the present essay, this explanation is redolent with aliveness and a grasp of the world in liminal terms and as ongoing process. On excessive substantiality, the notion of things and hypostatization is discussed at length in the next chapter.

Abstract naming contrasted to growing intimacy recalls hours and days and years of living with the monkeys. Living next door to them was indeed a reprieve from human isolation. As engaged practitioners, sharing in their world and their naming—a very personalized relationship, a doing and knowing rather than a saying—taught us to rest in the indeterminate aspect of things, reminded time and again that there was so much more to learn and understand about the monkeys, both as individuals and as a group. The Daoist differentiates between a distinguishing activity and the resulting distinctions and understands the relation between them. To understand naming fluidly is to understand it as integrator-operator. This is what is achieved by "abjuring the temptation to fix what is referenced."

Beyond the difference between distinguishing and distinction, there is the difference between a distinction and a distinction named. Not making this key differentiation is what allows Spencer-Brown to fuse together the ground and the upper scheme. So, the "nameless" falls into two categories—the not (yet) distinguished and the distinction not (yet) named. Take the initial whiff of a smell. This is a first stage. The second stage is the deliberate attempt to distinguish it, followed by the third, which is to attach a name to it. However, we tend to collapse the latter stages into one. This obscures the nature of our cognitional activity and leads to misplaced concreteness.

Regarding the first stage, recall []→[ ] in relation to EI, which expresses the drift inherent to A=A that allows for the unfolding of anticipation. It is the vector from a current integrator-operator [] to another [ ]. Just as EI, distinction is always in motion. Development and repetition are inherent to each other, as given by A=A, A≠A. Recall that []→[ ] stands for [A=A]→[A = A]. So, the name Helen refers to a person throughout her development, which is a sequence of integrator-operators. This is true for both Helen Keller and Helen of Troy. And just as every stage of the life of each is an integrator and an operator, so too their names are integrator-operators for cultural pointers—'Helen of Troy' for insights into love and war and 'Helen Keller' to insights about language acquisition. The deeper significance of this use of names beyond logic into metaphor and allegory, so common that we may overlook it, is what the paragraph on Daoism points to.

---

task is to experience that world as 'objectively' as possible. From the Daoist perspective, the problem begins when we insist that the 'objective world' is a world made up of objects—namely, concrete, unchangeable things that we encounter as over against and independent of us; things which announce themselves to us by asserting 'I object!' For the Daoist, the objective world cannot be objective in this sense because it is a constantly transforming flow of events or processes that belie the sorts of discriminations that would permit a final inventory of the furniture of the world. Paradoxically, for the Daoist the objective world is objectless. Sages envision a world of changing events that they can, for whatever reason, choose to freeze momentarily into a distinct pattern of discrimination, but that they recognize, when they see clearly, as being beyond such distinctions" 43.

There is a paradox in naming, one that is already present in EI. Call it the *paradox of labeling*. A specific EI arrived at elides EI as underlying process, and thereby the [subject] that carries it out. This is how we are built, as organisms. In general, adverting to the present moment tends to be at the expense of adverting to the process it belongs within. It takes much skill to maintain the two in focus, which is to advert to both continuity and discontinuity. This elision of liminality is what leads to substantialism. For one thing, it elides the inherent interdependence of individuals in self-regulating systems. Elided along with this is the paradox of self-regulating systems that, interdependently, form a larger self-regulating system. So, also elided, are the paradoxes inherent to the relationships between self-regulating wholes within larger self-regulating wholes. This, which is the nature of @@A, is a major topic taken up in detail in Chap. 9.

For now, the point is that labeling potentially leads to Whitehead's fallacy of misplaced concreteness. We first transpose from liminality to substance by rendering each element as an inherently independent in-itself, thereby granting them ontological independence. This reduction from *con-crecere* to abstract logic (@@A➔@@L) is the reduction from mutual to linear causation. Yet, nonetheless, labeling, as with individual instances of EI, is indispensable to aliveness. Thus, inquiry into aliveness must involve adverting to, and making sense of, how we make sense in the context of this paradox. Quite literally, to grasp our nature we must go against our nature.

To grasp the nature of [subjects] we must loosen our natural tendency towards either/or in the *A or $\bar{A}$* of @@L and reframe in terms of both/and in the *A and $\bar{A}$* of @@A. Doing so is to abjure the sharpness of clear and distinct ideas in @@L to dis-elide and recover the interrelating [subject] with all its inherent haziness, ambiguity, paradox.

Naming, at the bifurcation between @@L and @@A, points to two fundamental conundrums that take up this investigation: given that our nature is @@A, why do we do @@L? And given that we do @@L, how do we discern our @@A nature? Moreover, how do we express what we discern about @@A—in terms of @@L or @@A? Regarding the first, we have seen the complementarity of @@A and @@L in the transition from analog to digital as, in the course of distinguishing, the *verb*, distinctions are arrived at—*nouns*. Regarding the second question, part of what we do as @@L is to produce material products, only to then be tempted to make sense of ourselves, and of living beings in general, *as products*. Myths of creation do just that when they tell the story of the advent of humans as the fashioning of a product by a Creator—a great Producer.

In evolution, and even in history, @@L emerged but an instant ago. With EI as its remote precursor, distinctions, and more so named distinctions, carry misplaced concreteness. Paradoxically, naming stalls the process of []➔[ ]. Being inherently nonprocessual, naming obscures the continuity-discontinuity intrinsic to the developmental. (This, by the way, is why '[subject]' and '[object]' are bracketed throughout this examination.) It can come as no surprise that long anteceded @@L hold the key to account for @@A in its own terms, as the *Daodejing* so compellingly shows.

Given the current mathematical ideal of knowledge, returning to @@A to survey aliveness in its own terms requires us to turn to ourselves, like naming-un-naming-renaming Daoist subjects. Or, as Lonergan proposed, to inquiry that reaches for a virtually unconditioned, a provisional objective truth that holds until new, further questions can be asked. This differentiates and ushers in the play of @@A and @@L. This aligns with the key point that abstract distinction as perfect continence arises from distinguishing as a scheme of recurrence, which involves the ב of [subjects], the primacy of *con-crecere* oscillation in aliveness in the functional dynamics of joinings.

The problems inherent to naming, then, along with its tertiary status, point to *interiority* as a general method for inquiry into aliveness. Interiority refers to a methodological approach that systematically controls the making of meaning by rigorously including in its purview the knowing subject involved in the process of inquiry. Inevitably, this powerfully contributes to the dis-elision of the knowing subject. Interiority calls for injunction, because it is an invitation to the subject to examine the process of his or her own cognitional—distinctional—activity. The Daoist "doing and a knowing rather than just saying" points to injunctions such as "look down the microscope." We will take this up in the next section. Clearly, arriving at one's own experience gives rise to the possibility of drawing analogies—including analogies to oneself—or *isomorphism*. For instance, the isomorphism between distinction, EI, naming and inquiry, follows from the injunction: "Think of distinction as a deeply compact version of EI, and of naming as a highly differentiated version of EI." Such injunctions are invitations to thought experiments based on data of consciousness.

Modern philosophy began its turn to the subject, which is to say, its turn to data of consciousness, with Descartes' *I think, therefore I am*. Continued by Kant, this turn continues to this day. Lonergan's insight into insight through self-appropriation, as we are invited to do in his book *Insight,* is a key contribution to this systematization. Today, the social sciences and the humanities in general, including their teaching-learning methods, are primed to move away from the dyad theory/practice into the tripartite ontological and epistemological structure theory/practice/interiority.

This brief review suggests that this turn to the subject is the future of zoology and biology generally. Given everything that has been said about biomorphism and anthropomorphism, the foregoing reflection on names and naming points to a method of inquiry into aliveness that must include injunction, self-appropriation, and isomorphism. @@L became strongly rooted in Western culture as a result of the astonishing successes of mechanistic thinking. It cannot be surprising that contemporary processual thinking to supersede it finds roots and converges with non-western practices of interiority such as Daoism and Buddhism.[15]

[15] As it goes to press, this author has become aware of striking convergences between this book and Buddhism as presented in Eihei Dōgen's *Shobogenzo*. For excellent commentaries on this 13th classic see Hakuun Yasutani, *Flowers Fall* (Shambhala, Boston 1996) and Shohaku Okumura, *Realizing Genjokoan* (Wisdom Publications, Somerville, MA 2010).

Insights and Their Expression

Covertly, if not overtly, the epistemological structure theory/practice/interiority has long been part of the method of the sciences. We cannot leave out of this reflection on mathematics its most intimate relation to a subject, namely, mathematical creation. This too is implicit in the mathematical ideal. Spencer-Brown writes:

> A theorem is no more proved by logic and computation than a sonnet is written by grammar and rhetoric, or than a sonata is composed by harmony and counterpoint, or than a picture is painted by balance and perspective. Logic and computation, grammar and rhetoric, harmony and counterpoint, balance and perspective, can be seen in the work *after* it is created, but these forms are, in the final analysis parasitic on, they have no existence apart from, the creativity of the work itself. . . . Thus, the relation of logic to mathematics is seen to be that of an applied science to its pure ground, and all applied science is seen as drawing sustenance from the process of creation with which it can combine to give structure, but which it cannot appropriate.[16]

Lonergan referred to the process of creation as *insight*. Insight is the heart of our own cognitional activity, as we can advert to it for ourselves. Insight into insight is the foundation of Lonergan's metaphysics because this metaphysics is imminent to that cognitional activity. Our self-appropriation of our own cognitional activity, then, provides a foundation for an incisive critique of the oversight of insight in alternative metaphysics. This suggests that Kant's remark in Chap. 1 tells us more about what is overt in philosophy, yet which nonetheless also applies to mathematics, since here too the struggle from insight to expression involves grappling with definitions. Though proofs have no existence apart from insight, their rigorous expression calls for many additional insights. That a proof does not appropriate the insight is obvious enough from the mere fact that various proofs may be found of a single insight. All proofs draw sustenance from the insight itself as they furnish contingent structures to demonstrate necessity. Keith Devlin makes a similar point on notation:

> These days, mathematics books tend to be awash with symbols, but mathematical notation no more *is* mathematics than musical notation *is* music. The page of sheet music *represents* a piece of music; the music itself is what you get when the notes on the page are sung or performed on a musical instrument. It is in its performance that the music comes alive and becomes part of our experience; the music exists not on the printed page, but in our minds. The same is true for mathematics; the symbols on the page are just a representation of the mathematics. When read by a competent performer (in this case, someone trained in mathematics), the symbols on the printed page come alive—the mathematics lives and breathes in the mind of the reader like some abstract symphony.[17]

However, Spencer-Brown pointed out that these symbols are not in fact representations:

> It may be helpful at this stage to realize that the primary form of mathematical communication is not description, but injunction. In this respect it is comparable with practical art

[16] Spencer-Brown, *Laws of Form*, 102.

[17] Devlin, *The Language of Mathematics*, 5.

> forms like cookery, in which the taste of the cake, although literally indescribable, can be conveyed to a reader in the form of a set of injunctions called recipe. Music is a similar art form, the composer does not even attempt to describe the set of sounds he has in mind, but writes down a set of commands which, if they are obeyed by the reader, can result in a reproduction to the reader, of the composer's original experience… even natural science appears to be more dependent upon injunction then we are usually prepared to admit. The professional initiation of the man of science consists not so much in reading the proper textbooks, as in obeying injunctions such as 'look down the microscope'.[18]

This speaks directly to the method of inquiry into aliveness. At the core are injunctions and comparisons, all of which fall under the heading of what I refer to as *isomorphism*.

Lonergan explained at length in *Insight* that knowing is not taking a good look. What matters when the scientist looks down the microscope is not what she sees but what she understands, her insights. Oversight of insight in metaphysics performatively contradicts what we advert to through self-appropriation in our own cognitional activity. We scan the pieces of a jigsaw puzzle and have insights into possible assemblies. An insight is tested by attempting the assembly, which is verified by the judgment "yes this is where it belongs."

This too is how we learn about ourselves, our aliveness, and the nature of aliveness in general. It goes well beyond looking. This lies at the core of the isomorphism between judgment and distinction. This is why it is a powerful heuristic structure to keep in mind that our cognitional process is a highly differentiated version of distinguishing and distinguishing a deeply compact version of our cognitional activity. The challenge in any discipline is to have insights, arrive at judgments, and then to express them. Thus, the intent of this research is to express insights and judgments about distinguishing and distinction as these pertain to distinguishing distinctions such that readers can arrive at these kinds of insights for themselves.

As Devlin points out, the relation between mathematics and music runs very deep:

> In fact, for most of the two and a half thousand years of Western civilization, starting with the ancient Greeks, mathematics and music were regarded as two sides of the same coin; both were thought to provide insight into the order of the universe. It was only with the rise of the scientific method in the seventeenth century that the two started to go their separate ways.[19]

Mathematics is 'seen' and music is heard. It is no coincidence that *The Monkey Sanctuary*, described by Williams as a "study in interliving," was founded by musicians. This "interliving" was a very concrete and deliberately nurtured liminality. Thus, our approach to caring for and learning about the monkeys had to be rigorous by practical necessity. We had to know exactly what had to be done, and when, and how, but we could not aim for clear and distinct ideas in Descartes' sense. We played things by ear as jazz musicians do, what I dubbed an *artistic approach* as opposed to the *scientific approach*. Indeed, these two require each other. Just as we learned music by working on our musicianship, so we learned monkeys by adverting to our

[18] Spencer-Brown, *Laws of Form,* 77–78.

[19] Devlin, *The Language of Mathematics*, 6.

'primateship' as we tended for the monkey colony. Full involvement of the whole knowing subject is as indispensable to music as it is for monkeys. This is also true for mathematics and science, albeit the demand of the elision of the knowing subject in the expression of insights. The core difference is that, unlike a mathematical form, a microscope, or a guitar, when a monkey hugs you, the arms of nature reach out to embrace you.

Convergence of Platonism and Phenomenology

We conclude this chapter by taking a step back to gain some perspective. This may be done by comparing two radically different notions of number and showing were they converge. By obscuring the knowing subject in mathematical formalisms through naming, in the rationalist interpretation of mathematics, numbers are real in a platonic sense. Shapiro provides the following example of this interpretation:

> When the ordinary arithmetician counts a pair of shoes, each shoe is a unit, but the two shoes are not the same shape or even the same size. In contrast, when the philosopher counts 'two,' she refers to a pair of units that are the same in every way. For the philosopher, natural numbers are collections of pure units, which are indistinguishable from one another (*Republic*, 425; *Sophist*, 245).[20]

The difference between these two views of numbers may be expressed in terms of homogeneity. For arithmeticians, the name 'shoe' refers to a homogeneity that admits neglect of a large amount of concrete residue. This is what allows shoes to be counted as tokens of a type. In contrast, for the philosopher—Plato—a number refers to units that are homogeneous and require no arbitrary neglect of concrete residue. This could only be because they are made up of abstract collections of pure units that are indistinguishable from one another. Thus, for Plato and the rationalists, numbers belong to an ideational world in which absolute homogeneity obtains, unlike the actual concrete world we inhabit, which requires ב. For Plato, this is the sense in which numbers are Real and do not require living subjects. In contrast, according to the phenomenological account presented here, ב is the condition for the possibility of successive degrees of abstraction that eventually give birth to ideas that go by names such as 'shoe', or 'six' or 'number'.

We have seen how the elision of the subject confers to mathematics the status of ideal of knowledge and the claim to objectivity. Interestingly, Galileo's "Mathematics is the language in which God has written the universe" supplies natural laws with an omniscient divine [Subject] as guarantor of repetition that otherwise could only be provided by living biological [subjects]. Such a transcendent [Subject] allows logic to be independent from living [subjects]. But then, their construction of their world is no longer radically intertwined to their aliveness. In such a conception it is likely

[20] Shapiro, *Thinking About Mathematics*, 58–59.

that metabolism will be separated from cognition, instead of it being its early manifestation. This, then, is at the root of the classic body/mind split.

The failure to recognize this in a secular age, in the absence of an omniscient guarantor of repetition, continues to allow science to pursue mathematical-like truths and platonic forms that, knowingly or not, elide their origin in the distinguishing activity of living [subjects]. The present inquiry into distinction, and its consequences in our understanding of mathematics, amount to a subversion of the title of Stephen Hawking's book, renaming it *[Subjects] Created the Integers*—whether God can be counted among them or not—because the condition for their possibility is ב.[21]

To elide [subjects] is to elide the spirallic being indispensable for ב. This makes the mathematical ideal of knowledge a naïve realism.

As was shown with regards to *LoF*, the explicit, though equivocal, and somewhat blurred acknowledgment of the [subject] in the very last line of the formalism, marks the transition from mathematics to metamathematics. Beyond what *LoF* is mathematically, and whether Spencer-Brown was aware of it or not, this establishes a key link to phenomenology. At the very least it amounts to framing the mathematical formalism within a phenomenology. The two statements—"logic is nothing more than the properties of the *act* of distinction", and, *LoF* "aims to clarify the laws serving as a basis for the formation of forms, [which are] foremost about the marking of a difference . . ."—both point to the fundamental nature of [subjects].[22] However, though fundamentally phenomenological, the two statements together indicate the meeting point between Platonism and phenomenology.

Surprisingly, the foregoing reflections on *LoF* suggest how the irreducible opposition between Platonism and phenomenology can be overcome. It turns out that radical Platonism and radical phenomenology meet. What would be the foundational proposition of a radical Platonism? There is distinction. So, distinction is Real. The proper content of distinction is Real in the Platonic sense. From the proper content of distinction follows a non-numerical mathematics, from which follow oscillating values and thence number, space and time, all properties of distinction. If distinction is Real, these properties are Real. So, a coherent radical Platonism follows from the foundational proposition that distinction is Real.

What would be the foundational proposition of a radical phenomenology? That living distinctions are real. The same follows here as what follows for the radical Platonist: a non-numerical mathematics, from which follow oscillating values and thence number, space, and time. These are all properties of distinction. If there are living distinctions, the world appears, and the properties of distinction are real. So, a coherent radical phenomenology follows from the foundational proposition that living distinctions are real. A radical Platonism and a radical phenomenology converge on the same foundation: distinction, Real or living. The key difference

[21] Hawking, *God Created the Integers*.

[22] "We see now that the first distinction, the mark, and the observer are not only interchangeable, but, in form, identical." *LoF*, 63. The statements are from Kauffman, "Peirce's Existential Graphs," 61 (emphasis added), and Bakken, "George Spencer-Brown," 482.

between them is that this radical rationalism is the upper scheme @@L (logic), while the radical phenomenology is the ground scheme.

This brings us to Kant's noumena. The ground scheme, or @@A (aliveness), that makes living beings distinctions, is the foundation of appearance. The world only appears because we are ourselves distinctions. This is the naturalized version of Spencer-Brown's "The universe is simply what would appear if it could."[23] The appearance of the world confirms that we are distinctions. That distinguishing distinctions emerged, points with certainty to a moment the remote past, sometime after the Big Bang, when the universe ceased to be homogeneous. Long before appearance, this gave rise to differences that eventually allowed life to emerge. This is the point of the statement above about *LoF* clarifying the laws that underlie the formation of forms and the marking of differences. Difference unfolds logic, and we can thereby make sense of the universe as it was before there were [subjects] it could appear to. The condition for its appearance is ב, differences that make a difference for a [subject]. Mathematics lends itself well to understand differences as it unfolds in logic. Even to understand many aspects of living being. It is suitable for the proper content of abstract distinction. However, it falls short when it comes to grasping the origin of appearance, the proper content of aliveness, @@A. As we have seen with the ground and upper scheme, the condition for the possibility of @@L is @@A—joinings, relationships between distinctions that are for themselves, actual, living distinguishing distinctions. Thus, the condition for the possibility of @@L is a We, a plurality of distinguishing distinctions. Eventually, I/We led to I/World.

In sum, regardless of a Platonic or phenomenal interpretation of the nature of mathematics, this is not in itself an obstacle to inquiry into living being. On the contrary, as has been shown, distinctions, abstract and living, are inherently mathematical. The obstacle is the exigency of the mathematical ideal to collapsing, as *LoF* did, the upper and ground scheme, thereby to attempt to makes sense of @@A in terms of @@L. Such a collapsing pertains to the platonic framing of mathematics. A pure phenomenological approach involves an equivalent collapse if it attempts the opposite reduction, to make sense of @@L exclusively in terms of @@A.

It has been shown that @@L and @@A are *both* radically incommensurate *and* indispensable to each other. Substantialism, largely pertaining to @@L, and liminalism, largely pertaining to @@A, are likewise incommensurate and indispensable to each other. It is in the context of these two binary poles that we must understand how living beings relate to the world. Or, said simply and biologically, this is how we must understand functional closure.

[23] Spencer-Brown, *Laws of Form,* viii.

I/World

A radical phenomenology sets out the objective conditions for the activity of [subjects]. The proper content of distinction points to their fundamental nature. This addresses at its core the problem of anthropomorphism—a key obstacle in the development of biology.

The anthropomorphic fallacy amounts to relativism in biology. It is the *a priori* assumption that nothing can be known with certainty about the lived experience of other [subjects]. True, we cannot know another [subject's] experience—though our mirror neurons have over millions of years sharpened our skill at doing just that—we know with certainty that all [subjects] do and must organize their experience through acts of distinguishing, and what follows from these—distinctions, oscillation, number, space, and time. We are intimate with the proper content of distinction, both as embodied distinctions and in our distinguishing activity. We directly experience the indispensable and ever-changing borrowed content of distinction. And, by self-appropriation, we can learn about the conditions for the possibility of distinguishing and all that follows from them. In short, by self-appropriation we can advert to the primordial sense in which aliveness is mathematical, how organisms are and remain distinguishing distinctions.

Being distinctions and distinguishing, we organisms differentiate between ourselves, each other, and the world. That is, we differentiate [I]/[not-I]. An [I] is any instance, however differentiated or compact, of a distinguishing distinction for itself. Distinguishing maintains the distinction [I]/[not-I] and, doing so, the form of the [I] includes what as independent from it, the [not-I]. Thus, to maintain itself distinct from the [not-I], the [I] must elide part of itself as the source of the [not I]. For example, though we feel cold through our skin, we attribute the cold to the environment. Unlike a toothache, we do this by, to a certain degree, eliding the sensation as being our own. When we stumble and suddenly advert consciously to the pull of gravity, we elide this sensation too as being our own and attribute it to the environment. It follows from this that, to maintain itself distinct from the world, an [I] must partially elide its biomorphism-w as its own. Thus, a condition for the possibility of a [subject] is a partial elision of itself. It takes such a partial elision for such a [subject] to be *itself* and *for itself*. To be, a [subject] must elide its own construction of the [not I], the world. This is significant enough in its own right, and clearly belongs within a general phenomenology of aliveness.[24] As we shall see later in relation to identity-unity-whole, this elision of biomorphism-world (Bw) is made possible by an elision of biomorphism-aliveness (Ba). [Subjects] must partially elide themselves to be [subjects].

This is analogous to the elision of the subject in a mathematical formalism. In both cases, distinguishing, the verb, is elided through hypostatization, by nouns. A ready example is Alpha. She responds to ever-earlier EIs, but her own EI as heuristic

[24] Interestingly, because Buddhist philosophy is a phenomenology, the dualism I/not-I is considered a delusion (see for instance, "Ego" in *The Shambala Dictionary of Buddhism and Zen*, 62).

structure is well beyond her horizon. Indeed, this is what takes place within all organisms. Throughout, by EI, distinguishing (verb) is elided through hypostasis of particular EIs, distinctions (noun).

Such elisions of ourselves as [subject] occur at all levels. At the social level, it is others who knock us out from such elisions—our blind spots. Indeed, we most sharply come face to face with ourselves when we come face to face with others. Relationships, their difficulty, and inherent paradoxes, deepen our understanding of ourselves and sharpen our anticipativeness. As we shall later see, social relationships are a highly differentiated instance of relationship, as such. We can thereby learn about the nature of relationship, including the deeply compact instances, by examining them and observing how we understand them. Note that, by eliding the subject, a mathematical formalism produces a universe onto itself, a kind of solipsism, one impossible in principle for a living distinction, whose condition of possibility is the We. If this seems possible, it is only by reduction of @@A to @@L.

We are, others are, the world is, deeply mathematical. Yet, to understand relation, we must get to its foundation in relationship, its origins in joint, mutually coproduced I/We/World. This is the leap from the *a* and *b* of logical conjunctions (not *a*, *a* or *b*, *a* and *b*, *a* implies *b*), to the *a* and *b* of relationships. When they denote organisms and relationships, I shall refer to them with capital letters: not *A*, *A* or *B*, *A* and *B*, *A* implies *B*. Recall that this is a shift, not only from the abstract to the *concrēscere,* but a transposition from the substantial to the liminal. As the lower scheme is condition for the possibility of the upper one, so *A* and *B* is the condition for the possibility of *a* and *b*. To get beyond mechanical thinking, we must shift the priority in our fundamental framing, learning to treat relationships as the condition for the possibility of relations, and the liminal as condition for the possibility of the substantial.

Chapter 7
Four Kinds of Things

So far we have been dodging the question, What is a thing? Now that question must be faced. –Bernard Lonergan. (Lonergan, *Insight*, 270.)

If things shone with a brightness proportional to their complexity, then galaxies and stars would be dim light bulbs, while just your brain alone would shine out like a quasar beacon visible across the whole Universe. We are special, cosmically speaking, and we can feel genuinely good about that. But let's not get too arrogant about this, because an ant or a cockroach or your pet goldfish shines almost as brightly as you do. –Mark Whittle. (Whittle, *Cosmology*, 33.)

Things

In this chapter we take up the long due discussion of the four different kinds of things. We have begun to delve into @@A (aliveness) and shown that the nature of organisms is intimately related to how they are organized, how they organize the world, how the interrelate and, therefore, to organization as such. This makes them a particular kind of thing. Thus, we must differentiate between different kinds of things. Kinds of things are kinds of being. What are kinds of being? Which are the kinds of things? As we shall see, once clarified, there is much to learn about the nature of living being from the notion of thing. Joinings point to skin and membranes, to how living beings remain distinguishing distinctions. Like water that remains beyond the horizon of fish, being alive ourselves, we do not grasp just how thoroughly we understand the world in living terms. In what follows, then, we reflect on the notion of thing as an aid to help us take a step back from ourselves and our own nature.

Let us begin with the key term *thing*. Consider the following quotes:

> There really is no such thing as art. There are only artists.[1]

[1] Gombrich, *The Story of Art*, 15.

© The Author(s), under exclusive license to Springer Nature Switzerland AG 2023

D. C. Mayer-Foulkes, *The Nature of Living Being*, Biosemiotics 26,
https://doi.org/10.1007/978-3-031-24789-7_7

> "History" is not a person apart, using man as a means for *its own* aims; history is *nothing but* the activity of man pursuing his aim.[2]

> In our view philosophy does not exist. In whatever way we consider it, this shadow of science, this Gray Eminence of humanity is only a hypostatized abstraction. Actually, there are philosophies. Or rather—for you would never at the same time find more than one living philosophy—under certain well-defined circumstances a philosophy is developed for the purpose of giving expression to the general movement of the society.[3]

From these quotes it would appear that it is problematic to regard art, history, and philosophy as things, but not so artists and the individual actors of history. Notice Sartre's use of a key term:

> *hypostatize*: to attribute real identity to (a concept).[4]

> *hypostatize*: "We must beware of hypostatizing the 'market' as a real entity, a maker of inexorable decisions."[5]

The term 'hypostatize' is often used interchangeably with 'reify,' meaning '*thingify.*' Underlying all these issues is the notion of the real. What is a real entity? What is a real identity? Therefore, which entities are real? Which identities are real? What is real? For Gombrich and Marx, individual people are real because they are real entities and real identities. Not so art, history, or philosophy. This converges with the proposition that living beings are *proper* distinctions because they are distinctions for themselves. This is suggestive but, as we shall see, misleading.

Lonergan's Chapter 8 of *Insight,* entitled "Things," begins with the epigraph to this chapter. He immediately adds, "The reader may have wondered, to the point of impatience and annoyance, why we did not begin from the simple and obvious notion of the thing." This notion, "is neither as simple nor as obvious as it seems . . . it cannot be treated until there are assembled the elements to be synthesized."[6] He pointed to two difficulties with the notion of thing—one on the side of the subject and the other on the side of the object. He has addressed the first in the previous seven chapters of his book by clarifying the notion of insight.

This is not unlike my own clarification of the notions of distinction and distinguishing by [subjects], starting on the side of the subject by differentiating abstract distinction (the noun) from the ongoing acts of functional closure of distinguishing distinctions (the verb). Moreover, it was shown in Chap. 5 that Lonergan's notion of judgment is a species of distinction. Recall that Lonergan used the term *insight* to name both the moment of "ah-ha!" in the process of understanding, and the full process of cognitional activity of which this moment is a part.

---

[2] Marx and Engels, *The Holy Family,* 126.

[3] Sartre, *Search for a Method*, 3.

[4] Merriam-Webster.com *Dictionary*, s.v. "hypostatize," accessed April 5, 2021, https://www.merriam-webster.com/dictionary/hypostatize

[5] *Oxford English Dictionary* online, example sentence.

[6] Lonergan, *Insight*, 292–293.

The fundamental synthetic act of distinguishing is constitutive of [subjects] and gives rise to a lineage of differentiated instances such as Uexkull's *Umwelt*, Gibson's *affordances* and Lonergan's *things*. This is why inquiry into distinguishing is inquiry into the difficulty on the side of the subject. Isomorphism follows as a method because the proper content makes all instances of distinguishing analogous, no matter how compact or differentiated they may be. In establishing bet structure, all are thereby anticipative. Moreover, notions of sense-making by which living being is understood are heuristic structures that center on distinction. Thus, inquiry into things is at once inquiry into distinction—though it is a highly differentiated instance—and, therefore, itself inquiry into aliveness. This recalls Lonergan's injunctions for self-appropriation to understand the process of our understanding. This amounts to inquiry into inquiry, which is at once inquiry into [subjects], inquiry into distinction-distinguishing, and inquiry into method.

What, then, do we advert to in our use of the notion *thing*? What is the difficulty on the side of the subject? Lonergan pointed out that we must

> distinguish effectively between things and 'bodies,' between the intelligible unities to be grasped when one is in the intellectual pattern of experience and, on the other hand, the highly convincing instances of the 'already out there now real' that are unquestioned and unquestionable not only for animals but also for the general bias of commons sense.[7]

This is the very difference between abstract distinction and concrete distinguishing.

Consider two contrasting notions of automobile. On the one hand, think of images used in marketing: a slick shiny red object with doors, beautifully clean windshields, mirrors, handles and fittings, gliding on four slick black rubber wheels. On the other hand, an automobile intellectually understood and rationally affirmed is an object extended in space with duration in time and subject to change that is a synthesis of possibilities for driving, riding, carrying loads, maintenance, and so on, that, thereby, involves laws of traffic, norms of civility, principles of mechanics, design, engineering, along with the underlying laws and probabilities of physics and chemistry. The first is automobile as 'body;' the second is an automobile as a thing intellectually understood and rationally affirmed. One entry point to the fullness and complexity of automobile as thing rather than a 'body' is the injunction: think about your driving. Access your intimate knowledge of what an *automobile affords* you, the vast *Umwelt* it allows you to inhabit. What is 'real' about an automobile is only initially and very partially apprehended through the senses. Accessing your experience and understanding of what can be *done* with it, transposes automobile from 'body' to *thing*. 'Body' is isomorphic to distinction, the noun, and *thing* to distinguishing, the process or verb involving a subject.

Like 'bodies,' as Lonergan points out, distinctions as perfect continence are "highly convincing instances of the 'already out there now real' that are unquestioned and unquestionable not only for animals but also for the general bias of

[7] Lonergan, 293.

commonsense."[8] In both, the knowing subject is elided. In contrast, *thing* and concrete distinction-distinguishing are schemes of recurrence.

Having clarified the notion of insight, Lonergan's proposition is that *thing* is a kind of insight. It is not an insight that pertains to a specific conjugate form such as physical, chemical, or physiological conjugate forms. As was noted in Chap. 1, there is no logical progression between these conjugate forms, or the empirical sciences would not be distinct and autonomous. However, correlations of all these conjugate forms may be verified in a single thing. In this sense, the insight *thing* is a different kind of insight. It is the insight into a *unity-identity-whole*.[9] Recall the example from the first pages of Chap. 1 of the successive higher viewpoints to be verified in the woman jumping up and down because she had won the lottery.

> [T]he notion of a thing is grounded in an insight that grasps, not relations between data, but a unity, identity, whole in data; and this unity is grasped, not by considering data from an abstractive viewpoint, but by taking them in their concrete individuality and in the totality of their aspects. For, if the reader will turn his mind to any object he names a thing, he will find that object to be a unity to which belongs every aspect of every datum within the unity. Thus, the dog Fido is a unity, and to Fido is ascribed a totality of data, whether of color or shape, sound or odor, feeling or movement. Moreover, from this grasp of unity in a concrete totality of data there follow the various characteristics of things. Thus, things are conceived as concrete intelligible unities that are extended in space, permanent in time, and yet subject to change.[10]

The insight *thing*, or *unity-identity-whole*, is a highly differentiated distinction that results from highly differentiated distinguishing. An *E. coli* bacterium makes sense of the world by means of linear distinctions, not in terms of perfect continence, let alone 'bodies.' The result of its distinguishing activity is better captured in the notion of *Umwelt*: "the interactive unity of the organism and the world sensed by it."[11] Clearly, forness is at work in the distinguishing of a bacterium, though, given its limited capabilities, it remains controversial to refer to purposes, let alone intentions. Even Gibson's notion of *affordance* is a stretch for a bacterium because the term refers to "what the environment provides or furnishes the animal . . . . It is relational and characterizes the suitability of the environment to the observer, and so depends on their current *intentions* and their capabilities."[12] Gibson coined the term for visual and other forms of perception of animals equipped with specialized sense organs. The proper content of distinction is the reason why the foregoing notions all belong to a single lineage, while the differences between them is in the borrowed content.[13]

---

[8] Lonergan, 293.

[9] Lonergan, 271.

[10] Lonergan, 271.

[11] Favareau, *Readings in Biosemiotics*, 42.

[12] "Affordance."

[13] Though all belong to a single lineage, one must, as Favareau pointed out, "keep in mind the various milieus in which these terms originated. For instance, Uexküll's 'ego-tone' we would now call

The environment, and particular objects, present affordances according to whom they are for. A stick, or a piece of cloth, affords a large number of possibilities to us, compared to what they afford a dog. Affordances, and not 'bodies,' better account for a cat's exploration of the environment. We take for granted that a box is a 'body' into which other 'bodies' can be placed. For a cat, however, a box is a rise to sit on, a platform to leap from, a chamber to snuggle in, a surface to sharpen nails with, a site to stalk from. Shifting between a cat's notions and ours, is not unlike the difficulty shifting from box as 'body' to box as a thing, a *unity-identity-whole* of which we grasp, not relations between data, but a concrete individuality in the totality of its aspects. Adverting to our shift from box as a 'body' out there now, to box as a thing intellectually understood and rationally affirmed, we can self-appropriate what we as subjects bring to this synthetic notion, and thereby wonder about the synthetic notions of other [subjects]. For different [subjects] a box is a different [object].

The lineage of synthetic acts, all of which involve [subjects] and their own forness, is a heuristic structure for meticulous empirical research. The foothold for inquiry is isomorphism, the concatenation of the proper and borrowed content of distinctions. Implicit in this heuristic is a key question: What do I do, as subject, when inquiring into other [subjects] and, therefore, into the general notion of [subject]? What do I do when I inquire into my, and their [objects]?

Just as the injunction "think about your driving" short-circuits the notion of automobile as a 'body', and points to automobile as thing, a similar injunction points to how we make sense of other [subjects] as they make sense of [objects]. Watch a video of a monkey riding a bicycle, a parrot roller skating, a dog skateboarding, or a cat on a surfboard. First, notice you do not merely "see images" as you do when looking at an abstract painting or at the lines on the palm of your hand. Watching these acts elicit a kinesthetic response, body to body. A detailed physiological account of *how* this occurs might involve mirror neurons, say. While such *hows* are indispensable, underlying such lower conjugate forms is a *what*, and that is what we are interested in here. Namely, these events are *bodily-intelligibly grasped and affirmed.*[14] So, our own body-insights are mobilized whereby the proficiency of a parrot's roller-skating, or a cat's surfboarding, is grasped and affirmed. Our whole body as body-subjects mutely affirms, "Yes, that is exactly how it is done!" We can do this because neither the [object] driven nor the [subject] driving it are 'bodies' but things. Indeed, two kinds of things are involved, the driver, a [subject], a thing for itself as we are, and a device we are familiar with. Forness akin to ours is more easily involved if the activity with the [object] resembles our own—unlike, say, a cat sharpening its nails with a box. We grasp what these [subjects] grasp about their [object], not by regarding them or the object as 'bodies' but as [subjects] engaging with affordances provided by these [objects].

---

qualia and his 'meaning-carrier' properties of objects are precursors of Gibson's sensori-motor affordances..." Favareau, *Readings in Biosemiotics*, 89.

[14] I use 'grasp' for body intelligibility analogously to 'understand' for abstract intelligibility.

Here is another injunction: Watch the video, "White Blood Cell Chases Bacteria."[15] Notice yourself experiencing a kinesthetic grasp of swimming, chasing, navigating, evading, turning, escaping, reaching, capturing—even when the protagonists in this drama are a single white cell and a bacterium. You have a powerful sense, and with good reason, that what occurs with, and between these two [subjects], is deeply intelligible. This is the activity of [body-subjects], each a *unity-identity-whole* being-at-work-staying-itself.

Because distinguishing involves the synthetic intertwining of [body-subjects] and [objects], the difficulty of things on the side of the subject is inevitably linked to the difficulty on the side of the object. So, let us turn to the side of the object.

There are things on the side of the [subject] that are not things for themselves on the side of the object. Such is the case of art, history, and philosophy. Not being [subjects], things for themselves, they are only [objects] as things for us. This is in sharp contrast to Lonergan's example of Fido the dog, who is both a thing *for us* and a thing *for himself as [subject]*. For us, Fido is an object that is a [subject]. Later we shall discuss, in terms of anticipation, how we identify fellow living beings, no doubt a knack indispensable to our survival.

What is the relation between the notion of thing, and how we relate to living being? We need not look far to answer this question because we can advert to this in ourselves. Most obviously, we can only make sense of so much at a time; otherwise, we are in overload, and nothing makes sense. This point, which the reader can advert to here and now, will be the starting point of a discussion below. However, further preliminaries are necessary before that.

Different Kinds of Things

Lonergan uses Fido the dog as his example of a thing. This brings us to the difficulty on the side of the object. According to Lonergan, the greatest difficulty with the notion of thing is on the side of the subject, differentiating between 'body' and thing. But this difficulty is compounded when the object is itself a [subject], for we must now grasp not only the difference between 'body' and thing, but also the differences between kinds of things. This makes Fido the best and the worse possible example of a thing.

As a living being, a distinction that distinguishes, Fido is a thing for itself, a [body-subject] that is thereby undoubtedly a *unity-identity-whole*. Being-at-work-staying himself, Fido hardly needs us to be a concrete intelligible unity extended in space, with duration in time and subject to change.[16] So, there is little difficulty on the side of us, as subjects, making the case that Fido is a thing. But this brings its own complications. As a living kind of thing, an account of Fido as a [body-subject]

[15] "White Blood Cell Chases Bacteria."

[16] For Lonergan's "permanence in time" I substitute "duration," which I believe is more appropriate.

that is a *unity-identity-whole* being-at-work-staying-itself, must necessarily include his own making sense of the world. Fido is a [body-subject] relating to [objects], and this is inherent to a living kind of thing.

What of things not alive—such as a stick or a piece of cloth? These are not *unity-identity-wholes* on the side of the object. Unlike a stick or a piece of cloth, we cannot split Fido in two without him ceasing to be the kind of thing he is. But we can break up a rock or rip a cloth, and they remain the kind of thing they are. *Prima facie*, this makes them very different kinds of things to living beings. Fido's uniqueness as a thing is therefore also what makes him the worst possible example of a thing. Lonergan stated that "things possess properties that are subject to laws and to probabilities."[17] But this is true of tiny quarks and super-clusters of galaxies and anything else in between, whether alive or not. Lonergan also offers the Moon as an example of a thing. Again, a *thing* is an insight, but one different to an insight into explanatory conjugates. Unlike abstract explanatory intelligibilities, things are concrete intelligible unities. To regard something—anything—not as a 'body,' but as a concrete intelligible unity, is to regard it as a thing. Because we can regard Fido as a thing, it follows that we can regard a bacterium as a thing, and, likewise, we can regard a white blood cell as a thing. But this lack of difficulty refers to the side of the subject, not to the side of the object.

This brings us to two important points. First, though Lonergan speaks of kinds of things, he does so in a somewhat different sense than is done here. For him, kinds of things correspond to the different conjugates—physical, chemical, physiological—on the side of the subject. This brings up a key question, the second point, and a crucial one:

> Once things are recognized to be of different kinds, there arises the obvious question whether there are things within things. Are electrons things within atoms, atoms things within compounds, compounds things within cells, cells things within animals, animals things within men?[18]

Lonergan answers that there are no things within things. If conjugates of a lower order exist in things of a higher genus, these are not only defined by the lower conjugates:

> the thing is an intelligible unity grasped in the totality of the data. It follows that if any datum pertains to a thing, every aspect of the datum pertains to that thing. Hence, no datum can pertain to two or more things, for if in all its aspects it pertains to one thing, there is no respect in which it can pertain to any other.[19]

The first point, then, is that, for Lonergan, *kinds of things* refers to objects that correspond to specific explanatory conjugate forms. There are no things within things because, given an insight into a unity that grasps the totality of the data, all lower conjugate forms are subsumed under the highest conjugate form. This is a powerful

---

[17] Lonergan, *Insight*, 271.

[18] Lonergan, 283.

[19] Lonergan, 283.

conceptual resource whose outstanding importance is fundamental to our forthcoming discussion. Yet this is still the side of the subject.

Kinds of things, as used here, refers to the various things that in one way or another involve distinguishing activity. It was shown that there is no formal difference between the distinctions drawn by a T cell and those drawn by a group of astronomers. If, as Lonergan said, a thing is an intelligible unity grasped in the totality of the data, and if every aspect of the datum pertains to that thing, it follows that, if distinguishing takes place, or ever took place, this cannot be left out of the account of a thing. This is so for products and byproducts because distinguishing is inherent to both. In radical contrast to living beings, products and byproducts, distinguishing activity is not inherent to the nonliving—certainly not in the same sense. Except for the vanishingly infinitesimal mass of things that involve distinguishing, the nonliving accounts for the whole mass of the universe. Recall the epigraph to this chapter by Mark Whittle. Cosmically speaking, even the simplest distinguishing distinction is inconceivably special. This pertains to things on the side of the object and not on the side of the subject.

The nonliving flow processes that began with the Big Bang constitute physical and chemical things—on the side of the object—that long preceded living being and will long persist after we are gone. But our notions about these, such as the beginning and the end of the universe, are our own linear distinctions—the side of the subject. This is so even given the obvious fact that a rotating universe, condition for the possibility of ב, long preceded living being. The emergence of life is itself a distinction by living being, twice over. On the side of the subject, it is a distinction within the inquiry into life; on the side of the object, it is the advent of distinguishing with the advent of distinguishing distinctions, the living kind of thing. On the side of the subject, we cannot but make sense of the nonliving biomorphically, in spirallic terms. The ongoingness of the nonliving is our own attribution. As was shown in Chap. 3, in relation to topics such as biogenesis, it is indispensable not to elide ourselves as subjects—Lonergan's 'side of the subject'—when conceiving the nonliving. Making our biomorphism explicit, acknowledges the continuity-discontinuity between living distinguishing and the nonliving. These might be referred to as *nonliving 'distinctional processes'*. Consider this example:

> Without Jupiter and Saturn orbiting out past Earth, life may not have been able to gain a foothold on our planet, new simulations suggest. The two gas giants likely helped stabilize the solar system, protecting Earth and the other interior, rocky planets from frequent run-ins with big, fast-moving objects, researchers said. In other words, giant planets appear to have a giant impact on giant impacts.[20]

Being explicitly biomorphic, this amounts to saying that, just as girls skipping a rope must *get in* and *come out* without being hit by the swinging rope, so Jupiter and Saturn's orbits form a barrier that amounts to a closure. Both are biomorphically understood 'distinctional processes'. According to NASA's Tom Barclay, "If you don't have giant planets in your system, you have a very, very different planetary

[20] Lewin, "Life on Earth."

system . . . ."[21] Still, this 'distinctional process', like a geyser, is for us. Thinking otherwise is, quite literally, to treat Jupiter, Saturn, and their orbits as 'bodies,' if not as actual bodies. But they belong to the nonliving, in stark contrast to the three distinguishing kinds of things—byproducts, products, and living beings. Understanding them must include their distinguishing, the verb, the fundamental activity inherent to *them*. In sharp contrast, the nonliving is understood in terms distinction, the noun, that results from our distinguishing activity. We saw how crucial it is to keep this in mind in accounts of biogenesis.

Byproducts

Distinguishing distinctions were characterized as Ā→A→/→Ā. To repeat, an outside component (Ā) crosses a semi permeable boundary to become a component (A) integrated into a system-process in functional closure (/), from which it will eventually again cross the boundary to cease being part of the system (Ā).[22] This is what allows living beings to remain systematically far from thermodynamic control. Byproducts, the second Ā, can be vastly different from the first Ā. The first is an input and the second an output. The two together are constitutive of the functional closure at work in living being.

Byproducts are outcomes of the process of living. As we breathe, every in-breath is an input and every outbreath an output, a byproduct. Because we retain oxygen and expel carbon dioxide, compared to the air around us, our out-breaths are in a highly improbable or negentropic state. The same is true of all metabolic processes, starting from those of the simplest cells. Because byproducts are footprints of aliveness, beyond physical and chemical explanatory conjugates, their intelligibility as negentropic states calls for distinguishing as their highest explanatory conjugate. This pertains to the living beings they were once associated with.

Outputs are a defining feature of metabolism and, beyond that, a defining feature of the environment. The most outstanding example of a byproduct is the atmosphere itself:

> Before photosynthesis evolved, Earth's atmosphere had no free oxygen (O_2). Photosynthetic prokaryotic organisms that produced O_2 *as a waste product* lived long before the first build-up of free oxygen in the atmosphere, perhaps as early as 3.5 billion years ago.[23]

---

[21] As quoted in Lewin, "Life on Earth."

[22] From Chap. 3, section "What is an *Itself*?"

[23] "Geological History of Oxygen." It is a complex history: "The oxygen [these organisms] produced would have been rapidly removed from the atmosphere by weathering of reducing minerals, most notably iron. This rusting led to the deposition of iron oxide on the ocean floor, forming banded iron formations. ... Oxygen only began to persist in the atmosphere in small quantities about 50 million years before the start of the Great Oxygenation Event. This mass oxygenation of the atmosphere resulted in rapid buildup of free oxygen." On this topic, see the section Chap. 9, Microbial Partnerhips.

Not yet fully understood, the Great Oxygenation Event, outcome of byproducts of Cyanobacteria resulted in a highly improbable state that is the cornerstone of the *Gaia hypothesis*:

> The very low entropy associated with living systems suggests a model configuration of extreme improbability compared with the expected and predictable steady state abiological background. If highly improbable molecular distributions extend beyond the boundaries of living entities so as to include also their planetary environment, then the environment and life taken together can be considered to constitute a single larger entity. It is this operating system of life and its environment which we are calling Gaia. Effects due to the presence of life are not limited to chemical change. If life determines the abundance of a polyatomic gas such as CO_2 then the planetary radiation balance will be altered. An effect such as this could be part of a system of planetary thermostasis.[24]

Thus, the improbable state of the biosphere is due to the byproducts of living beings. So, byproducts may be defined as *highly improbable molecular distributions that extend beyond the boundaries of living beings*. These boundaries obtain in space and in time. The corpses of organisms fit this definition because an improbable molecular distribution extends beyond the lifetime of an organism. All organic outputs are in improbable states because, like an outbreath, they are remnants of what was *selected* by the organism, as previously pointed out. This remains true when one organism consumes another. That output too is in a highly improbable state compared to the nonliving.

Even actual footprints are highly improbable designs at odds with their surroundings. Natural designs on a beach, say, result from optimized flow of water and air. But a footprint is in stark discontinuity with these. Therefore, *footprint* is a good analogy for byproducts, and not only metaphorically.

Elephant footprints, for instance, create habitats for tiny aquatic creatures of diverse communities of invertebrates.[25] This makes them a keystone species in their environment:

> The African elephants' ability to locate underground water and dig pools helps provide many other species with water during droughts. African elephants may destroy trees and shrubs allowing for grasses to grow that other species eat. In the process, African elephants, and perhaps particularly the forest elephant, create an environment conducive to regeneration of healthy forest ecosystems. Management of elephants in the wild means maintaining a balance between elephants and the ecosystem.[26]

Likewise, oxygen, the footprint of cyanobacteria, was the condition for our emergence, and the ongoing footprint of trillions of organisms that maintain Gaia.

It is sometimes difficult to differentiate between byproducts and the nonliving. What kind of thing are the beautiful white cliffs of Dover? These massive cliffs are ancient underwater deposits of calcareous skeletons of Coccolithophorids. So, the cliffs are byproducts. They are part of the marine carbon cycle, a small piece of the

---

[24] Lovelock, "Regulation of Carbon Dioxide."

[25] Offord, "Elephant Footprints."

[26] Elephant TAG/SSP Key Messages (https://assets.speakcdn.com/assets/2332/elephant-conservation-messages.pdf), Message 4.

puzzle of how living beings contribute to the long-term stabilization of Gaia.[27] Significantly, this cycle involves two kinds of organisms, diatoms and coccolithophorids that together constitute a two-component regulatory system for the concentration of CO_2 in the atmosphere. Such systems—and there are many of them—are the fundamental building blocks, not only of metabolism, but of the environment that sustain them and which they themselves sustain. The white cliffs of Dover also show why unambiguous examples of the nonliving are hard to come by. This is why, throughout this essay, the instances given of the nonliving are the Moon, the Sun, and galaxies. With unique exceptions, such as lava and volcanic rocks, which are nonliving, we are surrounded by the three distinguishing kinds of things. Even the temperature and salinity of oceans is affected by living beings, as shown by concern over human-made global warming.

A controversy around the Gaia hypothesis is whether she is an organism.[28] This controversy dissolves as soon as by-products are acknowledged as a kind of thing. The exquisite and vast network of two-element regulatory systems that constitute her do not make Gaia an organism. Gaia has no outer membrane, is not in functional closure, and does not anticipate. Like all byproducts, Gaia is in a highly improbable state maintained beyond thermodynamic control by the interplay of a vast array of byproducts. Thus, though not a living thing, Gaia is neither nonliving nor a product but a byproduct.

Greenhouse gasses, for instance, are certainly not a body and hardly a 'body'. However, thy are highly intelligible as a thing; the kind of thing that is a byproduct.

Living beings, then, transmit organization through their products and byproducts. Organization pertains to all three. Just as Gaia demonstrates the effect of the organization inherent to byproducts, the organization inherent to products will be discussed in the following section. The claim could be made that, because organisms are part of Gaia, she herself is alive. For instance, if human beings deflected the path of a catastrophically destructive asteroid, this might qualify as an early rudimentary instance of Gaia's functional closure. But this, as we shall see, would not make Gaia a living being but a product.[29]

In sum, the intelligibility of byproducts, outputs of living being, calls for distinguishing as the highest explanatory conjugate. Though humble, they have an unfathomable massive effect on life as a whole. From the oxygen atmosphere that was a remote condition for our own emergence, to the human effects on the biosphere that marks the Anthropocene Age, byproducts are fundamental to life as we know it and may well determine its demise.[30]

---

[27] See for instance, "The Marine Carbon Cycle" available at http://earthguide.ucsd.edu/virtualmuseum/climatechange1/06_3.shtml

[28] See for instance https://www.newscientist.com/round-up/gaia/

[29] Recently it has been speculated that the universe could itself be a product. See Zeeya, *A Big Bang in a Little Room.*

[30] Two news items from 2018 pertain to our own products and byproducts and their effect on the biosphere: First, humans have wiped out 83% of mammal and plant life on earth. The Wiseman Institute of Science with the California Institute of Technology has conducted the first comprehen-

Products

If one had to characterize the Anthropocene Age with a single word, that word be *products*. Products are ubiquitous in our lives. Look about you; products are everywhere. Except for living beings and glimpses of stars, the sun, the moon, and a few rocks, everything in your purview is a product. Except for the atmosphere, byproducts mostly remain outside our horizon, either because we do not conceive them as such, or because they are deliberately concealed. Products exceed and long antecede the emergence of tools—but this depends on definitions. This is how Shumaker, Walkup and Beck defined tool use in their magnificent compendium *Animal Tool Behavior*:

> The external employment of an unattached or manipulable attached environmental object to alter more efficiently the form, position, or condition of another object, another organism, or the user itself, when the user holds and directly manipulates the tool doing or prior to use and is responsible for the proper and effective orientation of the tool.[31]

As they explain, this definition excludes 'architectural' constructions such as nests, dens, burrows, traps, bowers, hives, dams, food caches. Nonetheless, they consider Constructions, as they name them, to be products: "the exquisite, cumulated product of thousands of cases of tool *manufacture*." According to their definition "a Construction is not a tool because, once completed, it is not held or directly manipulated in its entirety. The Construction does not involve tool manufacture, because the final product is not a tool." They note that this is only one set of behaviors excluded from their study as borderline cases which are "too numerous to review exhaustively."[32] Yet most, if not all of these, are products such as paintings and objects of play that are seemingly functionless. One instance that also does not fit their definition of tool is a spider monkey infant that crosses an arboreal gap by walking across its mother while she holds to branches on each side.[33] And yet, these are extremely significant products, as we shall later see.

---

sive estimate of the impact of humanity on the total biomass of the planet. Humans comprise just .01% of the world's living creatures. By comparison, all marine life totals 1% of life on earth. Only 4% of the animal population is now considered wild. The remaining 96% of mammals are livestock and humans. Of the bird population, only 30% are now said to be free. Plants are the most dominant organisms on the planet, comprising 80% of life on Earth (see https://www.weizmann-usa.org/news-media/in-the-news/humans-just-001-of-all-life-but-have-destroyed-83-of-wild-mammals). Second, Antarctica is melting three times as fast as a decade ago. The continent's rate of ice loss is speeding up, which is contributing even more to rising sea levels. "Between 60 and 90 percent of the world's fresh water is frozen in the ice sheets of Antarctica, a continent roughly the size of the United States and Mexico combined. If all that ice melted, it would be enough to raise the world's sea levels by roughly 200 feet. While that won't happen overnight, Antarctica is indeed melting . . . ." (see https://mobile.nytimes.com/2018/06/13/climate/antarctica-ice-melting-faster.html)

[31] Schumaker, Walkup, and Beck, *Animal Tool Behavior*, 5.

[32] Schumaker, Walkup, and Beck, 204–205.

[33] Schumaker, Walkup, and Beck, 2.

As the layout of their book shows, use of tools—and therefore products—in the animal world is far more prevalent than was once supposed. This is shown by the chapters and their ordering alone: *Invertebrates*; *Fish, Amphibians, Reptiles, Birds*; *Non-Primate Mammals*; *Prosimians and Monkeys* and only then *Apes*.

> A wasp uses a pebble as a hammer. An octopus carries around a coconut shell to hide in. A shrike impales its prey on a sharp thorn . . . . Polar bears hurl great chunks of ice down on something like a walrus to kill it . . . . New Caledonian crows make hooks out of twigs to catch prey. Wild chimpanzees make wooden spears for hunting. A dolphin found a piece of tile and took it down to the bottom of their pool and used it to scrape algae off the bottom and then ate the algae.[34]

Living beings produce products to establish bet structure or A=A recurrently. Through their use, living beings *arbitrarily neglect concrete residues in order to recurrently differentiate between differences that make a difference and those that do not.* This points to function, which is part of the highest conjugate form in the intelligibility of products as things. It may seem obvious, but a spiders' web is only a web so long as it serves to catch the right insects. A bird's nest is only a nest so long as it holds her eggs and chicks. A wolf's lair is only a lair so long as it is warm and keeps out rain and predators. A beaver's home is only a home so long as it serves this function. Whether it is a dam for us and remains so after the beaver family has moved on is not its concern. A screwdriver is a screwdriver so long as it drives screws. A hammer establishes what may be hammered.

Products, like all things and all distinctions, are intelligible as schemes of recurrence. Materially built and maintained by a living being *for* something, they are verified and refined by iteration. The motto "If your only tool is a hammer, everything appears as a nail" is a way of saying that hammers have A=A built into them, just as automatic heating systems do. All tools are designed *for* a defined set of affordances. Thus, products are produced by living beings to carry out specific schemes of recurrence, any of which involves A=A. This often involves dense and complex nestings of A=A. No product, however, maintains *itself.* The A=A of a product is neither determined *by itself* nor is *for itself.* It is determined by the living being who produced it and it serves a purpose *for* them. Products are like prosthetics, extensions like artificial limbs to recurrently aid organisms to establish bet structure.

That tools and symbols are both products is hinted at by the symbolic use of tools. There are instances of human and animal tools being used to represent something else or to provide a change in psychological state.[35] This is what we do with coins or paper bills. Some monkeys and apes in captivity have learnt to trade tokens for various foods. More intriguing are symbolic tools that affect emotions, like a child's special stuffed animal or blanket. These are much more than tools and toys. Representing comfort or security, they are used to feel better. Once thought to be

[34] Hamilton, "Myth Busting." The webpage for the link to the recording and a transcript of the interview includes a startling instance of a New Caledonian Crow bending a tool into shape.

[35] Schumaker, Walkup, and Beck, *Animal Tool Behavior,* 16.

uniquely human, there is growing evidence of animal use of symbolic tools. Great apes in times of stress or sadness do so, for instance a female who has lost an infant: "That female would create something that researchers called a doll and then [she] treated it exactly as she had treated her infant that had recently died." The significance of these instances is still being debated, showing the sea change since the 1960s when scientists first realized that tools where not exclusively human.[36]

The notion of product is not unproblematic and raises difficulties that call for reflection. For one thing, like tools, the boundaries of products can be hazy. Take for instance the bolus, the rounded mass of chewed food prepared for swallowing. When does a sandwich we eat cease to be a product? Strictly, chewing prepares it to be recognized as food by the digestive system, else it will be rejected. This is A=A. Food must be recognized both at the table and to be assimilated. Digestion is, in fact, a process of ongoing recognition by organisms. Because of their similarity, we also refer to "assimilation" in learning. This is how concepts become part of the furnishing of our minds. Both are versions of $\bar{A} \rightarrow A \rightarrow / \rightarrow \bar{A}$, involving the functional closure of cells on many levels.

Living beings not only produce products, but they also make products of themselves. This is the case of the mother spider monkey who holds onto branches on either side of a gap for her infant to cross. Doing so, she produces herself as a product. Similarly, ants produce themselves as a bridge over a stream of water for a long column of ants to march across. An athlete, no less than a musician, practices intensively to produce themselves as such. As we become skilled at the use of a tool, the tool shapes us. A violin becomes an extension of a violinist as much as the violinist of the violin. According to Sandra and Matthew Balkeslee, in their book *The Body Has a Mind of its Own*, when we wear a hat, it becomes an extension of our own body so, we cannot help but crouch when walking through a low door. Likewise, when driving a car. The car becomes an extension of our body, we 'feel' it as we navigate it, for instance, in tight spots. A car is intelligible as embodied, as part of us, far more like a body than a 'body' already out there now. Should another car bump into us, our upset is about the experience that *we* ourselves have been hit. It follows from this that the more effective the use of some products is, the hazier the boundary between it and the user. Think of clothes and dental implants.

Just as we can turn ourselves into products, so we can turn ourselves, or parts of ourselves, into by-products. We shave off our beard and cut off our hair. A guitarist's broken nail is a by-product. We can turn ourselves into corpses by committing suicide. So-called economic externalities are mostly by-products of our products. Dying from our own pollution makes us by-products, the result of not being able to get away from our own by-products. We are back to Popa's suggestion that motility in living beings emerged to avoid poisoning ourselves with our own waste.[37]

Strictly, it cannot be said that products are never alive. Think of cattle, chicken, and fish raised for food. The moral considerations arising from living beings used as

[36] Hamilton, "Myth Busting."

[37] See Chap. 3, section "Autopoiesis with or without cognition."

products result, precisely, from their being turned into a different kind of thing, a thing *for* us, beyond their being things for themselves. In Chap. 2, it was asked: "What is an E. coli bacterium *for*?" The only possible answer is *for itself*. It is an end in itself. This convergence with Kant's notion of a kingdom of ends is deeply suggestive, pointing as it does to the origin of value in forness. Morally and epistemologically, it complicates things vastly that the norm in nature is that virtually all organisms are used *for* something by other organisms. This brings up a grey area between living being and products. For instance, by natural selection, a strategy for plant pollination is to *produce* flowers that are affordances *for* bees. The mutual benefit between plant and bee hinges on the long-established evolutionary interrelationship between them. The same is true of the dispersal of seeds by producing fruit for birds and other animals. Given how freely flowers and fruits are offered up and how relatively disconnected they are to the rest of the living physiology of a plant, it is not unreasonable to think of them as products or their precursors.

Michael Pollan suggested that apples, tulips, marihuana, and potatoes are wildly successful because they have harnessed humans to propagate themselves throughout the planet.[38] From a strictly Darwinian point of view, we are their tools. Such synergies are the rule in nature, rather than the exception. Organisms flourish together—the topic of joinings soon to be treated. Whether it makes sense to consider these products or not, the key point is that organisms represent affordances for each other: food, shelter, a place to lay eggs, a means for propagation, and so on. The epistemological point is that, whereas in some instances the nature of products is uncontroversial, in others, such as products that are alive, their nature as products depends on interrelationships between living beings.

This brings us to a class of products that is particularly important for our purposes: organizations. Living beings form lasting organizations that are not themselves alive. Just as we make products of ourselves and of other living beings, so too we belong to organizations that are products. We take up roles in organizations that allow them to flourish and succeed. In this strict sense, all social organizations are products—from marriage and family to friendships and institutions, from states to alliances between states. Colonies of ants and troops of gorillas are products. These relatively durable organizations are successful, among other things, because they lead to the production of further products. This is why they can collectively build bridges, rafts, and shelters, sometimes with their own bodies.[39]

It is highly noteworthy that ethics as a whole is grounded in social products. This goes back to the intrinsically organized nature of organisms, which eventually finds expression in social organization.

> Altruistic behavior, whether in kin relationships or towards other members of a group, and on occasion even towards members of different species, is common throughout the animal kingdom, particularly in species with *complex social structures*. For example, vampire bats regularly regurgitate blood and donate it to other members of their group who have failed to

[38] Pollan, *The Botany of Desire.*

[39] See for instance http://www.smithsonianmag.com/ist/?next=/science-nature/watch-fire-ants-use-their-bodies-to-form-living-architecture-180947854/

> feed that night, ensuring they do not starve. In numerous bird species, a breeding pair receives help in raising its young from other 'helper' birds, who protect the nest from predators and help to feed the fledglings. Vervet monkeys give alarm calls to warn fellow monkeys of the presence of predators, even though in doing so they attract attention to themselves, increasing their personal chance of being attacked. In social insect colonies (ants, wasps, bees, and termites), sterile workers devote their whole lives to caring for the queen, constructing, and protecting the nest, foraging for food, and tending the larvae. Such behaviour is maximally altruistic: sterile workers obviously do not leave any offspring of their own—so have personal fitness of zero—but their actions greatly assist the reproductive efforts of the queen.[40]

The key point is that social structures are products and that *qua* ethical behavior sustains them. To repeat, it is crucial to note that ethics as a whole is grounded in social products. We will return to the significance of this in the final chapters.

Are concepts products? Concepts are produced and they too are meant to remain available as tools for the future. Like food produced, they too must be assimilated. This, after all, is the purpose of schooling. The question is crucial to our topic because, if concepts are products, and if it is not possible to fully understand one kind of thing in terms of another, then it is not possible to fully understand living being in terms of products. It follows that it is not possible to fully understand aliveness in terms of concepts. But then, how to fully understand aliveness? This question is perhaps best reframed as: What is it *to understand* when it comes to aliveness?

Numbers are excellent instances of concepts. Like any other product, concepts have A=A built into them, even if the knowing subject who produced them is elided. This is also true of material tools, such as a hammer. That numbers are a kind of prosthetic whose producer is elided would explain why they, and concepts generally, may come to be regarded as autonomous, immutable, independently self-sustaining Platonic forms.

To consider the nature and formation of numbers let us follow their history in Keith Devlin's appropriately entitled section "The Token Advance," which evokes my own account of ב in terms of token/type.[41] Numbers first appeared as concrete tokens, tangible vouchers, or emblems. Their history is the slow but relentless progression from a concrete to an abstract notion of token. We all recapitulate this process when we learn numbers. Indeed, their formation is itself a token for the process of formation of all concepts:

> Numbers—that is to say, whole numbers—arise from the recognition of patterns in the world around us: the pattern of 'oneness,' the pattern of 'twoness,' the pattern of threeness,' and so on . . . . The counting numbers 1, 2, 3, . . . are a way of capturing and describing those patterns. The patterns captured by numbers are abstract, and so are the numbers used to describe them.[42]

---

[40] "Biological Altruism" (emphasis added).

[41] Devlin, *The Language of Mathematics*, 15–16.

[42] Devlin, 15–16.

As we saw, pattern is repetition, 'again,' the A=A in the axioms of *LoF*. By age five, we expect children to acquire the concept of number, a cognitive leap that took humankind many thousands of years. Counting numbers marks

> the final step in the process of recognizing the *pattern* 'numbers of a given collection.' The pattern is completely abstract—indeed, so abstract that it is virtually impossible to talk about it except in terms of the abstract numbers themselves. Try explaining what is meant by a collection of twenty-five objects without referring to the *number 25*.[43]

It has been shown that, being distinguishing distinctions, organisms are inherently mathematical.[44] Studies in developmental psychology and anthropology show that the capacity for high-level abstraction is acquired with difficulty, over time, and is directly dependent on cultural context. Numbers for counting originate as, and remain cultural artifacts. Indeed, in the proper context, even animals can be taught numbers and other concepts through human instruction. One shining example is Alex the gray parrot.[45]

The earliest human-made artifacts believed to serve for counting are notched bones dating back to around 35,000 BCE. The first known systems of enumeration that preserve information about what has been counted date back to about 8000 BCE in the Middle East, sometime after the advent of agriculture. These consist of small, carefully crafted clay shapes between one and three centimeters across, such as spheres, disks, cones, tetrahedrons, ovoids, cylinders, triangles, rectangles, and so on. Representing the kind of object being counted, these tokens provided a means to keep track of stock and for planning and bartering. Use of these tokens spread but remained unchanged until around 3000 B.C., when, with the growth of cities and, thereby, the complexity of Sumerian society, the temple institution and government appears to have led to more elaborate tokens. These now represented manufactured objects such as garments, metalwork, jars of oil, and loafs of bread. These tokens were strung together or stored in clay envelopes along with the objects they enumerated. But sealed clay containers must be broken to examine their content. So, accountants took to impressing the tokens on the soft clay exterior before closing them. With all the requisite information now available in the outer markings, the stage was set for dispensing with the tokens. So, after a few generations, clay tablets emerged with impressed marks. Sumerian accountants had thus substituted physical counting devices with written *numerals*. For a time, the outer markings on the envelopes had indicated what was in them, therefore representing the tokens in the

---

[43] Devlin, 14.

[44] A striking recent finding: "Bees can solve seemingly clever counting tasks with very small numbers of nerve cells in their brains." Given our argument that any such capacity is founded on distinguishing, the following statement is particularly striking: "Careful examination of the *actual inspection strategies used by animals* might reveal that they often employ active scanning behaviours as shortcuts to simplify complex visual pattern discrimination tasks. Hopefully, our work will inspire others to look more closely not just at what cognitive tasks animals can solve, but also at how they are solving them." ("Science News," *Science Daily*, December 21, 2018. https://www.sciencedaily.com/releases/2018/12/181221123718.htm. Emphasis added).

[45] Pepperberg, *Alex & Me*.

envelope while the tokens themselves stood for the real-world objects in the clay jars. In a significant cultural-cognitive leap of abstraction this redundancy was overcome, no doubt the result of endlessly repeated and redundant verification of the match between mark and tokens, and tokens and objects. Again: iteration is how all tools are verified and refined.

This path from material to abstract tokens is itself a token of the transition from material to abstract products. By the way, this is a perfect instance of integrator/operator. As integrator, a product represents the leading edge of what is understood; as operator, it serves as the starting point for further refinement. Exactly when the adoption of symbolic representation led to the explicit recognition of the number concept is hard to pinpoint, just as it is difficult to pinpoint when a child makes this cognitive leap.[46]

Their history conclusively shows that numbers are products and suggests how concepts develop by stages. If lumps of clay are products, so are scratches in clay. These mere shapes are products, whether they are engraved in stone or in our minds. The same is true of shapes such as triangles, circles, letters, and any combination thereof. Historically, developmentally, and conceptually, every refinement of these products leads to others. In many cases, a pre-requisite for such development is writing, another product. The history of numbers shows that *the immaterial nature of concepts is irrelevant to their character as products*. The long historical process of abstraction and introjection of concepts is today reproduced through schooling. Schooling is all about products, the prosthetics that furnish our minds.

This touches on the evolution of learning, from offspring imitating adults common in mammals, through the deliberate teaching in primates, to the Gutenberg revolution and the Internet. Culture is predominantly about products, starting from society itself. The details of their emergence, evolution, and development aside, one thing is certain: vast amounts of energy are expended creating products and transmitting skills in their use. Many products, like numbers, are layers of edifices that have long sedimented. This is as true of mathematics and language as of any system of symbols. All emerged through a long process of transformation from concrete to abstract. Etymology often points to this basic cultural vector. A symbol, for instance, from the ancient Greek *symbolon* (from *symballein*, to put together) was a product: "a sign or token by which one infers, an object such as a coin or a bone which two persons broke between them, each keeping one piece." These were reunited later as a means of recognition, for instance when letters or messages where delivered.[47]

Products *result from* repetition, are made *for* repetition, and *result in* further repetition. They are intrinsically spirallic, endlessly recurring with endless variation. Enlarging on the example above, when, very likely reproducing a proverb, Abraham Maslow wrote, "I suppose it is tempting, if the only tool you have is a hammer, to

[46] Devlin, *The Language of Mathematics*, 18.

[47] Liddell and Scott, *Greek-English Lexicon*, 759. As has been shown, distinguishing and EI show that, inherent to metabolism, is semiosis, the subject matter of the nascent field of biosemiotics. A key question, then, is the differentiation, as well as the deep roots in biosemiosis, of symbols such as numbers, which are products and the fruit of cultural evolution not exclusive to human beings.

treat everything as if it were a nail,"[48] he turned a tool that alludes to a tool into a new tool—a proverb about hammers into a pedagogical devise.[49] The hammer, the proverb, and the pedagogical device—all three are products shaped by a cultural context with a different kind of material. But all do A=A and all are designed, built, and maintained as prosthetics. The link between these products illustrates their enormous range and the ceaseless recombining by which new products are produced.

The significance of differentiating between the four kinds of things cannot be overestimated. If concepts are products, and if it is not possible to understand one kind of thing in terms of another, then it is not possible to understand living being in terms of products. It follows that it is not possible to understand aliveness in terms of concepts. This is why the question of understanding aliveness was reframed as, "What is it to understand when it pertains to aliveness?" Recall the point about following injunctions. We do not understand a chocolate cake, a symphony, or a mathematical formalism through concepts. We follow injunctions—recipes, scores—to illustrate these for ourselves. This is why injunctions and comparative study must be at the foundation of method for biology.

To understand aliveness by analogy to products is mechanism, whether the product is a clock, a steam engine, or a computer. Mechanistic thinking rejects what cannot be accounted for in mechanical terms, even to the point of banning it from the real. Something similar applies to concepts. To understand aliveness by means of concepts amounts to conceptualism. This is not to say that concepts—mathematics, for instance—do not have a part to play in understanding living being. Far from it, as this analysis itself shows. But it does suggest that concepts must be properly situated alongside the non-conceptual. So, one learns about aliveness from the verb, the doing of mathematics, and not merely from its finished products, the noun. This comparison with math, cooking and music is itself an instance of isomorphism.

A metaphor for conceptualism is helpful. Imagine a tribe of people unacquainted with movies being shown, before projecting it, the actual 35 mm roll of a film about them (this is back in the twentieth century). Scrutinizing the images frame by frame they discern thousands of tiny still images of themselves, their animals, their village. Viewing them does not lead them to imagine the motion picture, let alone understand how these numerous nearly identical frames produce the illusion of movement, and even less that this illusion is created in their own minds. Naturally, they are tempted to study each frame to discern movement. However, if anything, this makes it harder still to grasp that the movement is not in the frames themselves but produced by their rapid succession. This is analogous to the abstracting inherent to conceptualism. Close attention to the single still frames elides the indispensable role played by the subject in the synthetic act that produces the experience of movement. For their part, injunctions such as, *watch a movie, watch a movie in slow motion, watch a movie frame-by-frame*, lead to experiences that bring

[48] Maslow, *The Psychology of Science*, 15–16. For the history of this thought see https://quoteinvestigator.com/2014/05/08/hammer-nail/#note-8840-8

[49] Maslow, *Toward a Psychology of Being*.

understanding. As with the electric bell, *LoF*'s imaginary Boolean values account for the frame/no-frame oscillation that produces the illusion of movement. Still, sophisticated as this is logically, it is phenomenologically insufficient to account for our experience of motion. Such an account must involve the various conjugate forms that pertain to aliveness, all the way up to the highest, the distinctional. This analogy, then, suggest why concepts or logic alone are insufficient to account for lived experience, particularly since lived experience always involves movement. It shows how injunctions bring intelligibility to such experiences, and to the very possibility of experience, as such.

In sum, products as a kind of thing, are most likely to be considered 'bodies'. Like the automobile example we began with, this is how products are conceived and built. Mostly material, one justifiably conceives products mechanically. Even organizations, which are not material, are typically conceived and built as mechanisms.

The ubiquity of products makes mechanical intelligibility itself ubiquitous. The production of products consists mostly in integrating discrete units for a given purpose. Such integration is easily conflated with organization. Thus, products are regarded as bodies and bodies as mechanical products. As the use of tools throughout the animal kingdom suggests, once they appear, products are constitutive of aliveness. So pervasive and intimate is our relation to products that, to *biomorphism-a* (Ba), our form as organisms, and *biomorphism-w* (Bw), our construction of the world, might be added the sub-category *biomorphism-p* (Bp), our formation of, and by products. Bp refers to the pervasive relation of organisms and the physical environment that is mediated by products, thereby intelligible in mechanical terms.

It is almost impossible to resist the temptations of products. Prosthetics, more than like hats, they are often much more alike dental implants. Briefly aware of them while they are adjusted, their very function is to vanish as they blend in with the rest of our teeth—including previous implants. Concepts are prosthetics in precisely this way. And, just as we cannot help regard the two-headlight physiognomy of cars as faces, so we cannot help but ascribe a being to products that hardly differs from our own.

The Nonliving

Our experience of the physical environment includes the nonliving. The metaphysical relation between living being and the nonliving goes in two directions. Ontologically, living being emerged from the nonliving; epistemologically, the nonliving is known according to the spirallic nature of living being.

To reflect on biomorphism as a whole, and specifically Bw, which includes the nonliving, we must briefly venture into physics, specifically the physics of time. Both our origins and our experience are tied to time, and time hinges on our thermodynamic nature, which itself follows from our nature as distinguishing distinctions.

Carlo Rovelli explains that it all starts with directionality: "Heat is the fundamental reason why there is directionality in the universe. There is detectable difference between the past and future only when there is the flow of heat."[50] Negentropy is required for experience and not only because it allows organization. To experience is to *be* in time.

> Heat is linked to probability; and probability in turn is linked to the fact that our interactions with the rest of the world do not register the fine details of reality. The flow of time emerges thus from physics, but not in the context of an exact description of things as they are. It emerges, rather, in the context of statistics and of thermodynamics. This may hold the key to the enigma of time. The "present" does not exist in an objective sense any more then "here" exists objectively, but the microscopic interactions within the world prompt the emergence of temporal phenomena within the system (for instance, ourselves) that interact only through [a selection of variables in a] medium of a myriad of variables.[51]

Managing their own heat, then, living beings are small bubbles of negentropy briefly swimming against the vast river of entropy, tiny wavelets of causation as we interact with the world. Directionality allows bet structure to be established and living being to be anticipative.

> Our memory and our consciousness are built on these statistical phenomena. For a hypothetically supersensible being, there would be no 'flowing' of time: the universe would be a single block of past, present, and future. But due to the limitations of our [embodiment] we [advert] to only a blurred [apprehension] of the world and live in time. Borrowing words from my Italian editor, "what's not apparent is much vaster than what's apparent." From this limited, blurred focus we get our perception of the passage of time.[52]

So, at the heart of thermodynamics is the probabilistic nature of temperature. We take for granted that improbable behaviors will not take place and probable ones will. We would never expect a balloon to "fly out of the window, circle the lighthouse down there in the distance, and then return to land on my hand, at the point where it was released." We take for granted that a cold teaspoon placed in hot steam will heat up. The key question is this: "What does what we know and do not know have to do with the laws that govern the world?" To this question there is a subtle answer in terms of the differences that make a difference for us:

> Teaspoon and balloon behave as they must, following the laws of physics in complete independence from what we know or don't know about them. The predictability or unpredictability of their behavior does not pertain to their precise condition; it pertains to the limited

[50] Rovelli, *Seven Brief Lessons on Physics*, 62.

[51] Rovelli, 62 (modification in brackets). The original reads: ". . . that interact only through the medium of a myriad of variables." The author himself said something very similar, as shown in what follows shortly.

[52] Rovelli, 62–63 (modifications in brackets). The original reads: "But due to the limitations of our consciousness we perceive only a blurred vision of the world . . . ." Rovelli concludes: that there is still much to be understood. "Time sits at the center of the tangle of problems raised by the intersection of gravity, quantum mechanics, and thermodynamics" about which we still know little. "If there is something that we are perhaps beginning to understand about quantum gravity that combines two of the three pieces of the puzzle, we do not yet have a theory capable of drawing together all three pieces of our fundamental knowledge of the world."

set of their properties with which we interact. *This* set of properties depends on *our* specific way of interacting with the teaspoon or the balloon. Probability does not refer to the evolution of matter in itself. It relates to the evolution of those specific quantities we interact with. Once again, the profoundly relational nature of the concepts we use to organize the world emerges.[53]

Entropy is defined as that which we do *not* interact with—the differences that do *not* make a difference to us, the distinctions we do *not* make. According to Rovelli, this is what Ludwig Boltzmann understood and succeeded at expressing mathematically about the notion first coined by Sadi Carnot: "Entropy is precisely the quantity that counts *how many* are the different configurations that our blurred vision does *not* distinguish between."[54] And this, no less, is how our linear sense of time emerges:

> If I could take into account all the details of the exact, microscopic state of the world, would the characteristic aspects of the flowing of time disappear? Yes. If I observe the microscopic state of things, then the difference between past and future vanishes. . . . In the elementary grammar of things, there is no distinction between "cause" and "effect."[55]

Therefore, probability pertains to the differences that make a difference to living beings. It is according to what we can and do apprehend that we interact with the world. This is because we are in functional closure by self-regulating our own heat, and this results in a directionality that is relatively autonomous. Our ignorance, remedied by probability, *has a concrete effect on the world*. The relational probability that results in our interactions with differences that make a difference to us causes effects on concrete physical bodies, and this is how organisms permanently set up 'initial conditions' that are intrinsically unknowable. The point of was made beautifully in seventeenth century Spain:

> Porque ese cielo azul que todos vemos,
> ni es cielo ni es azul. ¡Lástima grande
> que no sea verdad tanta belleza![56]

It is also captured in a famous koan:

> The wind was flapping a temple flag, and two monks started an argument. One said the flag moved, the other said the wind moved; they argued back and forth but could not reach a conclusion. The Sixth Ancestor, Hui-neng, said, "It is not the wind that moves, it is not the flag that moves; it is your mind that moves." The two monks were dumb struck.[57]

---

[53] Rovelli, 56–57 (emphasis in original).

[54] Rovelli, *The Order of Time*, 32.

[55] Rovelli, 32–33.

[56] "Because this blue sky that we all see/ is neither sky nor blue. What great pity/ that so much beauty is untrue." Attributed to Bartolomé Juan Leonardo de Argensola or possibly his brother Lupercio Leonardo de Argensola ("Bartolomé y Lupercio Leonardo de Argensola").

[57] Wumen Huihai, *Mumokan, Case 29. The Sixth Ancestor's 'Your Mind Moves.'* The Mumokan is a collection of 48 Chan (Zen) koans compiled by the Chinese Chan master Wumen Hui-hai (1183–1260, Mumon in Japanese). Hui-neng (China, 638–713) is one of the key figures in Chan Buddhism and traditionally regarded as its Sixth and Last Patriarch. The fountainhead is the teaching of Shakyamuni Buddha (circa 563 BCE to 483 BCE). The paradoxical name of this work,

This brings us back to the origin of mind in our embodiment as distinctions:

> The skin of a living organism cuts off an outside from an inside. Not so the circumference of a circle in a plane.

Mind (Bw) is born of embodiment (Ba): functional closure, membranes (*membrains*, as we shall soon see). This is the heart of the paradox of boundaries—observers (Ba-Bw) are not merely objects but [body-subjects] whose distinguishing (Ba), produced distinctions (Bw). The river does not move, our minds do; movies do not move, our minds does; a drawn circle does not cut off an outside from an inside, our mind does it. Clouds do not touch each other, living beings do. This was discussed at length in Chap. 3.[58] Along with the discussions of the dangers of our own biomorphism when understanding biogenesis, everything presented there is consistent with Rovelli's account, and forms part of this discussion of the nature of the nonliving.

The only boundaries are those of living beings, distinguishing distinctions. Bw in relation to the nonliving, then, is radically different in relation to living beings, by-products, and products—the three distinguishing kinds of things. Treating the nonliving as one of these three gratuitously ascribes attributes that do not pertain to it—the fallacy of misplaced concreteness.

In sum, Rovelli shows that not only logic, but also physics "is nothing more than the properties of the act of distinction."[59] Given our thermodynamic nature, how we know the world follows from the fact that we are spirallic distinguishing distinctions. The world can only appear to a distinction that distinguishes. We are what we are because we are limited, our limits allow appearance, and appearance is necessarily limited.

Spencer-Brown said: "The universe is simply what would appear if it could. Its laws are the laws of the possible, called by Shakyamuni the links of conditioned coproduction…"[60] Living being and the world mutually co-produce. Appearance is made possible by the fundamental, primary, most important equation: 2. For 2=ב.

Joinings: Empirical Evidence in Biology

The discussion so far of the four kinds of things has treated the fundamental difference between living being and the other three kinds of things. As has been shown, the fundamental character of living being may be expressed with the single term *joinings*. Still, just what is meant by joinings has so far been mostly discussed in abstract terms. So, we turn now to five sets of empirical findings to show what it

Gateless Gate, suggests 'closed opening' or closed-open, and also recalls Boolean imaginary values and the functional closure of living beings.

[58] See section "What is an *Itself*?"

[59] Kauffman, "Peirce's Existential Graphs," 61.

[60] Spencer-Brown, *Laws of Form* (Limited ed.), viii.

means in practical biological terms that organisms are joinings. These findings pertain to the autonomy of cells, environments, individuals as environments, endosymbiosis, major evolutionary transitions, and sexual reproduction.

Autonomous Cells

Watch again the video *White Blood Cell Chases Bacteria*.[61] This white blood cell is anything but 'passive' or 'vegetative.' On the contrary, it displays an almost shocking degree of autonomy as it relentlessly pursues the bacterium. It is a predator pursuing its prey, a cat or a dog playing with a ball.[62] Aristotle coined the notion of the 'vegetative' as a halfway house between the animate and the inanimate. He wrote: "Now that which is ensouled seems to differ from that which is without a soul in two ways most of all: in motion and in sensing."[63] Because this would make plants inanimate, notwithstanding the fact that they reproduce, he had to reconsider. So, he opted for a low-level soul, one that only allowed them to reproduce. Thus, not inanimate, plants for Aristotle were not quite alive.[64]

Watching the scene between the white blood cell and the bacterium, one cannot but experience an overpowering shift in our sense of what individual cells can do. Their capacity for autonomous movement and sensing leaps out at us. Moreover, we cannot help but connect with the high-level metabolic quality of the event, for us expressed emotionally as anticipation and excitement of the chase. This connection, we shall see later, far from being anthropomorphic, is fully justified. This brings up profound questions. What is a self? Are the white cell and the bacterium selves? Are all cells and all bacteria selves? If we are made up of them, are they us, part of our self? Are they a 'they' for us or are they us? Who or what is a *they*, and who or what is *me*? What am I? What is an 'I'?

With Aristotle, we take for granted that motion and sensation make us autonomous. We do not doubt that we self-regulate. But what of the motion and self-regulation of this white blood cell? It could very well be (in) us. Indeed, so could the chased bacterium be (in) us. Can there be any doubt that both are [body-subjects] in their own right—and by extension so are all cells and all bacteria? This brief scene hits us with the realization that we are what we are because we are made up of 3.2 trillion cells—every one a [body-subject].[65]

Like us, all [body-subjects] have the hallmark of agency, even if, like all agency, it is relatively circumscribed. Had he seen this video, Aristotle would have been

---

[61] "White Blood Cell Chases Bacteria."

[62] Indeed, it is remarkably like a turtle chasing a ball (as can be seen in another video found at https://www.youtube.com/watch?v=81720bVYf9Q)

[63] Aristotle, *On the Soul*, Book 1, Chap. 2, 403 b24.

[64] Mancuso and Viola, *Brilliant Green*, 13.

[65] Add to those ten times the number of microbial cells that live within us, as discussed in the next section.

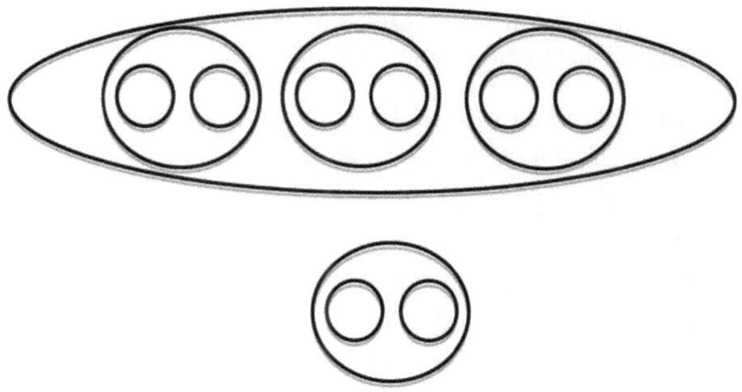

Fig. 7.1 The basic nature of [body-subjects]

stunned. His notion of the vegetative would have been shattered as much as our own notion is that autonomy and agency are restricted to us as persons. It follows inexorably that, as [body-subjects], we constitute vast complex environments made up of [body-subjects]. This nesting, which appears to be the condition for our being [body-subjects], is summarized as follows (Fig. 7.1).

We are back at A@@ (aliveness). To @@A in the sense of two-element regulatory system, we must now add the sense of relationship as exemplified by the white cell and a bacterium. These relationships within an organism pertain to all [body-subjects] and therefore to living being as such. This too is part of the proper content of @@A. Notice that this is so, both on the side of the object, as just shown, and now to the insight *living thing* on the side of the subject. If living beings are constituted by joinings, then these joinings pertain to them—us—as objects of inquiry as much as they pertain to us as the subjects who inquire. Despite the difficult questions it raises, both together, and the further empirical evidence that follows, tells us how and why we became the kind of *thing* that we are.

Living Takes Place in Environments

Joinings are inherent to the nature and history of life. The volcanic pool scenario suggests that, for biogenesis, a collective had to undergo chemical cycles of increasing complexity. As the volcanic pool scenario suggested, ongoing oscillations in heat and chemical energy through cycles of dry, wet, and moist conditions allowed lipids in close proximity to link together and form ever more complex molecular chains. In turn, these led to protocells packing together to form a progenote. Thus, whether tightly or loosely coupled, living beings are joinings constituted in environments and as environments. According to Emeran Mayer (no relation to the author),

> For about 3 billion years, bacteria where the sole living inhabitants on Earth. They occupy every tranche of land, air and water, driving chemical reactions setting the conditions for the evolution of multicellular life. Slowly, through trial and error across the vastness of time,

> they invented the complex and robust feedback systems, including the most efficient "language" that to this day supports all life on earth.[66]

Tightly packed in these three sentences is the foundation of joinings and the proper content of @@A as the foundation of biology to which this investigation points. Three billion years of distinguishing activity—the proper conjugate form to understand living being—led to the emergence of vast networks of two-component regulating—feedback—systems, with simple distinguishing distinctions upon which ever more complex distinguishing distinctions are built. This is the foundational relation Ba-Bw. If we take the Santiago school's framing of metabolism = cognition, this goes well beyond, to environment = cognition, or an ecology of mind. The environment as a whole is distinctional. The emergence of more complex organisms is therefore the process of nesting of distinguishing distinctions within higher level distinguishing distinctions. Growing empirical evidence supports Mayer's points. For instance, researchers have found that "the signaling molecules that our endocrine system and brain use to communicate probably originated in microbes." This includes insulin and "ancient microbial versions of many gut peptides and hormones, including noradrenaline, endorphins, and serotonin, and their receptors." Indeed, it has been proposed "that these signaling molecules represent the words of a universal biological language" used by frogs, plants, and microbes in our intestines as well.[67]

So, built from the ground up are systems of distinctions. Since these are in relationships, this implies systems of signs. It repays to follow Mayer's account because it dovetails perfectly with the unfolding of EI while also showing that life is synonymous with joinings. About four billion years ago life emerged in the form of single-celled microorganisms, the archaea. For the next three billion years, only microbes inhabited the earth. Trillions of them, more than the stars in our galaxy, a billion species of different shapes, sizes, colors, behaviors, floating silently in a marine universe. Through natural selection, signaling molecules, receptor molecules and decoding mechanisms evolved by which they could communicate.

> About 500 million years ago, the first primitive multicellular marine animals began to evolve in the ocean, and some marine microbes took up residence inside their digestive systems . . . .

Gradually, the animals and microbes developed a symbiotic relationship, and the microbes found ways to transfer vital genetic information to their host animals. This information provided host animals with a range of molecules they were lacking but which the microbes had learned to manufacture during billions of years of trial and error. Some of these molecules became the neurotransmitters, hormones, gut peptides, cytokines, and other types of signaling molecules our bodies used today.

Over millions of years, as primitive marine animals evolved into more complex creatures, they developed simple nervous systems in the form of nerve networks

[66] Blaser, *Missing Microbes*, quoted in Mayer, *The Mind-Gut Connection*, 18.

[67] Mayer, *The Mind-Gut Connection*, 78.

surrounding their primitive guts, not very different from the networks of the enteric nervous system that surround our gut today. The nerve networks in these creatures used some of the genetic instructions they received from microbes to produce signaling chemicals, which allowed neurons to pass messages to each other and instruct muscle cells to contract. These were the precursors of our human neurotransmitters.

Amazingly, these simple nerve networks and the signaling molecules enabled the primitive animals of millions of years ago to respond to ingested food in a similar, programmed away as our guts do today.[68]

Underlying a vast array of emergent *how's* for signaling is a single *what*: EI. Indeed, the reciprocal EI between organisms. Microbes and their relationships co-emerge. Notice that, metaphysically, this is a co-emergence of substance and liminality. With this co-emergence EI unfolds. New differences make a difference, and this leads to the unfolding of distinctions, both as Ba and as Bw. So, at the heart of the play Ba-Bw are interrelating [body-subjects], joinings.

This amounts to an astonishing shift in our view of life. A 3.4-billion-year soup of distinguishing activity, vast trial-and-error experimentation in EI, results in ongoing joinings that leads to what, strictly speaking, may only be fuzzily differentiated into single cells, multicellular organisms, environments. The empirical evidence shows that clear and distinct differentiations between the distinguishing distinctions that participate in joinings simply do not hold up. This is so even if, or rather, precisely because, they are each an in-itself for-itself. They truly can only be in-themselves and for-themselves because they are for each other. This is the primordial meaning of *A* and *B* as origin of the logical conjunctions. What do they mean in this context? What is *A* and *B* in act in this primary sense?

This is our purpose. To ask questions about living being in its own terms, about their constitution, interrelationships, the nature of individuality, agency, and so on.

We Are Environments

The full significance of the fact that complex organisms are environments has only recently begun to be understood. It underlies today's growing medical interest in the human microbiome, defined by the National Human Genome Research Institute as "the collective genomes of the microbes (composed of bacteria, bacteriophage, fungi, protozoa and viruses) that live inside and on the human body."[69] According to

---

[68] Mayer, 90–91.

[69] Yang, "The Human Microbiome Project." According to Yang, "We have about 10 times as many microbial cells as human cells. So, to study the human as a 'superorganism,' composed of both non-human and human cells, in 2007 the National Institutes of Health (NIH) launched the Human Microbiome Project (HMP) as a conceptual extension of the Human Genome Project." For an account of the enormous complexity of merely defining the term see Ursell et al., "Defining the Human Microbiome."

Rob Knight, "microbial genes far, far outnumber human genes, and most microbes live in your gut . . . like a large-scale ecosystem in which different groupings of microbial species can carry out the same functions."[70] Analyzing the microbiome has become a hot topic because the technology to study it has recently improved dramatically and because it can aid in a wide variety of disease treatment and prevention. Most strikingly, our microbiome is a critical factor in our *individuality*:

> If you consider the amount of money that's been spent on human genetic research and the limited success in solving hard problems like obesity, it makes sense to look not just at the human genome, where we're all over 99 percent the same as each other and we can't make any changes, but also at the microbiome, where we can be 90 percent different from one another, and where we have all changed it profoundly during our lifetimes via processes we're just beginning to understand . . . . We're mostly talking about diseases associated with digestion and the immune system. Microbes are involved to a much greater degree than you might expect in diseases involving the neurological system, like multiple sclerosis, and psychological conditions like depression. We're trying to figure how well we can diagnose depression and autism by looking at the microbiome.[71]

Not only are we environments; we are individuals in no small part *because we are unique environments*. Thus, A@@ is the condition for the possibility, not only of our individuality but also of our *uniqueness*. The experimental results are stunning:

> There is no question that the emerging science of gut-brain communication has been one of the most fascinating topics for scientists and the media for the last few years. Who would have ever believed that simply transferring fecal pellets containing gut microbiota from an "extrovert" mouse could change the behavior of a "timid" mouse, making it behave more like the gregarious donor mouse? Or that doing a similar experiment transplanting stool and its microbes from an obese mouse with a voracious appetite would turn a lean mouse into the same overeating animal? Or that the ingestion of probiotic-enriched yogurt for four weeks in healthy human females could reduce their brains' response to negative emotional stimuli?[72]

What then is an individual? What is an environment? Can a clear differentiation be made between them? What is their relation? The current findings in biology turn these into vexing questions. To catch a glimpse of the implications of our emerging view of life, a single example should suffice. *Toxoplasma gondii*, a microorganism, reproduces only in the intestinal tracts of infected cats, but it can infiltrate the brain of any mammal, including humans. Ideally for toxoplasma, after cats excrete it, they will be ingested by rodents, eventually forming cysts throughout their bodies, particularly in their brain. When a cat eats an infected rodent, the ingested cysts again reproduce in the cat's intestinal tract. This is where things get interesting. Under normal circumstances rodents avoid cats. But rodents infected with toxoplasma, far from being instinctively fearful of them, begin to prefer areas that smell of cat urine. The parasite cyst targets the area of the rodent's brain responsible for triggering their fight-flight response. Moreover, while infected rats exhibit all the expected

[70] Mayer, *The Mind-Gut Connection*, 216.

[71] Robbins, "Interview with Rob Knight."

[72] Mayer, *The Mind-Gut Connection*, 26.

defensive behaviors towards any other predator, they become sexually attracted to a cat's odor—a fatal attraction.[73] By whose agency do these rodents end up as prey for cats? What is agency when organisms are understood as joinings, relationship—@@A?

According to Mayer, study of the microbiome has ushered in a new era in biology that involves deep questions, very much along the lines pursued here:

> Even though we are just at the very beginning of this rapidly unfolding journey of scientific discovery, we can no longer view ourselves as the only intelligent product of evolution, distinct from all other living creatures on the planet. Just as the Copernican Revolution in the sixteenth century fundamentally changed our understanding of the world's position in the solar system, and Darwin's revolutionary theory of evolution proposed in the nineteenth century has forever changed our position within the animal kingdom, the human microbiome science is forcing us again to reevaluate our position on earth. According to the new science of the microbiome, we humans are truly superorganisms, composed of closely interconnected human and microbial components, which are inseparable and dependent on each other for survival.[74]

In a nutshell, this is joinings or A@@. The integration, interrelations, and interrelationships between organisms, superorganisms, and environment are best understood in terms of A@@. The title of a recent post on the website of the National Human Genome Research Institute sums it up best: *The Human Microbiome Project: Extending the Definition of What Constitutes a Human.*[75] For our present purposes, the title can be generalized as *The Microbiome: Extending The Definition of What Constitutes Living Being.*

Endosymbiosis

Symbiosis has long been recognized in biology. The counterpart to the synchronic study of the microbiome is the diachronic hypothesis of endosymbiosis. The Modern Synthesis combined Darwin's natural selection acting on mutations and the underlying inheriting of genes to generate new adaptations and new species. But is that the only way new lineages and adaptations form? French marine biologists wrote a short paper in 1937 suggesting the division of cells into *pro* (before) *karyon* (seed, kerel, nucleus) and *eu* (true) *karyon*. This brings up Lynn Margulis' work.

> Virtually all biologists now agree that: "[the] basic divergence in cellular structure which separates the bacteria and the blue green algae from all other cellular organisms, probably represents the greatest single evolutionary discontinuity to be found in the present-day world."[76]

---

[73] Mayer, 85–86.

[74] Mayer, 19.

[75] Yang, "The Human Microbiome Project."

[76] Steiner, Doudoroff, and Adelberg, *The Microbial Wor*ld, cited in Margulis and Schwartz, *Five Kingdoms,* 9–10.

Eukaryotes are joinings. These are joinings on a huge scale, at the level of lineages. The young earth supported a vast population of many types of bacteria. Mitochondria and chloroplasts within modern cells likely originated when bacterial cells came to live inside larger cells:[77]

> Endosymbiotic theory proposes . . . organelles were once prokaryotic cells, living inside larger host cells. The prokaryotes may initially have been parasites or even an intended meal for the larger cell, somehow escaping digestion . . . . Multiple lines of evidence support the endosymbiotic theory. Endosymbiosis is observed elsewhere in biology. Mitochondria and chloroplasts have intriguing similarities in structure, reproduction, biochemistry, and genetic makeup to certain prokaryotes. The plain fact that mitochondria and chloroplasts have any genetic information of their own argues in favor of the theory.[78]

Endosymbiosis shows that joinings are the rule and not the exception:

> Symbiosis is ubiquitous among organisms throughout the tree of life, from the species level to the kingdom level, and even to the domain level. It is integral to evolution as cooperating organisms gain survival advantage by a quid pro quo between them. For example, you (and for that matter all herbivores and omnivores) could not digest your food without the exquisite symbiosis between your gut and the bacteria therein.[79]

The insight of endosymbiosis eventually led to the study of the microbiome. Together they open new horizons in biology. As if that were not enough, a recent article declares: "Bacterial Microbiome Move Over: The Gut Virome Makes Its Debut:"

> A new study reveals that eukaryotic viruses are able to both shape mucosal immunity and support intestinal homeostasis in mice. Specifically, infection with murine norovirus (MNV) appears able to replace the beneficial function of bacterial colonization in the gut. … The new findings are the first strong evidence that viruses in the gastrointestinal tract can help maintain health and heal a damaged gut. Before this study, there had been very little investigation of the viruses that colonize the gut.[80]

Major Evolutionary Transitions and Social Evolution

Over millions of years, ever more complex organisms have evolved. Multicellular organisms are made up of trillions of cells. Maynard Smith and Szathmáry proposed that evolution has been driven by a small number of major evolutionary transitions (METs). Among these, evolutionary transitions in individuality (ETIs) consist of groups of individuals that replicate independently cooperating to form new, more complex individuals.[81] According to Bourke, contemporary inquiry into social

---

[77] Challoner, *The Cell*, 55–56.

[78] Mikulecky, Gilman, and Peterson, *AP Biology for Dummies*, 188.

[79] "Endosymbiosis: The Appearance of the Eukaryotes."

[80] Pullen, "Bacterial Microbiome Move Over."

[81] West et al., "Major Evolutionaty Transitions." MTEs that are not ETIs include mass extinctions, invasions of land, the advent of the oxygen atmosphere, the emergence of the nervous system, etc. See Hanschen et al., "Individuality."

evolution has moved well beyond beehives and baboon troops. In *Principles of Social Evolution*, Bourke argues that inclusive fitness theory is the key to understanding these transitions.[82] He defines six of these—each one an instance of joining that marks a major transition in evolution: 1. Separate replicators [genes] to cell enclosing genome; 2. Separate unicells to symbiotic unicell [eukaryotes]; 3. Asexual unicells to sexual unicells; 4. Unicells to multicellular organism; 5. Multicellular organisms to societies of eusocial species; and 6. Separate species to interspecific mutualisms. Bourke shows that these pivotal moments in the history of life occur in three principal stages: social group formation, social group maintenance, and social group transformation. All follow the same principles of social evolution. The selfish interests of individuals are overcome as they form mutually dependent cooperative groups. This is inclusive fitness, or Hamilton's law:

> Since Darwin, the only fundamental change in our understanding of adaptation has been W. D. Hamilton's development of inclusive fitness theory (Hamilton, 1964). The traditional Darwinian view struggled to explain many cooperative social behaviors, with the most famous example being the sterile out worker caste in eusocial insect species, the ants, bees, wasps, and termites. Fisher (1930) realized that genes can spread not only through their impact on their own direct transmission (direct fitness) but also through their impact on the transmission of copies of the same allele in other individuals (indirect fitness), but he explicitly chose to neglect the latter effects in his derivation of the fundamental theorem. Hamilton (1964) incorporated indirect fitness into a genetical theory of social evolution and showed that the characters favored by natural selection are those which improve the individual's inclusive fitness, which is the sum of its direct and indirect effects.[83]

So, inclusive fitness is all about joinings as understood in classic Darwinian and Hamiltonian genetic terms. The inclusive fitness of our genes is not only what we inherit to our offspring directly, but also what is indirectly inherited by our sibling's offspring—our nieces and nephews—and to whose survival we contribute socially. We have a vested interest in them, so to speak. Maynard Smith referred to this as "kin selection" because, having a common ancestor, these genes are identical by descent. So, inclusive fitness does not account for all the reproductive success of our close relatives, only what we ourselves maximize by our behavior.[84]

Bourke claimed that Hamilton's law explains the advent of cooperation at various level of social organization because the same principles of inclusive fitness are at work. They underly ETIs because all share the three principal stages of social group formation, social group maintenance, and social group transformation. These stages occur in analogous ways at the "different hierarchical levels." Hamilton's framing of "social evolutionary theory in its expanded form offers an interpretation of the selective basis of life's basic organization and history more profound and satisfying than alternative frameworks."[85] It therefore "conceptually unifies a very

[82] Bourke, *Principles of Social Evolution*.

[83] West, El Mouden, and Gardner, "Sixteen Common Misconceptions," 233.

[84] West, El Mouden, and Gardner.

[85] Bourke, *Principles of Social Evolution*, 198.

large range of phenomena across all taxa, hierarchal levels, and times."[86] In this review we shall refer to this vast unification as the [social].

> Living things are organized in a hierarchy of levels. Genes group together in cells; cells group together in organisms; and organisms group together in societies. Even different species form partnerships.[87]Throughout life's history, '*major transitions*' have occurred in which previously independent units have formed groups that in time become individuals in their own right. Common to all such events is a process of social evolution.[87]Each transition occurs only if natural selection favors one unit *joining* with another in a new kind of group.[87]

In sum, Bourke "presents a fresh synthesis of the principles of social evolution that underlie the major transitions" to show that "social evolution lies at the heart of the history of life, shaping biological organization across different taxa at all hierarchical levels.[88] The [social]—joinings—is at the foundation of evolution.

Sexual Reproduction

What is the outcome of a human sperm and a human ovum joining together?[89] The reader may wish to puzzle over this riddle for a moment—a genuine question—this being a clue to the answer given below. The key to the success of sexual reproduction is the advent of offspring with unique DNA, the outcome of the joining of two strands of DNA, one from a father and the other from a mother. This is an astonishingly powerful adaptive strategy to combine lineages because it results in a huge variety and novelty of features to be offered up for natural selection. The origin of every individual born from sexual reproduction is a joining. Perhaps this instance of joining a is the one we are most familiar with.

Returning to the riddle, what is the outcome of this joining? Questions. Questions? Recall the innate drive of a two-year-old to ask questions. This is how their construction of the world enters a new conceptual phase, a crucial turning point in their socialization. Responding to their questions, we begin to pour the whole of our culture into them. These questions have made humans uniquely adaptive to the point that we adapt the environment to ourselves. How did questions emerge? Our capacity to explicitly interrogate our fellows is as good a marker of the dawn of the Anthropocene Age as any other. Questions lead to further questions. This capacity rests on the deep organic nesting of @@A that unfolds anticipation through relationships. Nurtured in ever deeper organized [social] complexity, it leads to ever deeper play of Ba-Bw. In a nutshell then, the [social] and the [cognitive] are conditions for the possibility of the unfolding of each other, which is the unfolding of distinguishing distinctions and their anticipation. Sexual reproduction involves a

[86] Bourke, 198–200.

[87] Bourke, abstract. Emphasis added.

[88] Bourke, abstract.

[89] Phillip McShane, personal communication.

couple in a relationship, @@A, which results in a joining, @@A, that is often protected for a time in some social context, @@A.

The multifaceted revolution taking place in biology ushers in a new era in our understanding of living being. The empirical evidence for joinings is overwhelming. Living beings are [body-subjects] interrelating with [body-subjects], and themselves constituted by a plurality of [body-subjects]. Joinings are the condition for the possibility of the living kind of thing, its emergence and organization. Living beings, and most conspicuously multicellular organisms, have been built from the bottom up as complex pluralities of autonomous units. The condition for this possibility was three billion years of molecular signaling by trillions of microbes establishing bet structure and anticipating through EI.

Chapter 8
[Subjects]

Organisms as Objects and as [Subjects]

What kind of thing is a living being? The empirical findings outlined in the previous chapter show why an accurate one-word answer to this question is *joinings*. Joinings are @@A, *con-crēscere,* the ground scheme of distinguishing and its inherent dynamics.

> ב➔Repetition, *Again*➔Axioms of *LoF*➔Calculus of distinction or "Boolean Arithmetic"➔**Cells as distinctions**➔Boolean Algebra or Logic.

This scheme is made possible by plurality. That this is so ontologically poses deep epistemological conundrums about the nature of living individuals and their interrelationships. Following our outline of major strands of the empirical evidence that living being takes the form of joinings, we can now consider some of these conundrums. To do so, we start from Lonergan's fundamental point that there are no things within things. It follows from this that, if a thing is to be fully intelligible, its entire datum must enter the account up to the highest explanatory conjugate form. Recall that Lonergan's point pertains to the notion of thing on the side of the subject. We start there and then move to a reflection on the nature of [subjects] on the side of the object.

Consider Fido again. Though he is a vastly complex multicellular organism, most of what can be said about him, with few exceptions, could be said about multicellular organisms generally, and even about single celled organisms. Imagine it is Fido's white blood cell that chases and engulfs a bacterium. For a moment, while the bacteria is still alive, this is a body within a body within a body. Because all three are bodies, or, properly speaking [body-subjects], what becomes of Lonergan's point that there are no things within things? To reflect on this, a careful differentiation must again be made between the side of the subject and the side of the object.

© The Author(s), under exclusive license to Springer Nature Switzerland AG 2023
D. C. Mayer-Foulkes, *The Nature of Living Being*, Biosemiotics 26,
https://doi.org/10.1007/978-3-031-24789-7_8

The side of the subject is our own process of understanding; the side of the object are living bodies, in this case Fido, a [body-subject] made up of [body-subjects].

For us as subjects, a thing is a concrete intelligible unity. That unity can be Fido, one of his white blood cells, or a bacterium in his bloodstream. This depends on which concrete intelligible unit we, on the side of the subject, happen to be adverting to. We can shift from one concrete intelligible *unity-identity-whole* to another—from the thing Fido to the thing white blood cell to the thing bacteria and onto the thing molecule and the thing atom. Indeed, this is what is entailed by the notion of shifting. Shifting necessarily implies there are at least two or more things. Because, on the side of the object, the highest explanatory conjugate is distinguishing by Fido, the white blood cell, and the bacterium, and moreover, because all three are things for themselves, the problem of things within things must be resolved on the side of the subject. The same applies to the nonliving, for instance, if Fido swallowed a marble. On the side of the subject, a thing is a kind of insight. This is why there are no things within things.

As inquiring subjects, we make sense only of one *thing* at a time, of one concrete intelligible *unity-identity-whole* at a time. Synthetic acts are done one at a time. This is their nature, a key point that speaks to the very nature of living being. Synthetic acts are done one at a time—this obtains both on the side of the subject and as [body-subjects] on the side of the object. It is so at each level and in each case in which distinguishing takes place in organisms constituted by a vast multitude of simultaneous distinguishing processes.

This is fundamental. Once living beings are understood in terms of distinguishing, Lonergan's point that there are no things within things on the side of the subject tells us something crucial about the nature of [body-subjects] on the side of the object. The proposition above was that a [body-subject] is a *unity-identity-whole* being-at-work-staying-itself, on the side of the subject; now it is that a [body-subject] is an identity-unity-whole being-at-work-staying-itself, on the side of the object, as was explained in relation to *things*. Note the similarities and differences between Lonergan's *unity-identity-whole* (in italics) that pertains to the side of the subject, and identity-unity-whole that pertains to the side of the object.

This is an outstandingly important point. It is rooted in the homogeneity of distinctions as discussed in the section *Distinction: A/Ā, A=A* that opens Chap. 2. What is true of distinction, the root synthetic act as was explained there, obtains in all synthetic acts. Here it is again:

> Inherent to any distinction A/Ā is the homogeneity of A on one side and the homogeneity of Ā on the other. That is, for a distinction to obtain, A must be homogenous in respect to the distinction, else a further distinction is introduced. Likewise, Ā must be homogenous, insofar as Ā is that which is not A, else again, a further distinction is introduced. Therefore, homogeneity, or A=A, must obtain for both A and Ā for the distinction A/Ā to obtain. This is not to say that further distinctions cannot be drawn in either A or Ā. They can. But if they are, then in these too A=A obtains, else they do not constitute distinctions.[1]

[1] See Chap. 2, section "Distinction: A/Ā, A=A."

This is exactly Lonergan's point. On the side of the subject, there are no things within things. So, if the distinction is *Fido*, homogeneity obtains; the concrete intelligible unity is Fido. If the distinction is this white blood cell, homogeneity obtains; the concrete intelligible unity is *this* white blood cell. If the distinction is this bacterium, homogeneity obtains; the concrete intelligible unity is *this* bacterium. Likewise, if the distinction is this musician or this melodic line, homogeneity obtains; the concrete intelligible unity is *this* musician or *this* melodic line. Given a distinction or a thing, homogeneity, or A=A obtains, the concrete intelligible unity of *this* distinction or thing. On the side of the subject there cannot be distinctions within distinctions. Thus, the homogeneity of distinction explains why there cannot be things within things.

Before continuing with what may be learned about the nature of [body-subjects] from this—call it the no-things-within-things principle—notice that it does not obtain when things are conceived as 'bodies.' An instance is when the axioms of *LoF* are conceived as perfect continence and illustrated with box diagrams. A box diagram can represent the sentence *In the car there is a seat, and in the seat, there are springs*. These are bodies within bodies, and it is to conceive things as real as already out there now. I shall return to this question later and ask why doing so is so effective, even if it fails under scrutiny.

No-things-within-things is an insight into ourselves on the side of the subject. But, being distinguishing distinctions, it is also an insight into ourselves as [body-subjects] on the side of the object. Our distinguishing activity, foundation of our aliveness, is the synthetic acts between subject and object. It straddles subject and object. Adverting to this activity, we advert to ourselves both as subject and as object. Thus, we are ourselves the bridge between the insight *thing*, specifically the insight *living thing*, on the side of the subject, and our nature as [body subjects], on the side of the object. In short, *living thing*, on the side of the subject, converges with [body-subject], on the side of the object.[2] This is highly significant because, as shall be shown, it opens the door for inquiry into [body-subject] *qua* subjects.

This is so because the fundamental nature of organisms, as distinguishing distinctions, obtains for any organism. Therefore, the single nature of synthetic acts applies to any instance of a synthetic act. This pertains to the proper content of @@A, for organisms as embodied distinctions; for distinguishing as verb; and even for the resulting distinctions, the noun.

It follows from homogeneity that, in the distinguishing activity of all [body-subjects], there are, *on the side of the subject*, no distinctions within distinctions. In general, there are no synthetic acts within synthetic acts. There are no linear distinctions within linear distinctions, no *Umwelten* within *Umwelten*, no affordances within affordances, and so forth. In fact, there are also no 'bodies' within 'bodies.' 'Bodies,' too, are distinctions on the side of the subject. Only the elision of the knowing subject lends the appearance that they are real 'bodies' already out there

[2] Left pending for now is discussion of how and why [body-subjects] differentiate other [body-subjects] from the rest of the world. This pertains to kinds of anticipation that correspond to kinds of things, a topic taken up in Chap. 10.

now, unaffected by the homogeneity inherent to distinguishing activity. So, for instance, adverting to the springs in the seat of a car, a subject may claim to advert to 'bodies,' unaware that here too she adverts to one thing at a time. She refers to each and to their composite, which could be named 'spring-seat-car.' The fact that she can and does advert to this distinction while regarding the rest of the car and the world beyond as homogeneous shows that this elision of herself is performatively contradictory.

This brings us back to products and their production from discrete units—this spring, this seat, this car. Yet this does not affect our distinguishing of them. The seat is a case in point. We treat it as a thing, as *unity-identity-whole* even though it was produced by combining dozens of components. Biomorphism-product (Bp) is the least elided. This is where the notion, indeed the bias, originates for clear and distinct ideas. The mathematical ideal of knowledge centers on products and takes the form of mechanistic thinking.

Think of a bottle of wine. As a real 'body' already out there now this seems quite straight-forward, but it soon becomes problematic. For instance, are the tags part of a bottle of wine? What of the price tag? What about the cork and the liquid inside and the government warning and statements about tax? Immediately there are whole/parts problems. As a thing, however, a bottle of wine is a product produced by human beings. Its full intelligibility is in terms of distinguishing. The liquid within contains alcohol, an intoxicating substance, which is an important distinction at the root of many of its features. Psychopharmacologist Ronald K. Siegel pointed out that "drug seeking and drug taking are biologically normal behaviors . . . . In a sense, the pursuit of intoxicating drugs [in animals] is the rule rather than the exception." His controversial conclusion is that "the pursuit of intoxication with drugs is a primary motivational force in organisms."[3] This distinction, then, is millions of years old and, palpably, fundamental to the intelligibility of a bottle of wine, along with various other features involving manufacture, taxation, commerce, and so on. A bottle of wine is a thing, an insight that includes insights into all these various elements. This—and not as bodies—is how [body- subjects] advert to the world. The fundamental principle obtains for any organism being-at-work-staying-itself.

This principle is: *a [body-subject] being-at-work staying itself involves a highest-level synthesis of synthetics acts*. The ongoing state of any [body-subject] results from its ongoing responses at any moment to the differences that make a difference *to it as a unit*. This response-as-a-unit in ongoing *entelecheia* moment by moment is what makes an organism (be for) *itself*. Regardless of how it is done, this is *what* an organism does to be *a self*. This, which may be referred to as *unity*, is accomplished through a nesting of distinctions. The general principle underlying unity will be taken up in the next section.

Let us consider the possibility that unity is contingent. Imagine that the flagella of an *E. Coli* bacterium rotate randomly, and that coordination between them is coincidental. Clearly, natural selection would favor slight increments in

[3] Siegel, *Intoxication*, 11, cited in Kotler and Wheal, *Stealing Fire*, 116.

coordination that would make chemotaxis slightly more effective. Unity would tend to be selected for. The same applies to any single-celled organism and to the cells of multicellular organisms. In all instances, unity amounts to optimization and is likely to be selected for. Regardless of complexity, the converse is also true: good design optimizes for unity. Moreover, living beings that we are, operating in unity, we tend to take unity for granted, both in ourselves and in other organisms.

Another principle of good design is *whatever is necessary but the least possible.* This applies to the number of monitoring systems (for instance, those that detect pH or temperature in *E. Coli*, as well as those that coordinate between higher and lower systems, and so on.) The same principle applies to levels of hierarchy, organs carrying out similar functions, and so forth. An optimal design also uses the necessary but least possible amount of energy and other resources. Likewise, it achieves indispensable coordination between systems but avoids what is not necessary. Optimizing for unity offers designs a better future prospect for efficient high-level performance, the chance of exploiting new possibilities. In a word, it offers better chances of survival.[4]

Thus, even assuming unity is contingent and not necessary, it is a foundational feature of well-designed [body-subjects], even more so because they are, by definition, anticipative. To anticipate as a unit is more optimal than not doing so. Unity, then, is the seed, in the most compact [body-subjects] we remotely descend from, of what we in ourselves discover to be our single consciousness. This feature of [body-subjects] on the side of the object, converges with our experience of ourselves as [body subjects]. Homogeneity obtains in both. On the side of the subject, if I make the distinction *my left leg*, homogeneity obtains; the concrete intelligible unity is my left leg. If the distinction is *my right thumb*, homogeneity obtains; the concrete intelligible unity is my right thumb. If the distinction is this thought or that melodic line I hear in my mind, homogeneity obtains; the concrete intelligible unity is *this thought* or *that melodic line I hear in my mind.* The concrete intelligible unity of *this* distinction or thing obtains in my experience of myself and my experience of objects. Notice that, on the side of the subject, we can advert to specific areas or parts of ourselves and regard them as a thing, even if on the side of the object, all these areas and parts are us, part of ourselves, as a thing.

The reader can experiment with this. To attend to one aspect of yourself as object is to regard it as a thing that eclipses all other aspects. Should another aspect arise in your attention, either you switch to it, or it fuses with what you had been adverting to, resulting in a new thing or distinction. In either case, homogeneity obtains.

[4]Unity may take many forms, and we must not take for granted our own form of unity. The unity of a tree, say, may be radically different to ours and yet still be unity. A flock of birds is made up of many [body-subjects]. Perhaps in trees, not being as centrally organized as animals are, their leaves are somewhat more like flocks. Mancuso and Viola say: "A tree is much more like a colony of bees or ants than an individual animal" (Brilliant Green, 36). Where to place trees in a continuum between a centrally organized [body-subject] and flocks of them is an important empirical question.

So, I may attend to my left hand, to my right hand, to both hands as one, and even to my attending to my hands.

This is a bridge between me as living thing, on the side of the subject, and me as [body-subject] on the side of the object, and hence to all [body-subjects]. This is the unity brought by me and all [body-subjects] to synthetic acts. That [body-subject], *for* whom the thing or the distinction shifts from left leg to right thumb to scent to melodic line, is a unit. I can advert to myself being alive with the insight *living thing*. That I may do this, on the side of the subject, affirms my unity as [body-subject] on the side of the object. They confirm each other—my adverting *in* myself with the insight *living thing* on the side of the subject and my adverting *about* myself as [body-subject] on the side of the object. And yet, being a thing, how do I manage to advert to a part of myself as a thing?

Identity-Unity-Whole

The unity of our consciousness is but a highly differentiated version of the protean nature of the attending-acting of [body-subjects] in *entelecheia. Consciousness* is the name we give to the highest-level synthesis of human synthetic acts. Though what we advert to self-consciously is a deeply nested and deeply recursive highest-level synthetic unity, it is nonetheless, at its foundation, a feature of the embodiment of all [body-subjects]. Unity is at work when we "think about our driving" or when we observe a monkey swinging, a cat running, a spider knitting a web, or an *E. coli* in chemotaxis. In all instances we see a display of the unity of [body-subjects] in *entelecheia*. Our adverting to this outward unity converges precisely with our insight into ourselves as *living thing* on the side of the subject, and as [body-subject] ourselves, on the side of the object. What depth of nesting and recursive unity merits the label *consciousness* is an arbitrary distinction, just like the distinction *planet*—though no doubt more complex. The name denotes an arbitrary stage in the unfolding of unity that is in the very nature of [body-subjects].

Now, my capacity to advert to one part of myself as a thing, makes one wonder how homogeneity actually applies on the side of the object. It was proposed that Fido, as a [body-subject] that distinguishes, is made up of a multitude of [body-subjects], such as his white blood cells that also distinguish. That being so, what to make of the claim of his unity as a [body-subject] that makes sense of only one distinction at a time? To answer this question, we must delve further into the nature of [body subjects]. This requires two further terms—wholeness and identity. For Lonergan, the insight *thing* consists of regarding something as a *unity-identity-whole*. While he did not assign a specific meaning to these terms, that is what I propose to do here.

There are two possibilities for Fido *on his side of the subject*. Either there is unity—a highest-level synthesis of synthetic acts—and therefore a hierarchy of synthetic acts in his distinguishing activity—or there is not, there is no hierarchy, and he does not experience unity. This second possibility seems unlikely, given what has been said about design, our own experience of him as unity, and our experience of our own unity. In contrast to people whose two brain hemispheres have been separated surgically, and who do indeed sometimes act like two people in a single body, nothing that Fido does would make us suppose that he does not experience himself as unity. On the other hand, suppose that Fido experiences unity as we do. His *making sense* of the world fits with the definition of adverting and acting according to the highest-level synthesis of synthetic acts. When Fido *makes sense*, just like us, he adverts only to a small subset of all the distinctions taking place at all levels in him, given that he is a highly complex [body-subject]. If this describes Fido as [body-subject], it also describes his [objects]. A single [object] or world appears to him because of his highest-level synthetic act. Not experiencing the world as unity there would be no hierarchy in Fido's distinguishing activity. His *making sense* would be by adverting to all the distinguishing taking place in him at once and at all levels. Lacking a highest-level synthetic act, multiple [objects] would appear before him. But is this feasible? Beyond this being a recipe for overload, rather than a centrally organized system, Fido would be a flock of single cells. Being centrally organized, Fido does not, indeed could not, make sense of everything at once. To make sense of the world, he must selectively attend and prioritize, moment by moment, to what is relevant. This is to be a unity.

What does Fido do? Again, we can advert to what we do. At this very moment, writing this sentence, I am not adverting to the distinguishing activity being carried out, moment by moment, by every one of the 3.72 trillion cells that conform me. Rather, I am *making sense* by adverting to my own highest-level synthesis of synthetic acts, given what I am doing. So, I am adverting and acting according to unity. My attention is on the composition of this sentence. I am hardly even aware of my fingers as I they tap the keyboard. As I turn my attention to myself, I now advert to the sense of my weight on my chair, my back slightly tensed, my hunching, my breathing, my mouth slightly clenched, a plane flying overhead—and I realize I must call my attention back to the writing of this sentence. This is a miniscule fraction of the distinguishing activity going on within me, which is the foreground of an overall deep and underlying sense of aliveness that forms an all-pervading background sense of wellbeing, with islands of ill-being when I am hungry, stiff, stressed, and so on.

Given my third party understanding of myself as a [body-subject], I surmise that the overall background sense of (myself as) unity is the outcome of all the distinguishing activity that is (within) me. Therefore, a corollary to unity is *wholeness*. I am a unit which, to conduct itself as such, must experience as unity only a fraction of itself. For humans, it is admissible to add a term to this last sentence, so it reads, "I am a unit which, to conduct itself as such, *consciously* experiences only a fraction of itself. . . ." The label for this highest-level synthesis aside, the key point is the necessary differential between unity and wholeness.

Unity involves nesting of distinctions, which in turn requires hierarchy. Lower-level systems are monitored by higher-level systems. Each monitors the differences that make a difference to them, their own diff-diffs. Lower-level systems cannot, by definition, monitor what make a difference to higher-level systems. And it makes no sense for higher-level systems to monitor what makes a difference to lower-level systems—assuming this is possible. Even if it were the case, then this monitoring would itself be monitored by a higher system. In short, additions or reductions of monitoring activity are irrelevant to the fundamental point that organized systems are hierarchically nested.

How is the relation between unity and whole set up? The notation in *LoF* is helpful here. When a higher-level system receives the signal Γ from a lower-level system, instead of the habitual , it means the lower-level system is malfunctioning. So, when it is ongoing, signals proper functioning—no news is good news. To function properly, a [body-subject] like Fido only needs to become aware when something is not right by actively feeling unwell, or Γ. This is the principle that underlies the now widely accepted *gate control theory of pain*. The signal for business-as-usual or 'all is well' is . This is replaced by Γ when something is amiss. The theory states that "gates" are closed to prevent pain sensation from traveling to the central nervous system if it remains within a certain threshold.[5] The opening of these gates, allowing pain to be adverted, is the transition from whole to unity.

If unity is the organism *making sense*, including making some sense of itself, as in the case of pain, wholeness is the totality of the distinguishing activity occurring in the organism at all levels. Unity and whole are necessary features of the hierarchy by which organisms function. The differential between them, along with the opaqueness between levels, and thereby of the system unto itself, is necessary for organisms to grow in complexity.

It follows from this that a further semi-opaqueness is indispensable for organisms to function. This brings us to the partial elision of our own biomorphism-a (Ba) mentioned in the section I/World in Chap. 6. Concretely, the highest-level system cannot monitor every next-to-higher-level system at once. At any one time, an organism monitors what is most relevant about the situation at hand. This calls for sequence. Think about your driving. You continuously monitor the road but not your hands on the steering wheel or your foot on the accelerator. You certainly do not monitor your body unless some signal switches from , to an intermittent Γ, Γ, Γ... for backache. As you listen to that interview on the radio, you are not even aware that you are monitoring the road. We often awake to the fact that we have arrived at our destination. If on departing, I should trip and fall in the parking lot on my way to the car, during the drive I will closely monitor the Γ, Γ, Γ, . . . of knee ache while doing my best to avoid it distracting me from monitoring my driving. In short, beyond unity is the ongoing shift of my attention. This is *Identity*. Whatever part of unity I monitor, it is me—*(my) identity*—who monitors. Just as the opaqueness between wholeness and unity is necessary and constitutive of both, so is the semi-opaqueness of unity to identity. We attend to a signal of pain when it reaches beyond whole to identity. To function optimally, identity, the highest-level system, shifts its monitoring from one to another next-to-highest-level system. This shifting is a

[5]For a brief overview of gate control theory, see https://en.wikipedia.org/wiki/Gate_control_theory

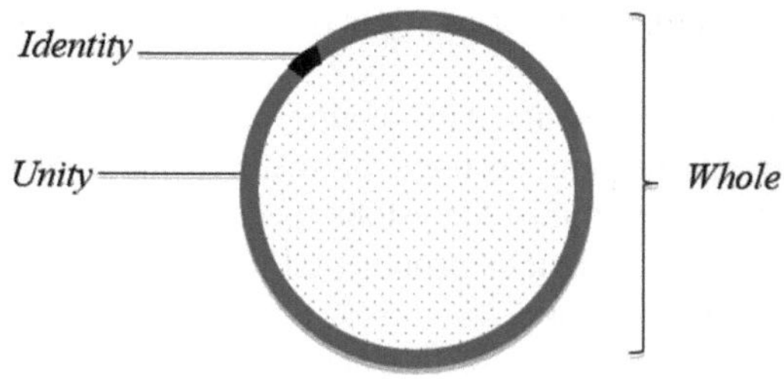

Fig. 8.1 Identity-unity-whole
Whole is all distinguishing activity of a system and its subsystems
Unity is all possible monitoring of *Whole*
Identity is the sequential shifting around the rim of *Unity*

switching from , to Γ in any of the monitored systems (see Fig. 8.1). There is discontinuity of what is monitored but continuity in the monitoring itself, unless that system is off (during sleep, for instance). This continuity-discontinuity obtains at all levels. At our most conscious level, this continuity-discontinuity sometimes feels like an ongoing crossing of portals. A lifetime ago I was sitting in traffic. In this lifetime I am sitting at home by the chimney. In a future lifetime I will be falling asleep in bed.

Identity-whole-unity, then, results from nested distinguishing and oscillating systems, and continuity-discontinuity is the foundation of their integration. Recall the electric bell. It rings by oscillation. For a system monitoring the bell—say, a person—this is a continuous sound. On a higher level, the bell rings and then goes quiet. It rings intermittently. This is a higher-level oscillation. On yet a higher level, a person monitoring the bell may report that it rang constantly in the early morning, then went quiet, and then rang constantly in the afternoon. On a higher level still, the report may be that it rang a lot on Monday, was mostly quiet Tuesday, rang a lot on Wednesday, and so on. This illustrates ever higher levels of continuity-discontinuity. This week compared to last week and next week, this month compared to last month and next month, and so on. One important corollary is the possibility of confusing levels. Someone who heard the bell ring constantly during the morning may wonder if it has gone quiet in the afternoon due to a malfunction. Another corollary is that what is optimal functioning at one level may register as malfunction at another. So, for instance, medical cures for some ailments aim at dialing down the immune system, because it's appropriate response turns out to be excessive, and thereby potentially lethal for the system. This is a risk inherent to nested systems that are relatively autonomous. The point is that identity functions with intermittence.

In sum, identity-whole-unity explains how [body-subjects] made up of [body-subjects] experience homogeneity by means of nested levels of continuity-discontinuity. Recall that this is another name for distinction. Remember how our unfortunate victim constructed his environment by nesting sets of continuity-discontinuity. Crucially, for our purposes, continuity-discontinuity points to self-reference and the imaginary Boolean values in *LoF*. These are disallowed by logic, so the fundamental intelligibility of living being in terms of identity-unity-wholes overtakes logic.

We are ourselves environments. This analogy goes both ways because, as has been shown, environments are nestings of distinguishing distinctions. So, for instance, it may be helpful to think of keystone species such as wolves and elephants as playing the role of unity in their environment. Likewise with the carbon cycle that is so fundamental to the biosphere.

But this also points to the limit of this analogy. Whereas [body-subjects] have *identity*, particular environments do not, nor does Gaia. This is why neither environments nor Gaia are [body-subjects]. This is in sharp contrast with living beings, [body-subjects] whose character as identity-unity-wholes explains their emergence and growing complexity, and the increasingly complex [objects] they make sense of.

In addition to self-appropriation, as the reader was invited to carry out above, can identity-unity-whole be verified empirically? Our muscles are structured and function as identity-unity-wholes, thereby providing us with a helpful image. Muscles remain beyond the horizon of our awareness until we feel pain, which becomes available at *unity,* and we advert to at *identity*. The pain originates at the level of *whole*, for instance as a result of a tear following strenuous exercise, and involves metabolic processes in fascicle, single muscle fiber and myofibril. This is what raises to the level of *unity* and becomes available to *identity* (Fig. 8.2).

Clearly, the functioning of muscles is readily understood in terms of identity-unity-whole. An instance of isomorphism is the injunction to seek intelligibility of other aspects of living being in terms of this fundamental heuristic. Similars are similarly understood.

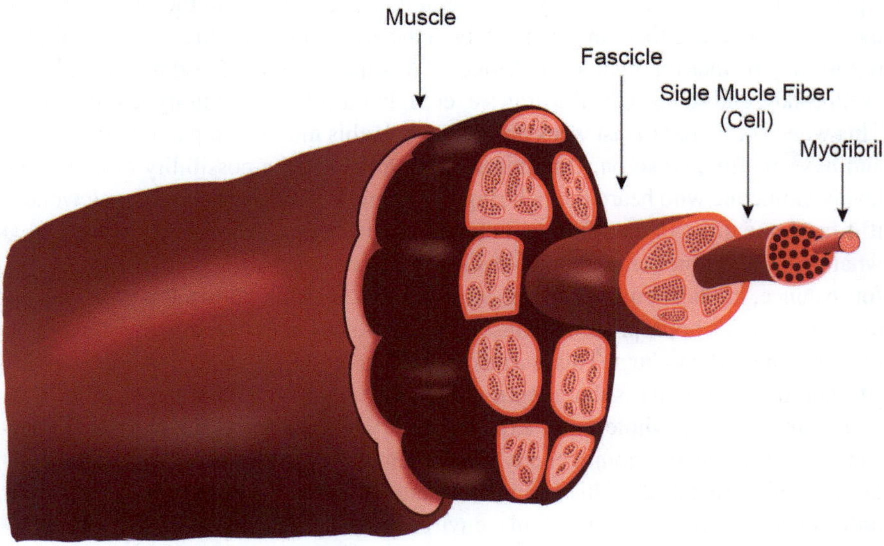

Fig. 8.2 Muscles levels or organization. (The author)

A moment's thought confirms that isomorphism itself is made possible by our own functioning as identity-unity-wholes. This is what allows us to shift our attention from one nested aspect or component of muscles to another, each as a thing, to then integrate them as a further thing, one that is ours, and to then keep this set of images in mind as a template to think, for instance, about our driving. This is the key underlying theme of our next topic, our integration as [body- subjects], which is our nature as things on the side of the object.

Skin and Membranes

One way to immediate advert to identity-unity-whole is through our skin. Briefly sit quietly and comfortably and close your eyes. Deliberately exhale all the air in your lungs, hold for a count of three, then inhale. Repeat this a few times and then breathe naturally. Now advert to the skin throughout your body. Link it to your breathing. Notice the sense of unity that arises. Dwell in this sense. If your awareness shifts to mental activity or data of sense organs, gently bring it back to the feel of your skin. Anything you are aware of belongs to unity. The shifting awareness is identity.

We are multicellular organisms. I shall return to this topic in the final chapters. Our focus here is on the condition for this possibility and its most fundamental consequences: skin. According to Nina Jablonsky, multicellular organisms began to emerge in the sea about 2.1 million years ago. First there were colonies of single cells, like contemporary choanoflagellates.[6] The cells facing the water had a second membrane and small tail-like structures that allowed the colony some movement. As these colonies grew larger and denser, surface cells could no longer absorb sufficient nutrients and oxygen to supply the entire colony. So, a simple mouth evolved that led to a pocket lined with specialized cells. This lining, a primitive type of epithelium, absorbed and distributed nutrients and oxygen within. This was the first pharynx, or primitive gullet. As organisms grew bigger, so did the problem of distribution. To meet their metabolic needs, different types of cells began to evolve—the first tissues with specialized functions such as sensory cells and nerve cells. Increased complexity within also called for better outer protection. Primitive invertebrates like sponges and jellyfish display undifferentiated outside protective layers, but in more advanced species such as *Caenorhabditis elegans*, a simple epidermis

[6]This paragraph follows the account in Jablonsky, *Skin*, 25–26. Findings on the age of multicellular organisms keep changing. Based on the work of a research team led by Professor Abderrazak El Albani of the Institut de chimie des milieux et matériaux in Poitiers and reported at iflscience.com, "The record for the oldest multicellular life was broken, nay smashed, 4 years ago with the finding of 2.1-billion-year-old fossils in Gabon." For a summary and link to the scientific paper see http://www.iflscience.com/plants-and-animals/oldest-multicellular-life-revealed-detail/. According to Elizabeth Pennisi—also mentioned in Chap. 6—multicellular life may have arisen much faster than previously thought. She said that "experiments have shown that in the test tube, single-celled life can evolve the beginnings of multicellularity in just a few hundred generations—an evolutionary instant." See Pennisi, "The Momentous Transition."

shields against potential disease-causing microorganisms while providing a chemical environment tuned to the needs of the inner tissues. Unlike our own skin because it is composed of a single layer, this epidermis of invertebrates is the forerunner of the skin of vertebrates.

Even this brief account follows the logic of identity-unity-whole. The distinguishing activity of every single cell in the colony begins to form a whole as unity begins to be achieved by the cells forming the rudimentary epidermis. At this early stage, the behavior of the colony may be likened to a flock, with all participants sharing similar EIs. With the primitive epithelium tuned to the needs of the inner tissues, the emergent unity begins to set the stage for the eventual emergence of identity. Here, too, is the logic of integrator-operator. At every stage in its emergence, the epithelium is an integrator from which new operators emerge. The same is true of other systems that provide unity, such as the nervous system.

This account of the evolution of skin eventually meets up with the account of anticipation in which the interaction of two layers of skin underlies the evolution of scales, teeth, hair, feathers, sweat glands, and even mammary glands—each an integrator-operator that played a key role in our evolution.[7] This touches on the living meaning of our koan

> The skin of a living organism cuts off an outside from an inside.
> Not so the circumference of a circle in a plane.

From single cells with membranes to layers of them in skin, followed by the emergence of various other tissues, it all began with lipid bilayers, the protagonists of biogenesis in the volcanic pool scenario and studied by Luisi in the lab. According to Jesper Hoffmeyer,

> human skin, as the outermost semiotic interface for an organic community of some 50 trillion individual living cells, is therefore an evolutionarily advanced example of a fundamental principle of life—i.e., the organizing principle of the membrane or interface, the semiotic bridge driven by the persistent asymmetry between the inside and the outside relations of biological organization and interaction. In its most basic manifestation, the operation of this semiotic bridge, and the inside-outside asymmetry that drives it, is a biological membrane composed of a simple lipid by-layer mixed with numerous protein molecules. Such membranes encapsulate each and every cell on Earth.[8]

Consistent with his point that the basic unit of life is the sign and not the molecule, Hoffmeyer examined skin in terms of semiosis as the highest conjugate form.

> From birth we humans are in fact skin more than we are anything else . . . . What enters the awareness of the newborn infant is only qualities or differences between qualities—grades of intensities of touch, taste, and smell. In a certain sense, then, the newborn child's skin is a type of brain, in the sense that it is the place where encounters with the world first freeze into the vague structurings of knowledge. And, not coincidentally, the skin and the brain both originate from the same germ material, i.e., the embryo's ectoderm layer.[9]

---

[7] Jablonsky, *Skin*, chapters 2 and 3.

[8] Hoffmeyer, *Biosemiotics*, 28–29, slightly altered.

[9] Hoffmeyer, *Biosemiotics*, 17.

To clarify: "The main function of the ectoderm is to form the central nervous system (brain and spinal cord)."[10] The nervous system itself is a paradigmatic case of identity-unity-whole. Its relation to skin and the digestive is extremely old and deep:

> A mouth is present in all animals and comprises an opening from the outside into the oral cavity and the beginnings of the digestive tract to allow eating... The mouth arose once during evolution. In all animals, the mouth forms from ectoderm and endoderm. A direct association of oral ectoderm and digestive endoderm is present even in triploblastic animals, and in chordates [which includes all vertebrates]... Further support for a single origin of the mouth is a conserved set of genes that form a 'mouth gene program'...[11]

The ectoderm, then, points to the locus of joinings, the seat of relationship, the cogeneration of [subjects] and the cogeneration [body-subject]-[object] or Ba-Bw. In the title of this chapter ("Surfaces within Surfaces") Hoffmeyer's explicitly points to nesting and identity-unity-whole. He alludes to the cultural bias that there is a precise locus of our personality, that it abides deep inside in the brain and, moreover, that everything pointed to by "I"

> is thought to fall outside the purview of natural science and is therefore considered imaginary by many philosophers (Churchland, 1986; Dennett, 1987; Churchland, 1991). Yet, the simple fact that our personhood presupposes our brain does not imply that our personhood is in the brain—and should we feel so compelled as to finally place our personhood in a definite biological locus, why not place it in the skin? After all, this is where we encounter the world around us and, in so many very obvious ways, the skin is the place where all the fun occurs.[12]

The skin is the largest and most diversified organ of the human body. It at once keeps the world away physically and present psychologically. It is, quite literally, how we remain ourselves and how we make sense by assigning meanings to sensations. "The self exists only insofar as that which is inside contains an intentionality toward or reference to that which is outside—an *aboutness*, as it has often been called."[13] This is forness, born with biogenesis and condition for Ba-Bw. Whether or not skin is the locus of personality, it certainly is a principal locus of unity, and, therefore, of identity—for humans, our sense of personhood. We shall return to this point below. Let us go back to the very foundations, the lipid bilayers that form the membranes of all cells, and the deepest foundation for our koan.

---

[10] *Biology Dictionary*, "Ectoderm."

[11] Chen et al., "Mouth Development."

[12] Hoffmeyer, *Biosemiotics*, 17–18, slightly altered. The references are to Patricia S. Churchland, *Neurophilosophy: Toward a Unified Science of the Mind-Brain* (Boston: MIT Press, 1989); Daniel C. Dennett, *The Intentional Stance* (Boston: MIT Press, 1989); and Paul M. Churchland, "Folk Psychology and the Explanation of Human Behavior," in John D. Greenwood, ed., *The Future of Psychology* (Cambridge: Cambridge University Press, 1991).

[13] Hoffmeyer, *Biosemiotics*, 26 (slightly altered).

Mem-brains

Membranes are the very heart—or rather, the very limen—of distinguishing distinctions. According to cell biologist Bruce Lipton, they are also the brains of cells. Embedded in the lipid bilayers are Integral Membrane Proteins or IMPs. These are basically of two kinds: receptor proteins and effector proteins. Receptors function as "molecular 'nano-antennas'" that respond to specific environmental signals. Some extend inward from the membrane to monitor within the cell, suggesting that identity-unity-whole starts with simple cells.

> When a receptor protein binds with an environmental signal, the resulting alteration in the protein's electrical charges causes the backbone to change shape in the protein adopts an "active" configuration. Cells possess a uniquely "tuned" receptor protein for every environmental signal that needs to be read.[14]

So, the interplay Ba-Bw starts with the earliest membranes, as discussed in relation to biogenesis, and Luisi and Bibtol. What these receptors respond to constitutes the organism's *Umwelt.* What Rovelli said in terms of thermodynamics applies at the most fundamental level of aliveness: "Probability does not refer to the evolution of matter in itself. It relates to the evolution of those specific quantities we interact with. [Thus,] the profoundly relational nature of the concepts we use to organize the world" began billions of years ago with the earliest membranes.[15]

Only very recently have scientists realized the importance of "hundreds of complex information pathways between the membrane's reception of environmental signals and the activation of the cell's behavior proteins. The study of signal transduction is catapulting the membrane to center stage.. .." [16] According to Lipton, some receptors respond to physical signals, while others, like tuning forks, respond to energy fields—light, sound or radio frequencies. The idea that only molecules impact a cell is outdated.

> The membrane's receptors are the equivalent of sensory nerves, and the effector proteins are the equipment of action-generating motor nerves. Together, the receptor-effector complex acts as a switch, translating environmental signals into cellular behavior.[17]

The most significant finding amounts to a radical paradigm-shift: *the cell's operations are primarily molded by its interaction with the environment, not by its genetic code.*[18] The DNA blueprints stored in the nucleus are remarkable molecules accumulated over 3 billion years of evolution. But they do not "control" the operation of the cell. Genes cannot preprogram an organism's life because survival requires the ability to adjust to ever-changing conditions. "The membrane's function of

[14] Lipton, *Biology of Belief,* 83.

[15] Rovelli, *Seven Brief Lessons*, 56–57.

[16] Lipton, *Biology of Belief,* 84.

[17] Lipton, 83–84.

[18] Lipton, 86 (emphasis in original).

interacting "intelligently" with the environment to produce behavior makes it the true brain of the cell."[19]

Typically, the nucleus of a cell is regarded as its "brain." A simple test is to compare what happens to a cell when the nucleus is removed and when its membrane is removed. An enucleated cell can survive for up to 2 or 3 months and in conditions that are anything like a brain-dead lump of cytoplasm.

> These cells actively ingest and metabolize food, maintain coordinated operation of their physiologic systems (respiration, digestion, excretion, motility, etc.), retain an ability to communicate with other cells, and are able to engage in appropriate responses to growth and protection-requiring environmental stimuli.[20]

In sharp contrast, if the membrane is destroyed, the cell dies, precisely as if our brain had been removed. Even if the membrane is left intact and only its receptor proteins are destroyed, the cell becomes "brain-dead" as it ceases to receive indispensable environmental signals. The same occurs if receptor cells are left intact and the effector proteins are immobilized.[21] This is why Lipton playfully refers to *mem-brains*. Their collective power results from this:

> By examining these basic units of perception, we have engaged in an ultimate reductionist exercise, taking the cell down to its fundamental nuts and bolts. In this regard, it is important to note that at any given time there are up to hundreds of thousands of such switches in a cell membrane. Consequently, the behavior of a cell cannot be determined by examining any individual switch. The behavior of a cell can only be understood by considering the activities of all the switches at any given time. That is a holistic—not reductionist—approach . . . .[22]

This holistic approach regards individual cells as identity-unity-wholes that form part of identity-unity-wholes, or [body-subjects] made up of [body-subjects].

Lipton recalls his shock in the early 80s on discovering the similarity between his own description of membranes and that of a computer chip: "*The membrane is a liquid crystal semiconductor with gates and channels. A chip is a crystal semiconductor with gates and channels*"[23] A few years later, an article in *Nature* reported on an experiment combining a cell membrane and a piece of gold foil demonstrating that "the cell membrane not only looks like a chip but also functions like one."[24] A biological cell membrane had been successfully turned into a digital-readout chip. This brings us squarely back to *LoF* and nested sets of switching systems oscillating between on and off. Single cells, tissues, organs, and organic systems—all are nested switching systems, identity-unity-wholes within identity-unity-wholes. The

---

[19] Lipton, 86. Though it may seem a trivial objection, as has been pointed out, the environment is not 'ever-changing' but cyclical in nature, with occasional and gradual changes. This investigation as a whole speaks to the fundamental importance of this point.

[20] Lipton, 65–66.

[21] Lipton, 86–87.

[22] Lipton, 87.

[23] Lipton, 90, emphasis in original.

[24] Lipton, 91 (reference is to B. A. Cornell et al., "A Biosensor That Uses Ion-Channel Switches," *Nature*, 387 (1997): 580–583.

very nature of cells, then, is the meeting point between @@L—oscillating Boolean values—and @@A—joinings. The evolutionary account of membranes points to the huge selective advantages of joinings:

> At the cellular level, the story of evolution is largely the story of maximizing the number of basic units of "intelligence," the membranes receptor/effector proteins. Cells became smarter by utilizing their outer membrane surface more efficiently and by expanding the surface area of their membranes so that more IMPs [Integral Membrane Proteins] could be packed in. In primitive prokaryotes, the IMPs carry out all of its fundamental physiologic functions including digesting, respiration and excretion. Later in evolution, portions of the membrane that carry out these physiologic functions go inside, forming the membranous organelles that are characteristic of eukaryotic cytoplasm. That leaves more membrane surface area available to increase the number of perception IMPs. In addition, the eukaryote is thousands of times bigger than the prokaryote resulting in a tremendous increase in membrane surface area, i.e. a whole lot more room for IMPs. The end result is more awareness, which translates to greater survivability.[25]

This is where Lipton's account of membranes and Jablonsky's account of skin meet. Cell membrane size reaches a critical limit after which it cannot get larger. This happened after 3 billion years of evolution during which single cells were the only organisms on the planet. Then "cells started banding together with other cells to form multicellular communities through which they share their awareness . . . ."[26] The preceding 3 billion years of signaling practice give way to new stages of identity-unity-whole, increasing the complexity of both [body-subjects], their [objects], and the interactions between them:

> The limbic system offered a major evolutionary advance through its ability to sense and coordinate the flow of behavior-regulating signals within the cellular community. As the internal signal system evolved, its greater efficiency enabled the brain to increase in size. Multicellular organisms gained increasingly more cells that were dedicated to responding to an ever-widening variety of external environmental signals. While individual cells can respond to simple sensory perception such as red, round, aromatic, and sweet, the extra brainpower available in multicellular animals enables them to combine those simple sensations into a higher level of complexity and perceive "apple."[27]

This extra brainpower allows organisms to advert to new classes of things. This is clear evidence for the correspondence between the notion of thing, on the side of the subject, and our nature as [body-subjects], on the side of the object. Billions of years of elementary distinguishing and ever deeper nestings of distinguishing led to ever higher levels of affordances until an [object] such as "apple" could emerge.

In sum, membranes make cells nodes of causation, as explained in relation to Autopoiesis. This is how cells distinguish. That being so, multicellular organisms are multiple loci of causation, and identity-unity-whole is how they are organized. All cells, and all systems of cells, form distinctions that distinguish. The transcendent role of membranes, in stark contrast to the cell nucleus, overwhelmingly calls for the shift from substantialism to liminalism. If cells are nodes of causation, it is

[25] Lipton, 87–88.

[26] Lipton, 88.

[27] Lipton, 132.

because membranes are loci of relation and relationship. Skin, and its evolution, provide overwhelming empirical evidence for the notion of Identity-Unity-Whole.

Identity-Unity-Whole and *Unity-Identity-Whole*

Organization within goes together with organization without. [Body-subjects], organized as identity-unity-wholes (Ba), organize the world (Bw) into [objects], that are *unity-identity-wholes*. That being so, it cannot be surprising that Lonergan's account of the organizing insight *thing*, on the side of the subject, closely matches our own organized nature as [body-subjects], on the side of the object. This is Lonergan again:

> . . . the notion of a thing is grounded in an insight that grasps, not relations between data, but a unity, identity, whole in data; and this unity is grasped, not by considering data from an abstractive viewpoint, but by taking them in their concrete individuality and in the totality of their aspects.[28]

To grasp relations among data, this data must first be apprehended—and that takes identity-unity-whole and, grasping that data as a thing—the unity, identity, and whole in the data—is also only possible by an identity-unity-whole. The distinguishing activity required for apprehension, whether we are aware of it or not, is whole; that which we can be aware of is unity; and in identity our apprehension shifts from one aspect of a thing to another as we assemble the totality of its aspects. Lonergan requests his reader to "turn his mind to any object he names a thing. . . ." [29] This injunction, explicitly directed to identity, is what allows [subjects] to construe *things*. Only by the play of unity and identity can we "grasp of unity in a concrete totality of data [from which] follow the various characteristics of things."[30]

So, only because we are identity-unity-wholes can *things* be "conceived as concrete intelligible unities that are extended in space, permanent in time, and yet subject to change."[31] Were it not for the [subject's] identity, which operates with memory and anticipation, there would be no experience of change. Our experience of [objects] would be of annihilation rather than transformation. The apprehension of change results from [body-subjects] projecting their own organized nature as identity-unity-whole, such that an [object's] features change, yet it remains the same, only apparently of its own accord.

In short, because distinguishing distinctions are organized, they organize the world. This is the coproduction Ba-Bw. According to an organism's level of complexity, this Bw ranges from elementary distinctions to *Umwelt* to affordances all the way to the production of products, whether material or conceptual, and onto

---

[28] Lonergan, *Insight*, 271.

[29] Lonergan.

[30] Lonergan.

[31] Lonergan.

thing as a highly differentiated placeholder to organize the world. Lonergan's account of *thing* on the side of the subject, then, accurately mirrors the nature of [body-subjects]. It turns out that [body-subjects] project their own organized nature onto [objects]. This is consistent with my claim all along; namely, that only a living being is a proper thing because it is *for itself*. It takes Ba-Bw for organisms to be organized, and we make sense by organizing—or projecting organization upon—the world. Therefore, organisms, proper distinctions, project their own character onto what they distinguish. If, on the side of the subject, a thing is a *unity-identity-whole*, it is because we are, on the side of the object, identity-unity-wholes.

Likewise, we cannot not project our own nature onto our products. Paradoxically, we produce products by organizing them, to then elide our role in their organization when we make sense of ourselves by analogy to products and their organization. The fashionable instance is to think about organisms in terms of hardware and software. Yet, no computer software is up to the task of producing the hardware by which it runs, and there are sound logical reasons to wonder whether this is possible in principle. Still, the analogy is parroted even as all organisms routinely make copies of themselves.

A key result of the interaction identity-unity is narrative. This is how memory leads to further anticipations, new ב's. Narrative unfolds anticipation, linking the future to the past. This is EI. Recall how Alpha retains memories of previous EIs and, thereby, anticipates future events. The progression of EIs she adverts to is, in effect, a [narrative]. As an identity-unity-whole, she, as a [body-subject], is an integrator-operator forming ongoing sequences of integrator-operators as she constructs herself and her world. The mutual coproduction Ba-Bw at work in learning and development is built in the ongoing play of continuity-discontinuity that constitutes [body-subject] and their [objects].

Human identity-unity-wholes achieve layered sets of narratives, both diachronic and synchronic. This is the matrix form of ב writ large. Diachronic narratives are time progressions (say Jane's life story, who was first a musician and later an entrepreneur). Synchronic narratives hop between diachronic narratives (Jane's joint celebrations today of her birthday, her concert debut, and her 10th year as a CEO). That both diachronic and synchronic narratives are intelligible to us points, first and foremost, to our spirallic nature, and the highly differentiated sense of identity it makes possible. Of course, such a layered sense of identity has evolved, it must be developed, and it must be learned. It is isomorphic with the process of differentiation and specialization of organs and physiological systems such as skin.

Identity-unity-whole, a heuristic structure that focuses on *what* rather than *how*, serves to frame sets of empirical data from different fields. Our autonomic nervous system, for instance, regulates functions without our conscious control such as heartbeat, digestion, and many others such as sneezing, coughing, and trembling. Our immune system, though permanently busy, goes mostly unnoticed. Both are instances of unity-whole that only exceptionally rise to the level of identity. This framing of organization invites isomorphism between the gate control theory of pain. It is a general principle of organization. So, a manager is effective to the extent that she does not need to know. To delegate is to not know what goes on in other layers or sectors of the organization unless it exceptionally requires attention. In organizations as in organisms, this opacity is the condition for the possibility of identity-unity-whole as organizing principle.

Opacity, along with identity-unity-whole, is a principle, a *what,* of organization. Inquiry into it by isomorphism serves as a heuristic from which follow questions and inferences about the nature of organization as such, as well as about particular organisms and organizations. The value of this heuristic may therefore be verified empirically, along with the general principle and heuristic of distinguishing-distinction as highest conjugate form.

Organisms and Organization

The problematic notion of thing was discussed. What are the things art, history, and philosophy? It must now be clear that these are all things on the side of the subject, not 'bodies' or bodies that are real because they already out there now, but instances of Lonergan's *unity-identity-whole* that are real because they are intellectually grasped and rationally affirmed. And what about the side of the object? There is a striking isomorphism between the vast networks of activity that art, history, and philosophy refer to, and networks of cells, whether in ecologies or in multicellular organisms. Like particular environments, or Gaia, their intelligibility must include [body-subjects], their interrelating and their knowing, but these are not themselves [body-subjects]. Art, history, and philosophy refer to forms of organization. This makes them products, perhaps blended with byproducts.

As has now been shown, distinguishing distinctions, or [body-subjects], are in-themselves and for-themselves by being for each other in joinings. It has been noted that this is the primordial meaning of *A* and *B*, or @@A (relationship, aliveness) that grounds *a* and *b*, or @@L (relations, logic), as expressed by the logical conjunctions. The following chapter continues to assemble elements to take up the question, What is *A* and *B* in this primordial sense?

To reach for this understanding, the differentiation first had to be made between the four kinds of things. This, in order to understand the kind of thing we are and how we make sense of the world. We have noted the overwhelming importance of byproducts, even if for the most part they go unnoticed. Reflection on the physics in the section on the nonliving has brought a radical confirmation of *Umwelt.* Our very experience of space and time follows from the fact that we are limited. Perhaps most important of all, we have begun to acknowledge our production of products, their temptation as templates for intelligibility, and the kinds of mistakes that follow from this. This is not as simple as it sounds.

Given the ubiquity of products and their nature as prosthetic—after all, language and mathematics are products—it is no abstraction to say that understanding aliveness must involve both the difference and complementarity between @@L and

@@A. We live our lives in this tension. We must make sense of aliveness in terms of this tension. Indeed, *making sense*, as such, is to work this tension, no less. The transition from distinguishing, the verb, to distinction, the noun, is the working of this tension. A very humble instance of this tension is an electric bell: do I hear a series of discrete sounds or a continuum? This simple question, as we have seen, takes us deep into metaphysics, because the condition for the possibility of repetition is ב, the fundamental character of life.

To paraphrase the koan about the flag, it is not a bell that sounds like a continuum; it is identity-unity-whole that does it. Perhaps the most important elision required for a [subject] to be a [subject], is the key function of identity-unity-whole: at all levels, shifting from analog to digital. As has already been pointed out, the digital is the proper content of distinction (the noun), whereas transitioning from analog to digital is the proper content of distinguishing (the verb). This raises a key point. In inquiry into living being, to elide this elision is to overlook the nature and key function of identity-unity-whole, with its indispensable opaqueness. This is to conceive organisms purely as identity, digitally, mechanically, only in terms of @@L. This was Spencer-Brown's elision of *again* in the axioms—his own elision of the phenomenological—and what led to the digital conception of distinction (the noun), as it obtains in products, to be transposed to distinguishing (the verb) and thence to [subjects]: "We see now that the first distinction, the mark, and the observer are not only interchangeable, but, in the form, identical."[32] The "clear and the distinct" of @@L pertains to the upper scheme. It elides the ground scheme of aliveness and thereby the opaqueness indispensable for its functioning. This opaqueness is inherent to the analogical, and part and parcel of the paradoxical nature of joinings. While oscillation is at the limit of @@L, self-reference subverts @@L as classically conceived. Perhaps this is why *LoF* remains marginalized by logicians despite its re-axiomatization of Boolean Algebra and the imaginary values it justifies. Still, even is @@L includes oscillation and self-reference, it cannot account for the opaqueness of identity-unity-whole and paradox that is the proper content of @@A. That is, the proper content of organization.

Having delved at length into the four kinds of things, and the need to differentiate between them, we can now turn to the functional dynamics of joinings, *con-crēscere*, organization, the proper content of @@A. What is organization? How are its functional dynamics to be understood? What is it to understand when it comes to organization? To seek answers to these questions is to investigate, in its own terms, the nature of aliveness, of living being. And thereby to seek the primordial sense of A and B.

---

[32] Spencer-Brown, *Laws of Form*, 76.

Kate Opens a Door

Kate was a wonderful vibrant monkey with beautiful expressive eyes. Every last hair shone on her. Like her mother Polly, she was full of fun and ingenuity. They both had the same merry *eeolk*! One day she surprised us: she had learned to open a door. She had figured out how to work the knob. It was obvious because she could throw the door open anytime she wished. But she did not imitate us. She did not do it with her hands. Instead, she hung vertically from the knob with her tail, head near the floor, resting hands and feet on the door. Then, holding onto the knob with her tail, with a single impetus away from it with all fours, her body flew up in an arch until she was parallel to the floor, her head reaching the height of the nob. At this point she turned the nob with her tail. The door flew open with Kate careening on it hanging by her tail and calling out a triumphant *eeolk*!

Charlie Jumping

Sitting on the floor of the ante-cage in Room 1 watching the monkeys, all of a sudden Charlie starts to jump up and down on the near shelf. He is 6 ft away, just across from me. He looks intently at something on the surface of the shelf, slightly above my eyelevel from where I am sitting. He crouches, looks intently, and then jumps. He stops, again looks intently, and then does some more jumping and some more looking. For the life of me, I cannot work out what he is doing. An insect? That's unlikely. Monkeys don't demonstrate at insects. If they demonstrate at something it is because they regard it as dangerous. So, they keep their distance, even if it is small. They certainly do not alternate between demonstrating and sitting quietly by. When he sits he does something with his hand, I cannot tell what. He goes through the whole cycle several times. Jumping up and down, looking intently, doing something quietly. Jumping up and down, looking intently, doing something quietly. I can no longer bear the suspense. But it is impossible to make sense of what he is doing unless I know what has captured his attention. I get up slowly, tentatively. I don't want to distract him. As I approach, he glances up at me but continues as before. I arrive by him as he is finishing the jumps. Though the wire mesh is between us, our faces almost touch each other. Looking keenly again, he picks at something. I cannot see anything! Then he jumps up and down again. What is this about? What is he after? Even now, I cannot see it. There is nothing there! Now I really am baffled. He stops jumping. He reaches with his hand and, as he does, just before he covers it, I see it: a tiny seed. Is that it? But then, why all the jumping? He tries to grip it several times and then jumps up and down. Still, there on the shelf, while he jumps up and down, is the seed. I see! He is trying to get hold of this tiny seed, but it is so small and slick that it slips between his fingers. After pinching at it delicately a few times his frustration mounts, becomes unbearable, and so he vents by jumping up and down. Then he stops and tries again. He fails and jumps. It's driving him nuts! He keeps at it. At last, he succeeds at prizing it up and nibbles it in an instant. Charlie's expressiveness is thrilling.

To this day, I wonder why he didn't simply lick the seed off the shelf or pick it up with his lips. Maybe his tongue could not reach. The monkey's hand thumbs are small and only laterally opposing. They can only grip sideways with them against the base of their index finger. Charlie could not pick up the seed as we would, with the tips of thumb and index finger. The best he can do is an awkward sideways pincer action as if we pressed our thumb against the mid-knuckle of our index finger. To reach further as he does this, he is actually twisting his wrist awkwardly to use the back of his thumb and finger. He can only do this when the whole of his index finger touches the shelf, which is what hides the seed from me. Because of their difficulty picking up small objects with their hands, monkeys use their lips or tongue. This is how they catch a spider. Frustrated as he was, and for reasons best known to himself, Charlie persisted in trying to pick up this slippery customer by hand. This tiny incident made me realize just how much our understanding of the monkeys depends on us being deeply and thoroughly familiar with the situations in which we observe them.

Chapter 9
Living Being

No
soy
contradicción
soy
No (Vicky Nizri[1])

Aliveness is in the We. *Entelecheia* is always already in the plural. In these two final chapters, it is shown that, from plurality and the nature of living beings as joinings, our ethical nature follows. To understand living being *qua* ethically is to understand it its own terms. The full arch of the argument is presented, even as several important topics must be glossed over.

A key point that was reached when treating the four kinds of things is that organizations are products. This relates closely to organization in organisms, which pertains to the proper content of @@A (aliveness). If @@L (logic) unfolds the properties of distinction, @@A unfolds the properties of the dynamics of plurality. If the foundation of @@L is repetition, the foundation of @@A is relationship. And relationships make up both organizations and organisms. The similarities and differences between them will be central to the discussion that follows. In both cases we have pluralities of individuals, along with their interrelationships. In both cases, this is the proper content of @@A. Recall that the term *plurality* was reserved for this instance of multiplicity: multiple individuals, actual distinguishing distinctions in functional closure. These pluralities may be of individual cells in a multicellular organism or in a primate group. At work in any plurality is *con-crēscere*—growing together—, the proper content of @@A. This will bring us to the primordial nature of *A and B*.

In the previous chapter, lines of research in contemporary biology were described that center on joinings. We now examines the dynamics of joinings, the form of relationships, and implications this has for inquiry into aliveness.

[1] Nizri, *Lilith*, 98. "No/ am I/ contradiction/ I am/ no" (trans. Daniel Mayer).

© The Author(s), under exclusive license to Springer Nature Switzerland AG 2023

D. C. Mayer-Foulkes, *The Nature of Living Being*, Biosemiotics 26,
https://doi.org/10.1007/978-3-031-24789-7_9

Boundaries

Inherent to plurality is the fundamental problem of boundaries. "Searching for a perfect all-encompassing definition of grouping," say Kraus and Ruxton (in this case about animal collectives), "can easily result in a sterile list of criteria that are hard to apply and often rather arbitrary."[2] We are back to the arbitrariness inherent to distinction, in this case establishing the homogeneity between members (A) and non-members (Ā) of a group. Animal group behavior is so diverse that, to make sense of grouping, they opt for a general framework. This points to a general problem inherent to all pluralities because "the division of animal species into group-living and solitary ones is largely artificial." In most species, association with conspecifics takes place at certain times and not others.[3]

> When a frequency distribution of group sizes is recorded, it often becomes apparent that singletons (or very small groups) are the most frequently observed ones. This is a paradox to which we will return.[4]

Plurality is inherently dynamic. The distinction group-living/non-group-living is somewhat arbitrary because group-living involves both synchronic and diachronic aspects, individuals associated both in space and in time. Diachronically belonging to a species, all organisms belong to a collective. Synchronically, the authors adopt the definition *elective group size* (Fig. 9.1).

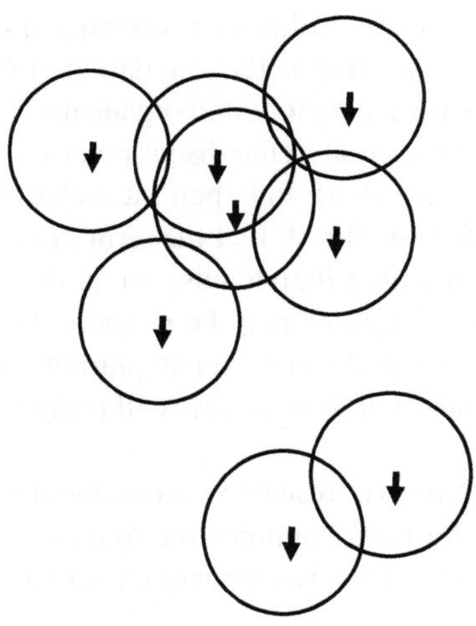

Fig. 9.1 Elective group size concept. Each individual, symbolized by an arrow) is the center of a circle. The radius represents the maximum distance of effective communication between individuals. . . . Group size is the number of individuals inside overlapping circles. (based on Krause and Ruxton, 3)

[2] Krause and Ruxton, *Living in Groups*, 2.

[3] Krause and Ruxton.

[4] Krause and Ruxton, 4.

This definition underscores the paradox of plurality. Members of a group may disperse beyond sight or earshot of each other, but they do not thereby cease to be a group. A definition of a group must accommodate fluctuations because membership—the ongoing process of grouping, ungrouping, and regrouping—is inherent to their intelligibility. Recording distribution of group sizes show that singletons or very small groups are the most frequent configuration. This is a paradox.[5]

What lies behind this paradox? Group-living individuals attract each other and seek group formation when they become separated. However, social attraction does not often involve physical contact. Approach is generally to within about a body length. In fact, direct physical contact tends to result in avoidance behavior. Such attraction/repulsion is observed over most taxonomic groups.[6]

Indeed, attraction/repulsion is inherent to joinings. These *two forces* are the heart of joinings and a key aspect of the proper content of @@A. They are inherent to the nature of relationship. These forces are expressed in multiple ways. One of these is trade-offs:

> ...[T]he inter-individual distance is a trade-off of repulsive and attractive forces between individuals that are, in turn, a reflection of the costs of close association with conspecifics (such as competition) and the benefits (such as dilution of predation risk).[7]

Clearly, a group is not a 'body' already out there now, but a thing intellectually grasped and rationally affirmed. Inherent to the thing *group*, then, and to *plurality*, is paradox. Paradox too, then, is part of the proper content of @@A, as expressed in the two forces and in trade-offs. Specific instances of these, therefore, are the borrowed or contingent content of @@A, as displayed in the following table of empirically identified reasons for animal groups, which in turn suggests the many possible reasons for joinings throughout the living world (Fig. 9.2).

Perhaps the most fundamental problem with joinings is that they remain outside our horizon, even when they are right under our noses. Western botany, with its recent discovery of kinship and friendship relationships among plants, long known to indigenous peoples, seems ever more like zoology.[8] Peter Wohlleben notes the interdependence between trees of the same species interconnected through root systems because "it appears that nutrient exchange and helping neighbors in times of need is the rule." From this, he concludes:

---

[5] Krause and Ruxton.

[6] Krause and Ruxton, 139.

[7] Krause and Ruxton, 140.

[8] See for instance, Robin Kimmerer, *Braiding Sweetgrass*.

Benefit	Species
Anti-predator	
Defense, communal	Lions
Defense against parasites	Horses
Encounter-dilution	Aquatic insect larvae
Many-eyes effect	Aquatic insects
Predator, confusing	Gulls
Predator, learning,	Insect larvae
Predator, swamping	Mayfly
Selfish herd effects	(Theory)
Foraging	
Enhancement, coarse-level, local	Geese
Enhancement, fine-level, local	Starlings
Group hunting	African wild dogs
Information center	Swallows
Information, public	Starlings
Mate choice	(Multiple)
Keeping warm	
Reduced heat loss	Mice
Protection from desiccation	Butterfly eggs
Reduced cost of transport	
In air	Pelicans
On water	Ducklings
Under water	Fish

Fig. 9.2 Summary potential benefits of grouping (based on Krause and Ruxton, 7)

> Forests are superorganisms with interconnections much like ant colonies. Plants—and that includes trees—are perfectly capable of distinguishing their own roots from the roots of other species and even from the roots of related individuals.[9]

Trees are social beings. They share food with conspecifics and at times with competitors. There are many advantages to plurality:

> A tree is not a forest. On its own, a tree cannot establish a consistent local climate. It is at the mercy of wind and weather. But together, many trees create an ecosystem that moderates extremes of heat and cold, stores a great deal of water, and generates a great deal of humidity. And in this protected environment trees can live to be very old. . . . Like the heard, they, too, look after their own, and they help their sick and weak back up onto their feet. They are even reluctant to abandon their dead.[10]

Thus, tree behavior forms part of the [social].

[9] Wohlleben, *The Hidden Life of Trees*, 3. See also the outstanding work on trees by Suzanne Simard and on plants by Monica Gagliano.

[10] Wohlleben, 3–4.

Microbial Partnerships

Another account of joinings right under our noses is presented by Marlow and Braakman in the article "Team Players: Microbial Partnerships That Rule the Planet." Their research touches on another aspect of the proper content of @@A (aliveness): division of labor. They point out that decades of research into microbes has focused on individual species and their molecular components. In the lab, single organisms have been isolated to determine their biochemistry and the functions of their genes.[11] Here is a prime example of how the epistemological elision of the [social] involves assumptions with practical consequences and profound implications in research. In this case, it has led to major gaps in our understanding because only a tiny fraction of microbes in the wild are isolated. Species that live in complex natural communities are intertwined in ways not easily replicated in the lab. The coexistence of a multitude of species playing complementary roles goes flatly against the assumption of microbial ecosystems in a struggle for resources with a winner-take-all. Metabolic activity observed in the lab rarely matches the real-world environment, if only because species that can be isolated tend to be more vigorous than those that cannot. These authors observed that "the whole was sometimes more, sometimes less, but always different from the sum of the parts."[12] Very recent advances in biomolecular sequencing and microscopic imaging confirm this:

> The latest findings indicate that collaboration is a critical driver of the biosphere: as individual organisms evolve to share energy, genetic information and metabolic duties, they unlock new ways of life and gain entry to previously inaccessible habitats.[13]

Marlow and Braakman researched the division of labor between Prochlorococcus and Pelagibacter, two microbes that recycle each other's waste extracting otherwise unused energy. This is what Bourke named *interspecific mutualisms*.[14] In a section entitled *The origin of teamwork* Marlow and Braakman arrive at "two startling conclusions . . . increasingly inescapable":

> First, bacteria and archaea are far more diverse than anyone had imagined—the number of branches on the tree of life has exploded. But perhaps more surprising, their genomes are suspiciously small; many do not have enough information to build a fully functional cell or to complete the metabolic transformations that convert food into energy. . . . The entire community has the capacity for a certain function, like nitrogen cycling. All the pieces are there, but to identify a single organism that has all the pieces in its own genome—that's really unusual. The newly discovered cells' genomes often lack the ability to make all the amino acids needed to build their proteins or the nucleotides for constructing their DNA, suggesting they acquire these building blocks from neighboring cells with a surplus.[15]

---

[11] Marlow and Braakman, "Team Players," 35.

[12] Marlow and Braakman.

[13] Marlow and Braakman.

[14] Marlow and Braakman, 37.

[15] Marlow and Braakman, 36, citing Laura Hug, professor of environmental microbiology, University of Waterloo.

In short, energy is extracted from the environment collectively:

> Individual cells perform certain chemical conversions and pass the product down the chain to other cells for subsequent reactions. Sharing cellular building blocks and energetic resources in this way both requires and enables cohabitation among diverse organisms.[16]

The radical importance of this collaboration over 3 billion years is astonishing:

> The Prochlorococcus-Pelagibacter partnership may have emerged out of just a few small genetic changes, but its long-term effects were enormous. When the ancestors of Prochlorococcus and Pelagibacter colonized the oceans between 600 million and 800 million years ago, the waters were still largely devoid of oxygen and rich in iron.[17]

Iron is indispensable for photosynthetic proteins that generate oxygen. But "it cannot dissolve and be wrangled into proteins when oxygen is around." Enter a partnership:

> This catch-22 would have kept photosynthetic organisms from expanding into the open ocean, where accessible iron would become scarce if they moved in and started making lots of oxygen. But Prochlorococcus's organic carbon waste products—fueled through growth alongside Pelagibacter—had a remarkable ability to bind iron, increasing its availability even in the presence of oxygen. Thus, we hypothesized that through the interplay between their organic waste and the critical iron, Prochlorococcus and Pelagibacter ultimately helped to pave the way for photosynthesis to oxygenate our planet's oceans. Life on Earth would never be the same.[18]

They conclude that "microbial partnerships turn out to be more common and influential than scientists could have ever imagined."[19]

Beyond their title, Marlow and Braakman refer to "collaboration," "cooperation," "ulterior motives," "roommates," and draw they direct parallels to human society:

> In much the same way that the global economy capitalizes on local strengths and the exchange of goods, groundwater and deep-sea microbial communities use division of labor to efficiently extract resources from sparce supplies, making harsh environment livable.[20]

To label 'anthropomorphic' such assertions of fundamental aspects of the [social] is to entirely miss the point. The deep isomorphism that obtains in all instances of plurality derives from the proper content of @@A. As was shown, Eukaryotic cells originated from a partnership. Except for Prokaryotes, such as some marine diatoms in plankton or a microbe that turns up in our gut, all life forms we are familiar with are Eukaryotes: plants, fungi, invertebrate animals, fish, amphibians, reptiles, birds,

---

[16] Marlow and Braakman.

[17] Marlow and Braakman, 39. Analogous problems occur in medicine. According to James Meigs, "In cancer and other fields, doctors have discovered that only rarely does a single drug serve as a silver bullet. But a combination of drugs tailored to the individual patient—a kind of silver shotgun blast—can be effective. Of course, the FDA approval process isn't set up to test such combination therapies" (Meigs, "Thank God for Big Pharma," 22).

[18] Marlow and Braakman, 39.

[19] Marlow and Braakman, 33.

[20] Marlow and Braakman, 36.

and mammals.[21] As we have already seen, molecular signaling over billions of years is what allowed them, and partnerships between them, to emerge. Thus, all Eukaryotes share metabolic features for teamwork and collaboration. If anything is anthropomorphic, it is to regard the aforementioned notions as such because, properly speaking, they are *eukaryotic* if not *biomorphic*. This is how our own molecular signaling originated, the very one underlying our own Ba-Bw. This is where our own grasp of the proper content of @@A springs from. Marlow and Braakman's empirical research and Bourke's principles, then, all point to the dynamics of joinings that are the proper content of @@A. These include the two forces of attraction and repulsion, trade-offs, and division of labor. All function as integrator-operators, and indeed, integrator-operator is itself part of the proper content of @@A.

Kinds of Wholes

Multicellularity poses a first class ontological/epistemological conundrum. A collective of [body-subjects] transitions to become a colony of [body-subjects] which in turn become a new higher-level [body-subject]. Physiological complexities aside, this involves two shifts between different kinds of wholes. Single unicells are identity-unity-wholes. Not so colonies, which are a collective of identity-unity-wholes. But some collectives then transition into higher-level multicellular organism, a new, higher-level identity-unity-whole. Any identity-unity-whole, be it a unicell or a multicellular organism, is a living being. Not so a colony. A living being is a whole *for itself* because biomorphism-a and biomorphism-w (Ba-Bw) co-create. A colony is not, it is a whole only for us, in our own Bw. One is a proper whole; the other is not. I shall refer to wholes that are made up of wholes as WW or organic wholes, and to collective wholes made up of Wholes as wW or social wholes.[22]

Multicellular organisms, then, emerge through two transitions, from organic wholes to social wholes, and then to a higher-level organic whole (WW➔wW➔WW). Both these transitions involve blurry and, thereby, problematic boundaries. But the second is vastly more problematic, both biologically and epistemologically, because it involves the emergence of a new higher-level identity-unity-whole. These are discussed below.

The proposition that the condition for the possibility of biogenesis was plurality, implies that life emerged as an instance of the second transition, from a social whole to an organic whole (wW➔WW). Strictly speaking, then, there has never been a singleton organic whole (W). A living whole *for itself*, a proper whole necessarily belongs to either an organic (WW) or a social whole (wW).

---

[21] Janke, *A Correlated History of Earth.*

[22] Stuart Kauffman refers to Kant, who "saw organisms as wholes. The whole existed by means of the parts; the parts existed because of and in order to sustain the whole" (*At Home in the Universe*, 37). This is a suggestive definition, but it applies to both kinds of wholes discussed here

To this we may add a third kind of whole that I shall refer to as a mechanical whole. That is, a nonliving whole made up of parts (wp). An organic whole is for itself (Ba-Bw), a social whole is for itself in a narrow sense to be discussed ("Ba"-Bw), and a mechanical whole is not for itself, it is only for a living being (Bw).

Social wholes (wW) and organic wholes (WW) are both instances of joinings, and this is what makes multicellularity a metaphysical conundrum. Joinings themselves are a conundrum, let alone the transition from one kind of joining to another. So, the current discussion will later lead to the relation between the [social] and identity-unity-whole, and to how this is at work in major transitions in evolution. These transitions involve nesting, as has been mentioned, which is yet another aspect of the proper content of @@A.

> The hierarchy of life is the central landscape of collectivity in the living world—eusocial societies composed of multicellular organisms, multicellular organisms composed of single (eukaryotic or prokaryotic) cells, single (eukaryotic) cells composed of (prokaryotic) cells, cells composed of gene networks, and gene networks composed of replicating genes.[23]

Evolutionary transition are diachronic accounts that correspond to synchronic account of nested identity-unity-wholes. According to Bourke, each of these transitions went through three principal stages: social group formation, social group maintenance, and social group transformation. Clearly, these stages speak to both emergence and development (diachronic) and to nesting (synchronic). Notice that the transition from one stage to the other occurs either by emergent probability, or through development (integrator-operator) or, most likely, by a complex blend of these two.

Humans as individuals are at a very mature second stage, and human society seems to be at an obnoxious teenager second stage, with very early signs of a potential third stage. This is why, in terms of social group maintenance, we embody two radically different kinds of joinings. We are each a multicellular individual made up of trillions of [body-subjects] organized into an identity-unity-whole. And we also belong to collectives of [body-subjects] that form organizations. Bourke's claims that both these kinds of joinings may be understood in terms of inclusive fitness shows how closely interrelated they are, even as they differ radically from each other.

Let us briefly turn to the transition into multicellularity. Choanoflagellates are a favorite organism to study this transition because these single-celled eukaryotes can join together into multicellular structures and later revert to being single-celled. Mostly ocean dwellers, they use flagella for locomotion (though unlike *E.coli's* these do not spin but oscillate). Most strikingly—and this underscores our common heritage—the genes they use for signaling to join together and separate are nearly identical to those used to signal by our own cells.

Multicellularity is a paradigmatic example of joining. Current research shows that it emerged far more easily than was once thought. According to Elizabeth Pennisi, "hints of multicellularity date back 3 billion years when impressions of

[23] Hanschen et al., "Individuality," 255.

what seem to be mats of microbes appear in the fossil record."[24] Only in the late Proterozoic (now called the Vendian or the Ediacaran), 650–540 million years ago, do the first macroscopic fossils of soft-bodied organisms emerge.[25] So, colonies of cells have existed during 80% of the time since life emerged, yet only in the last 20% of that time did the first sponges emerge. With the beginning of cell division of labor, endless possibilities emerged with specialized organs. Processing of food, the number one priority, became increasingly efficient, and, with less energy spent on feeding and eliminating waste, organisms grew larger, denser, more active, and survived in hitherto inaccessible, threatening, or deadly environments. Studies involving cladistics and comparative anatomy show the most primitive sponges were formed by colonies of cells like choanoflagellates. So, eventually a simple mouth evolved, a primitive epithelium with a pocket lined with specialized cells; an outer protection formed a simple epidermis conducive to an inner chemical environment and limited movement with the coordinated use of flagella. With new solutions for the distribution and absorption of nutrients and oxygen, and the elimination of waste, division of labor, which is and continues to be at work throughout, led to further specialized tissues and organs.[26]

> Division of labour is a defining feature of complexity at all levels of biological organization. If individuals specialize to perform certain tasks, more complex social groups can evolve. In the extreme, if the different individuals become dependent on the tasks performed by others, then and new 'higher level' individual may emerge. Examples include genes with different functions in a genome, cells that form distinct tissues in an animal and castes that carry out different tasks in social insect societies. Consequently, to understand why complex life has evolved, we must understand the evolution of division of labour.[27]

In short, division of labor is an integrator-operator and also part of the proper content of @@A. Cooper and West, who offer the first mathematical model of division of labor, point to the kinds of joinings in which it is most likely to emerge:

> (1) division of labour is usually—but not always—favoured by high efficiency benefits to specialization and low within-group conflict; and (2) natural selection favours extreme specialization, where some individuals are completely dependent on the helping behaviour of others.

That is, *high benefits with low conflict tend to favor extreme specialization with high dependence*. Trade-offs are part of the proper content of @@A. So, "division of labour can lead to mutual dependence between different individuals." Therefore, it can "drive major evolutionary transitions, such as those to multicellularity and eusociality."[28]

Briefly, let us consider the relation between identity-unity-whole and flow. Slime molds like *Physarum polycepthalum* are extremely simple organisms, a system of

---

[24] Pennisi, "The Momentous Transition."

[25] Waggoner, "Introduction to the Vendian Period."

[26] The foregoing account is based on Perry, "Irreducible Complexity?"

[27] Cooper and West, "Division of Labour," 1.

[28] Cooper and West, Abstract.

interlaced tubes. Yet, they solve complex optimization problems such as finding the shortest path through a maze. Formerly classified as fungi, they are now known to be eukaryotic organisms that can live as single cells and fuse into a vast single cell. This cell can spread throughout a maze and, with peristaltic movement, displace water and nutrients throughout the structure. The parts that come in contact with a food source

> release a substance that softens the tube network's gel-like walls, making them widen from their inherent internal pressure. The slime mold moves by expanding along wider tubes and pruning narrow ones—so the enlarged tubes effectively record past food locations, as they influence the organisms overall direction of growth even after the food is gone.[29]

A single look at a picture of slime mold (*Physarum polycepthalum)* shows that its vascularization follows the constructal law.[30] Find a picture online and notice how the design of this vascularization is identical, even in appearance, to river deltas in which riverbeds of previous flooding seasons "remember" the routes with the largest greatest loads of water. These designs also resemble trees, leaves, coral, lungs, our circulatory systems, nervous systems, and even networks of trails in a forest. The key point is that some of these are, or belong to, organic wholes (WW), while others are social wholes (wW). So, we can ask ourselves about the similarities and differences between them.

The key similarity is that both of these kinds of wholes are organized. This is why both can be generically referred to as *an organized whole* (/W). This cardinal similarity allows both to be accounted for in terms of division of labor and Hamilton's law.

The difference between them is that they are different kinds of things. An organism is an organized whole, a living being, whereas an organization, a social whole, is a product. Just as we humans make products of ourselves when we belong to organizations, so do other organisms, with individuals taking up various roles—division of labor. According to Bourke, unique cases in which the benefits from, and dependence on, division of labor are high, can result in major transitions, for instance, the transition from unicells to multicellularity. This is how and why organized wholes turn into social wholes and then into higher level organized wholes (WW→W→WW), a higher-level individual.

But this still begs the question that was posed about these products. What precisely makes a social whole (wW) be a product and an organic whole (WW) be an organism? The difference may be stated in terms of identity-unity-whole. Whereas an organic whole (WW) has identity because it is an identity-unity-whole, the same cannot be said of a social whole (wW). Think of a human organization. At all levels, from low-ranking work groups to the board of directors, individuals are formally assigned functions or roles. This is division of labor. In these roles they decide for, and speak on behalf of, their sector of the organization. Doing so, they carry out the

[29] Fischer, "Slime Memory," 19. For an actual experiment see https://youtu.be/7YWbY7kWesI (retrieved May 2021).

[30] See Chap. 3, section "Liposomal structures and flow."

function of identity. Thus, individuals *lend* their *own identity to* the organization and/or to its sectors. Distinctions, of which judgments are an instance, are drawn at the level of identity by the WW that draws them. An organization, or a sector of it, not being alive, does not draw distinctions, unlike the individuals or organic wholes (WW) who constitute it. So, individuals serve the function of identity on behalf of the social whole (wW). This tracks with the point in the section Identity-Unity-Whole and *Unity-Identity-Whole* discussed above. What is the condition for effective delegating—willing to not know what goes on in the organization? It is to place people—organic wholes (WW)—in positions of authority whose *judgment* can be trusted. Opacity is the condition for the possibility of organization, and this is what allows it to work.

Furthermore, this is why we differentiate between persons and their roles. Similarly, this is why there is no problem when cells in a given tissue are replaced. In an organization, a person *both* is *and* is not the role. This paradox, often stressful for individuals, makes organizations resilient, flexible, and adaptable because roles remain steady while individuals come and go. This is true of all cells, organizations, even species.[31] The principle applies to bodies as the individual remains while cells come and go. But in contrast to organizations (wW), in organisms (WW), identity remains. More accurately, identity-unity-whole remains while cells come and go. Thus, identity-unity-whole is the *what* of the aliveness of organisms, rather than the biochemistry, which is the *how*. In short, this is the fundamental difference between the two kinds of joinings: organic wholes (WW) and social wholes (wW).

Their similarity is of enormous heuristic value. But organic wholes being living beings, and social wholes being products, makes them radically different kinds of things. The fundamental difference between them clarifies the nature of both. And yet, their isomorphism also points out that, when drawing analogies to understand living being, there is far more to learn from comparing them to social organizations than to mechanical products.

The foregoing shows that identity-unity-whole is part of the proper content of @@A, precisely because it is in terms of this content that the two kinds of joinings may be differentiated. It further allows us to trace this back to the fundamental argument. An organic whole (WW) is a distinguishing distinction that can establish bet structure for itself. The 'for itself' of a new, higher-level multicellular organism, pertains to identity. That being so, much of what was said before about biogenesis could be repeated here. The emergence of an organic whole from a social whole that is not for itself and therefore not alive, is a kind of biogenesis. This is how life emerged in the first place. Here, too, the cyclical nature of the environment results in random modifications of the unicells and the colonies they form. At some point, when ב begins, the social whole—a colony—ceases to be, to become a distinguishing distinction for itself, a properly self-regulating anticipative unit. With this, the new higher-level individual is not only itself the object of *natural selection*, but

[31] Recall the discussion in Chap. 3, section "ב and the limitations of narrow Darwinism" on natural selection, or the fortuitous "establishing of bet structure" at the species level, which makes species somewhat alive.

also a [body-subject] that *selects* for itself. As was shown, 'select' in both senses together accounts for aliveness and the emergence of higher levels of anticipation in organic wholes (WW), whereas only the former obtains in social wholes (wW), such as species.

This key underlying difference points to a fundamental and vast difference in the heuristic structure required to investigate them. Organic wholes develop, not so social wholes. So, whereas organic wholes must be understood in terms of integrator-operator—along the lines of EI and the unfolding of anticipation—social wholes must be understood in terms of emergent probability—along the lines of biogenesis. If social wholes do appear to develop, this is because, as has been shown with organizations, individuals lend them their identity. This, by the way, is precisely the case with the three things art, history, and philosophy.

The similarities between the two kinds of organized wholes has long led to analogies being drawn between them. Notable instances are Plato's *Republic* and Aristotle's *Nicomachean Ethics*, which make sense of the state by analogies to the individual and to the family respectively. The method of both is isomorphism. Plato analogizes between an organic and a social whole, while Aristotle does between two social wholes, one that belongs within the other.

As was shown, a key function of identity-unity-whole during the process of distinguishing, is the transition from the analogic to the digital. Eliding this tends to result in treating organisms as digital, and organization generally as @@L when it comes to their formal structure. Organizations are often understood by reducing them to levels of authority, the structure that refers to the *identity* that individuals (living beings that are organic wholes) lend to organizations (products that are social wholes). Eliding the analogical nature of the organic wholes that make up the social whole easily slides into conceiving organization digitally, merely as a set of loci where distinctions are made on behalf of the organization. In fact, the analogical nature of organic wholes—real people in actual organizations—makes up the massive and mostly unspoken paradoxical social dynamics that constitute the organizations. In plain terms, collectives of individuals are highly emotionally charged. They like and dislike each other, help and trip each other up, gossip, confess and betray each other. Contemporary leadership theory and practice now recognizes that the failure to understand organizations like this is at the root of many leadership failures, often with dire consequences.[32] This is why the elision of the analog amounts to conceiving organizations in @@L or mechanical terms. It leaves aside the highly complex functional dynamics within and between the living individuals that conform them—@@A. Elided is the proper content of @@A, including the forces of attraction and repulsion, trade-offs, etc. Elided along with this is the play of analog/digital, transparency/ opaqueness (T/O), in short, the general paradoxical nature of @@A at work in social individuals.

---

[32] Following Heifetz and Laurie, one might say the function of authority is digital and deals with the clear and distinct, whereas that of leadership is analogical and deals with the hazy and ambiguous. Authority is about the known and known unknown, whereas leadership is about the unknown and unknown unknown. See *Leadership Without Easy Answers* and *Leadership on the Line*.

Given the nature of transitions, Bourke must deal with the question of individuals and their emergence. How does he do this? In fact, he does not differentiate between organic and social wholes. For him 'individual' applies to both:

> By 'individual' in this book I mean some stable, physically discrete entity that is composed of interdependent parts acting in a coordinated manner to achieve common goals and is typified by the very property of lacking a high degree of within-individual conflict. 'Physically discrete' here means that the parts of the individual are either physically joined to one another or tend to remain in close proximity. An individual in this sense could be the product of the evolution of individuality at any of the hierarchical levels of life. For example, it could be a single cell, a multicellular organism, or a eusocial society.[33]

He clarifies that his definition may, in some cases, not settle whether some entity qualifies as an individual or not, given that evolution is a process of continuous change.[34] Further, even if blurry edges or a mosaic of features are to be expected, still, he states that definitions in evolutionary biology are useful, even when they go beyond ordinary usage. Bourke notes that some authors widen the meaning of the term 'organism' while others reintroduce the term 'superorganism' for colonies of species of eusocial insects. "The key point is that all agree that there's a special property found in the natural world, here termed individuality, that requires an evolutionary explanation."[34] Indeed, "physically discrete entity" recalls the problems of defining a group. However, like a group, even an individual is a thing and not a body.

True, the two transitions WW➔wW (from organic wholes to social wholes) and wW➔WW (from social wholes to organic wholes) are blurry. So, "interdependent parts acting in a coordinated manner to achieve common goals" can suggest the flock behavior of a social whole (wW) before becoming an organic whole (WW). But the problem is neither blurriness nor shared features but the biological reality of individuals, in the normal sense, as tacitly acknowledged by his proposed alternate use. Thus, Bourke's definition has Hamilton's law in mind, both for WWs and wWs. But it obscures the key difference between a living identity-unity-whole and a product. What, in fact, marks the transition from a wW to a WW? Figure 9.3 (above) illustrates the two transitions into multicellularity following Bourke.

The first transition elicits the paradox of group membership discussed above. The second transition recalls our koan:

> The skin of a living organism cuts off an outside from an inside.
> Not so the circumference of a circle in a plane.

What does the circle in the figure represent? Is this an abstract distinction for us or the skin of an organism? In other words, the illustration is ambiguous precisely where it matters most. It avoids stating the difference between a colony and an individual. This is the very problem discussed at length in relation to biogenesis. This

[33] Bourke, *Principles of Social Evolution*, 8.

[34] Bourke, 8–9.

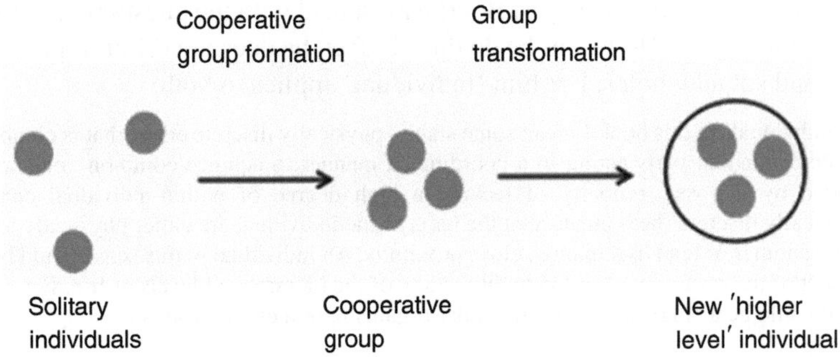

Fig. 9.3 Example of illustration of major evolutionary transitions (METs) in individuality (West et al., "Major Evolutionary Transitions," 10113. The final reference is to Bourke, *Principles of Social Evolution*)

too is a kind of biogenesis, so here too, we must not conflate our own biomorphism-world (Bw) with the biomorphism-aliveness (Ba) of our object of study. Is this object an object or a (new) subject?

Consider it from the point of view of the emerging subject, and therefore its emerging Ba-Bw. An outer layer that maintains a differential between inner and outer states turns a social whole into an organic whole (wW→WW). This is no abstract dotted circle; this is living skin. Even in sponges, the most primitive multicellular animal that hardly exhibits tissue-layer organization, different cell types perform differentiated functions—division of labor. Epithelial-like cells form an outer layer and enclose a jelly-like substance that performs various functions. Like an endoskeleton, it maintains a sponge's tubular morphology. The same is true of plants, whose organs (roots, stems, leaves) are made up of three kinds of tissue (dermal, vascular, and ground). Dermal tissue covers, protects, and controls gas exchange in stems and leaves and water absorption in roots.[35]

In multicellular organisms, then, the outcome of this initial division of labor is a major one: the differentiation of organs. Living skin maintains a differential between inner and outer states. This, very concretely, is how nestings can conform identity-unity-wholes. From random variations, natural selection arrives at optimally organized [body-subjects], identity-unity-wholes with autonomous organs, each an identity-unity-whole with their own supervening identity. In this framing, natural selection is for optimally organized nestings of identity-unity-wholes in layers, or as joinings, made possible by skin. And each of these layers—cells, tissues, organs, [body-subjects]—is an integrator-operator.[36] This is so for any organic whole. Beyond the biological, metaphysically, skin differentiates between two kinds of things, organic and social whole, organisms and products.

---

[35] "Organismal Biology," retrieved October 2020 from http://organismalbio.biosci.gatech.edu/growth-and-reproduction/plant-development-i-tissue-differentiation-and-function/

[36] See Chap. 4, "EI as Integrator-Operator." Everything explained there obtains here.

Like Bourke's 'individual,' my own term 'organic whole' ('WW') is not unproblematic. Given the plural nature of living beings, is there such a thing as a WW? What does it mean to say that a Whole is made up of Wholes? What does the process of integration of a new, larger, more complex Whole render what were previously Wholes? If the condition for their joining is, precisely, that they depend on each other, do they remain Wholes? Are they not now parts? Recall that, for Bourke, 'physically discrete' means that the *parts* of an individual are either physically *joined* to one another or tend to remain in close proximity—here he obviously has social wholes (wW) in mind. To avoid foreclosing the discussion merely by means of a definition, I shall use the term organic parts (/P). Properly speaking, a fully formed multicellular organism is an organic whole made up of organic parts (W/WP or WP). These organic parts were either once organic wholes themselves or they have become sufficiently autonomous to be considered organic wholes. Organ transplants, for instance, are possible only thanks to the relative autonomy of these organs. As we saw, choanoflagellates are organic wholes or organic parts (Ws or/Ps),depending on whether they currently belong to a w/WP or a W/WP. For most organisms, this transition is irreversible given the interdependence that results from division of labor. Organs as organic parts are integrator-operators, and their interdependence is itself a further integrator.

The key point, however, is that all of these—organic wholes (W/), organic wholes within an organic whole (/W) and organic parts (/P)—are not, do not make up 'bodies'. They are things, specifically living things that make up living bodies, *con-crēscere* [body-subjects]. How we make sense of them (Bw) does not take priority over how they are and how they make sense of the world (Ba-Bw). This goes back to the deep link between *unity-identity-whole* and identity-unity-whole.[37] We must make sense of them as *unity-identity-wholes* according to how they are identity-unity-wholes making sense of the world in terms of their own *unity-identity-wholes*. Of course, we do this based on how we are and do this ourselves.

Multicellularity, following the advent of Eukaryotes 1.6–2.1 billion years ago, was a key integrator-operator that set the stage for vast possibilities of sequences of operator-integrators. In contrast to the other two domains—Bacteria and Archaea, together known as Prokaryotes—Eukaryotes are true W/WPs. Their cells have a nucleus enclosed within a nuclear envelope—*skin*.[38,39] About 600 million years ago, the advent of multicellular Eukaryotes led to the Precambrian and Cambrian explosions that massively diversified life-forms. Though these are a tiny minority of organisms, their large size makes their collective biomass about equal to that of prokaryotes.[40] Bourke's notion of 'individual' aside, how to examine the actual acts of actual individuals at the level of *identity*? And how to refer to such inquiry? [Psychology] is a possibility, and we shall later see why it could be appropriate. But

[37] See the section "Identity-Unity-Whole and *Unity-Identity-Whole*" in Chap. 8.

[38] Woese, Kandler, and Wheelis, "Towards a Natural System."

[39] Zimmer, "Scientists Unveil."

[40] Whitman, Coleman, and Wiebe, "Prokaryotes."

deeper issues are at stake. As counterpart to Bourke's [social], I propose the [ethical]. This is how we finally wrench ourselves away from the abstract circles of @@L into @@A and the primordial meaning of *A and B*. Skin makes all the difference.

Liminality and Hymenomorphism

Skin makes distinguishing distinctions possible by allowing differentials to be maintained between inner and outer states. More precisely, skin, as membrane, allows *entelecheia* and is fundamental to biogenesis. Membranes make biomorphism-aliveness/biomorphism-world (Ba-Bw) possible, the distinguishing by which organisms organize themselves and the world. ב, done by membranes, is the oscillation captured by imaginary Boolean values. Let - - refer to the layered play of continuity/discontinuity that constitutes identity-unity-wholes. This icon captures continuity/discontinuity in the multifaceted synchronic and diachronic sense instantiated by membranes, surfaces with open or closed entries-exits that alternatively open and close. It thereby stands for everything membranes are and do that characterizes aliveness. Let - - be referred to as *liminality*.

Further, let *liminalism* and *hymenomorphism* refer to the fundamental metaphysical commitments that follow from liminality. The term is derived from Greek *morphē* ("form, shape; beauty, outward appearance") and Greek *hymen* ("'membrane, especially 'virginal membrane,' as the membrane par excellence; thin skin from the Proto Indian European *syu-men-, from root *syu- 'to bind, sew'"). Apparently, this root is not directly connected to Hymen, the god of marriage, but folk etymology supposes it does.[41] This is particularly apposite given that the condition for the possibility of liminality is joinings.

Hymenomorphism deliberately echoes Aristotle's *hylomorphism*:

> Central to Aristotle's four-causal account of explanatory adequacy are the notions of matter (hulê) and form (eidos or morphê). Together, they constitute one of his most fundamental philosophical commitments, to hylomorphism: Hylomorphism $=_{df}$ ordinary objects are composites of matter and form. The appeal in this definition to 'ordinary objects' requires reflection, but as a first approximation, it serves to rely on the sorts of examples Aristotle himself employs when motivating hylomorphism: statues and houses, horses, and humans. In general, we may focus on artefacts and familiar living beings. Hylomorphism holds that no such object is metaphysically simple, but rather comprises two distinct metaphysical

[41] See https://www.etymonline.com "form" and "hymen." In view of the opaqueness indispensable to the organization of living being, it is notable that *skin* derives from the Old Norse *skinn* "animal hide, fur," and in turn, its synonym *hide* comes from the "Old English *hydan* "to hide, conceal; preserve; hide oneself; bury a corpse," from West Germanic **hudjan* (source also of Middle Dutch, Middle Low German *huden*), from suffixed form of PIE **keudh-* (source also of Greek *keuthein* "to hide, conceal"), from root **(s)keu-* "to cover, conceal."

> elements, one formal and one material. Aristotle's hylomorphism was formulated originally to handle various puzzles about change.[42]

Though Aristotle is properly recognized as the first biologist, the differentiation of four kinds of things suggests that hylomorphism must be replaced by hymenomorphism.[43] His examples when reflecting on change and transformation, cause and effect, whole and parts, often refer to products. This is so, even as his differentiation between products from living beings is like my own: "artificial products... none of them has in itself the principle of its own production."[44] A lump of bronze is matter whose form transitions from lump to statue.[45] Transitions involve nesting: for instance, letters are matter that take the form of syllables; syllables are matter that take the form of words; clay takes the form of bricks; bricks take the form of a house.[46] For Aristotle, "animals are paradigm cases of natural existents, because they have a nature in the sense of an internal causal principle which explains how it comes into being and behaves."[47]

Living beings change in themselves because they are for themselves—this is Ba-Bw. Change of the other three kinds of things—including movement—depends directly on living beings: biomorphism-products (Bp) and Bw in the case of products and by-products and Bw in the case of the nonliving. This all follows from the phenomenological insight: "It is not the wind that moves, it is not the flag that moves; it is your mind that moves." Our koan is the foundation of hymenomorphism:

> The skin of a living organism cuts off an outside from an inside.
> Not so the circumference of a circle in a plane.

A discussion of hylomorphism versus hymenomorphism is beyond the scope of this study. I limit myself merely to pointing out why it arises. Fundamentally, metaphysics as conceived by Aristotle is grounded in @@L. His abstract distinctions, like Bourke's circumference, cannot overcome the problem of the disappearing circle of Fig. 2.4. With skin and everything that pertains to it, elided, @@A, the ground of @@L, is reduced to @@L. Hymenomorphism, on the contrary, begins from @@A as ground, the ground scheme previously discussed at length. To elide liminality, including the play analog/digital, is to elide the identity-unity-whole nature of organisms and reduce them to identity. ב, anticipation, analog/digital are elided, and with them the rest of the proper content of @@A. So, organisms are conceived digitally, mechanically, purely in terms of @@L. This elision of the phenomenological is none other than the elision of *again* in the axioms of *LoF*, now transposed to distinguishing and thence to [subjects]. Liminality, the nesting of skin within skin,

---

[42] See https://plato.stanford.edu/entries/aristotle/#Hyl (retrieved January 2021).

[43] See https://plato.stanford.edu/entries/aristotle-biology/ (retrieved January, 2021).

[44] *Physics* II.1.192b28.

[45] *Physics* II, 2–3.

[46] *Physics* II, 9.

[47] https://plato.stanford.edu/entries/aristotle-politics/supplement2.html, cites *Phys.* II.1.192b32–3 (retrieved January 2021).

points to the intentionality inherent to identity-unity-whole. In short, the excess of extension leads to the oversight of *intension*.

What is to be understood by intension? Recall that, throughout *LoF*, starting from the definition of distinction as perfect continence, and as expressed formally in symbols, distinction is conceived spatially. The diagram that calculates is spatial, as are electric circuits that are necessarily extended. In this regard, the covert use of "again" is that much more significant, because it is the one fundamental element that is not extended. Overtly, then, the calculus of distinctions is all about the noun, or *extension*, in Descartes' sense. Only covertly, with the use of *again*, and by drawing the equivalence of distinction and the observer, does *LoF* touch on distinguishing, the verb. Still, by positing imaginary Boolean values, *LoF* is sufficiently about process to link to what I am here naming *intension.* Distinguishing, in the sense of metabolism = cognition, may be referred to as [mind], Descartes' name for his second substance. In its stead, I prefer to pair *intension* with extension. This is the same Indo-European etymology *ten,* 'tend,' 'stretch,' but now juxtaposing the Greek *ex*, 'outside,' with its antonym, the Latin *in*, 'in,' 'within,' 'towards,' also a form of *un*, 'not.'[48] Thus, *intension* is the stretch or tendency inwards and towards 'not'.[49] The very expression *distinguishing distinction* alludes to intension/extension. Hymenomorphism refers to the indispensable play between intension and extension that allows the universe to appear, thanks to the calculating of living surfaces. The thrust of this exploration is that, to be about aliveness, and thereby also about *intension*, biology must start from the ground scheme @@A that is the condition for the possibility of @@L. Thus, @@A and @@L must be concatenated into a single conjugate form. Hymenomorphism can account for why the skin of a living organism cuts off an outside from an inside as a circumference of a circle in a plane cannot.

Aristotle's hylomorphism, the play matter/form grounded in @@L, metaphysics, is his first philosophy. How to name a first philosophy in which skinward plurality is the grounding condition for the possibility of [subjects] and their unfolding? Following Emmanuel Levinas, I shall refer to this first philosophy as *[ethics].* We thus arrive at the flesh of @@A—aliveness.

Ethics

What is [ethics]? A first point is that ethics is a subset of [ethics]. As in previous cases, bracketing a term generalizes its meaning to embrace all instances, from the most deeply compact to the most highly differentiated. The highly differentiated is what we typically understand to be ethics. As with all bracketed terms, the various instances clarify each other by isomorphism. It is Emanuel Levinas' unique notion

[48] Partridge, *Origins*.

[49] The previous and the next chapters show why opaqueness ('not') is indispensable to aliveness.

of ethics that I shall use to clarify what I mean by [ethics]. However, first a brief overview of the field is in order.

As Patrick Byrne explains, the four general approaches to ethics today are: (a) deontology, (b) consequentialism, (c) virtue ethics, and (d) the neuroscience/evolutionary approach. The key concept shared by all is autonomy and much of the discussion centers on the meaning of this term.[50] To understand joinings involves precisely understanding the autonomy of organic wholes (WW) within social wholes (wW). This generalization is what calls for the notion of [ethics]. Deontology is concerned with fairness in the procedural sense. It focuses on how decisions are made. Consequentialism grew out of utilitarianism and focuses on the results of decisions. Jeremy Bentham reduced results to pleasure and pain, upholding the idea of maximizing the former and minimizing the latter. Contemporary consequentialism concentrates on various fundamental goods, somewhat loosely defined, such as material well-being, contentment, and so forth. Virtue ethics, the third approach, is a recent revival of Plato's and Aristotle's approach in terms of courage, generosity, honesty, friendliness, self-restraint, and so forth. The underlying idea is development of ethical character. Fourth, and most recent, is the neuroscience and evolutionary approach that focuses on the neurophysiology and genetics of altruism.

Based on Lonergan's ethics, Byrne proposes an overall approach to bring these four into an integrated conversation. This also serves for inquiry into [ethics]. The four approaches focus on different things—how autonomous agents make decisions, their outcomes, character, and underlying physiology. If ethics is concerned with highly differentiated—mostly human—relationships, [ethics] is concerned with the proper content of joinings, the dynamics of relationships as such.

Take *tit-for-tat*, for example, the strategy in game theory first introduced by Anatol Rapoport in Robert Axelrod's computer tournaments in the 1980s.[51] In this simple strategy, an agent first cooperates and subsequently replicates an opponent's previous response. If the opponent is cooperative, so is the agent; if not, the agent is not. In these tournaments, various strategies were pitted against each other. Axelrod and W. D. Hamilton wrote the paper "The Evolution of Cooperation," which presented what would become the mathematics of reciprocal altruism in biology—the origins of Bourke's work on the [social]. Even this deeply compact instance of [ethics] integrates elements of the four approaches: (a) how each autonomous agent acts ("makes decisions"); (b) outcomes; (c) character (some agents are more aggressive, others less so); and (d) underlying physiology (in this case the algorithm and code of each agent). Because aliveness is plural from its inception, the proper content of @@A involves the [social] and the [ethical]. The [social], as shown by Bourke, is how Hamilton's law accounts for the dynamics of social wholes (wW) at all levels.[52] The [ethical], for its part, pertains to individuals made up by nestings of the [social]

---

[50] This paragraph draws on an interview with Dr. Patrick Byrne found at https://www.youtube.com/watch?v=Ed9KlxFY7b0 (retrieved September 2020).

[51] Axelrod and Hamilton, "The Evolution of Cooperation."

[52] Another notable attempt at integration is Jablonka and Lamb, *Evolution in Four Dimensions*.

yet enclosed by skin, which makes them organic wholes (WW). Contrary to Bourke, this refers to individuals in the normal sense of the term, thereby not eliding their nature as identity-unity-wholes. This in turn underscores the importance of relationships at all levels, and their paradoxical nature, as grasped by the four approaches.

What is meant by paradox? Take a child who responds to low expectations with poor performance. Her sense of her own identity determines her capability, and this involves unity and whole. Should this individual, in later life, respond to low expectations with high performance, this paradox will be intelligible only in terms of the development and interrelations between her and others. That is, interrelations between identity-unity-wholes. Needless to say, the boundary between the [ethical] and the [social] is hazy. Recall Fig. 9.3, which treats major evolutionary transitions in individuality. These transitions are framed in terms of the following six questions:

- What conditions favor the formation of cooperative groups?
- What conditions maintain cooperation during group transformation?
- What conditions favor division of labor?
- What conditions favor communication that coordinates cooperation of the group level?
- What conditions lead to negligible conflict within groups?
- What conditions favor mutual dependence?[53]

Cooperation, division of labor, communication, conflict, mutual dependence—all these topics straddle the [social] and the [ethical] because they refer to interactions between individuals. And yet, the paper from which this list is taken does not treat the individual. It is individuals that are anticipative and remain in *entelecheia*. The locus of *unity* and *identity*, it is individuals that carry out the transitions from analog to digital—even when the collective they form is a social group (wW). As was shown, human individuals (WW) ourselves, we draw distinctions—we make judgments—on behalf of organizations (wW). This is what makes us ethical agents, and what makes any individual [subject] participating in the [social] an [ethical] agent.

It can be argued that ethics can only refer to individuals who make conscious choices. This points to a type of judgment. Namely a judgment arrived at by an individual who can ask herself the self-reflective question, Have all pertinent questions been asked? But there are two objections to this.

The first objection is that this would imply that the *qua* ethical hinges on a very specific kind of borrowed content of distinctions, rather than on its proper content. However, borrowed content is what differentiates the ethical within the [ethical]. The proper content of [ethics] is what places all its instances in a single range, one of these is what we normally refer to as the ethical. The [ethical] refers to the proper content of living beings, what they are as embodied distinguishing distinctions that are made up of distinguishing distinctions.

[53] West et al., "Major Evolutionary Transitions," 10,113.

Second, that judgments in general are kinds of distinctions, itself points to the isomorphism one may expect in the full range of the [ethical]. If the ethical may be characterized by the question, Have all pertinent questions been asked? by isomorphism we can supposed that this is a highly differentiated instance of which the compact version may be identified as the [ethical]. And, even then, of all human beings who as such are considered *qua* ethical, only a very few ask ourselves that question. And if we do, we do so infrequently.

In short, it is the inherently plural nature of distinguishing distinctions that makes individuals [ethical]. The locus of their agency is identity, in identity-unity-whole, specifically when the individual is participating in social relationships. Therefore, skin, as the marker of individuality, points to when [ethics] is at work. As we shall see later, the unfolding of anticipation and of the individual go together, along with the growing differentiation between the [social] and the [ethical]. This again points to identity. It is at the level of identity that we anticipate. But the two aspects are present from the start, in the inherent nature of distinctions that distinguish. The six questions cited above show that biology tends towards the 'sociological' rather than the 'psychological.' In a remarkable statement, the authors unwittingly hit the nail on the head on what is elided by this:

> The evolution of cooperation requires two conditions. First, there is some ecological or efficiency benefit to cooperation. Second, there is some *mechanism* that leads to the benefits of cooperation being directed back to the operator and/or their relatives.[54]

Beyond its clear intent, no doubt justified, the term *mechanism* is notable. Just like their diagram, this key statement about individuals is framed in @@L terms just when @@A is especially called for. Benefits directed back to individuals benefit *them*, through skin, at the level of identity. Identity points to relationship, skin to skin. Relationships are the nature of joinings, another way of naming the proper content of @@A. We can now ask, What is a relationship? That is, What is the primordial sense of *A and B*? We now know that this is a question about the nature of the [ethical] From this unfold the four approaches named above.

Relationships

What are joinings? What are the conditions for their possibility? What sustains them? *Relationships*. Inquiry into aliveness (@@A) is inquiry into relationship. Organic and social wholes are relationships. [Ethics] is inherent to relationship and relationship to [ethics]. Inquiry into [ethics], then, is not inquiry into logic (@@L), abstract distinctions and relations as represented by circles, but inquiry into skin-to-skin, face-to-face, A and B. When you watch a white blood cell chase a bacterium, you are not watching a relation. You are watching a relationship.

[54] West et al., emphasis added.

As Lipton pointed out, at the cellular level, organisms maximize [intelligence] as the outer membrane surface expands. So, his notion of *mem-brains* converges with the idea that the skin of a newborn child is a type of brain, the locus of the earliest construction of knowledge, the more so because, as Hoffmeyer pointed out, skin and brain originate from the same ectoderm layer. Skin and membranes, the seat of forness and of multiple loci of causation, gradually unfold intensionality. Anticipation and intensionality unfold into ever more differentiated anticipation and intentionality. This too is the unfolding of relationship.

The general unfolding of Ba-Bw (biomorphism-aliveness—biomorphism-world) carries a growing capacity for anticipation that in turn sustains ever greater degrees of individuation—to which, at some point, we grant the title *person*. Skin, then, is the locus of autonomy, and, therefore, [ethics]. Autonomy is the *entelecheia*, the being-at-work-staying-itself, of distinguishing distinctions in plurality, in relationships. Since biogenesis, actual acts of distinction have been carried out in relationships, organic wholes (WW) interrelating within social wholes (wW). Even in this ontological sense, [ethics] pertains to the proper content of joinings, of @@A. It is there from the origin of life because it pertains to organization, to Ba-Bw, functional closure, forness, ב. [Ethics] is the dynamics of [body-subjects] interrelating, or [intersubjectivity]. This is the primordial sense of *A and B*. We shall soon see that, whereas the [social] is sociological, the [ethical] is historical.

There is much more to say about relationships, their nature, and dynamics. Before doing so, however, it is indispensable to add flesh, or rather, a face, to this notion of [ethics]. For this we must delve into the Other, which is the foundation of relationship and origin of [ethics].

Levinas' Ethics and [Ethics]

A precursor to the idea of [ethics] is Emanuel Levinas' notion of *ethics as first philosophy*.[55] Levinas developed his philosophy as

> a description and interpretation of the event of encountering another person. Giving rise to spontaneous acts of responsibility for others, the encounter unfolds, according to Levinas, at a precognitive level, thanks to what he called our embodied "sensibility."[56]

For Levinas, a phenomenology of intersubjective responsibility must be a 'first' philosophy "in the sense of interpretively reconstructing a level of experience precursive to both reflective activity and practical interests."[57] This clearly converges with a liminal and intensional understanding of [ethics] as inherent to joinings:

---

[55] Levinas, *Totality and Infinity*. In this section, I mostly follow William Large's accessible account in *Levinas' Totality and Infinity*.

[56] https://plato.stanford.edu/entries/levinas/index.html (retrieved June 2020).

[57] https://plato.stanford.edu/entries/levinas/index.html (retrieved June 2020).

> For Levinas, ethics begins not with the self but with the experience of the other. It is not I who decides to be ethical, rather I am forced to be ethical despite my wishes or intentions. Ethics has its source not in the will or freedom, but in the demand of the other. . . .[58]

Analogously, [ethics] does not begin with single organisms but with interrelationship. Indeed, as the intensional and part of the proper content of @@A, it is the ground for the emergence of individuals, and thereby 'my' wishes and intentions. External demands are constitutive of the *entelecheia* of all organisms, and, to that extent, they become differentiated from the internal. The source of [ethics], then, is the *con-crēscere* plurality of *entelecheia* of organisms in their co-evolution into [body-subjects]. These dynamics are the very definition of aliveness.

> Do our moral laws have their source in our reason, such that they would be true of everyone, or are our moral choices instead based upon our character that is socially and historically rooted?[59]

This may be transposed as the question: is the source of our [ethical] nature @@L or @@A?

> The big difference between Levinas' approach to ethics and the traditional way to think about morality is that the former begins with the experience of the other, whereas the latter starts with the self and works outwards.[60]

This is a transposition from substantialism to liminalism, in a different key, of Marlow and Braakman's revelation of the gaps in our understanding left by focusing on individual species of microbes; the realization that it is the interrelations of an entire community that allow it to carry out functions like nitrogen cycling.[61]

Levinas' work converges with systemic thinking, to which we shall soon turn. It also involves a critique of phenomenology for two key reasons worth noting. The first converges with the overall purpose of this research—to bring aliveness back into biology; the second allows the foundational problem of biology discussed in this analysis to be situated in a larger context.

> Phenomenology is the demand for philosophy to return to its roots and its beginnings . . . [to] the experiential. The world only is because it appears to someone. If the world did not appear to me as already meaningful then it would not be at all. The stone does not appear to the stone, or the supernova to the supernova. Things are only to the extent that they manifest themselves to someone.[62]

---

[58] Large, *Levinas' Totality and Infinity*, 2.

[59] Large. Notice in passing that the term *ethology* (Latin *ethologia*, from the Greek ἦθος, *ethos*) has been used both as the 'science of character' in the sense of morals (J. S. Mill) and as the study of animal behavior—first popularized by William Morton Wheele, American myrmecologist (a person who studies ants) in 1902. (see "Ethology.")

[60] Large, 2.

[61] Marlow and Braakman, *Team Players*, 36, quoting Laura Hug, professor of environmental microbiology at the University of Waterloo.

[62] Large, *Levinas' Totality and Infinity*, 2.

Things only appear because of joinings, plurality. Recall that it is a radical phenomenology, and not a radical Platonism, that can account for appearance. We are right back to the proper content of aliveness (@@A) as the ground of ב. [Ethics] brings about appearance. The paragraph underscores all that follows from inquiry into aliveness in terms of distinguishing distinctions—in broad strokes, the centrality of the experiential, thereby of self-appropriation, of isomorphism as method. So, from the [ethical] eventually the ethical unfolds. Which, as we now see, is another way of saying that [subjects] co-unfold with [objects] because we are organic wholes in social wholes. In plain terms, our discoveries in the world are grounded in our nature as distinctions, the condition of whose possibility is joinings, our social and ethical nature.

Consider this in terms of the Hegel-Lonergan nest of terms—implicit, explicit, abstract, alien, mediation, reconciliation. Starting from *again* in the axioms of *LoF* and arriving at ב, led us to joinings, which is implicitly [ethics]—relationship, the proper content of @@A. In this chapter we have arrived at [ethics] explicitly. This was at first abstract and must now seem alien. The reference to Levinas is an aid towards reconciliation. The starting point is that his is a phenomenology. It is grounded on relationship, but it is a phenomenology, nonetheless. And this points to one of the fundamental obstacles to reconciliation:

> One of the greatest opponents of phenomenology is naturalism. Naturalism is the belief that nature has a meaning in itself (stones and supernova exist independently of their appearing to someone) and that anything that exists can be explained by natural laws, which are in turn independent of the mind that thinks them, such that Newton's laws would still be true even if Newton or anyone else had never thought them.[63]

This was our starting point: the root of distinction, understanding, and therefore of natural laws, is ב. The fundamental part played by our limitations, and our arbitrary neglect of concrete residue was confirmed from physics by Rovelli. The laws of physics hold only *for* [body-subjects].

The problem with biology, then, is naturalism. This may seem a bizarre assertion, but everything said in this essay points to that. With the belief that things exist independently of their appearing to a [subject], why would biology concern itself with the problem of that appearing? Strictly guarding against anthropomorphism, it admits the subjectivity of [subjects]. But it does so by foreclosing inquiry, rather than addressing how to probe the subjectivity—emotional, cognitional—of [subjects].

Lonergan pointed out that in any philosophy one may distinguish between its cognitional theory and its pronouncements on metaphysics, ethics, theology. He called coherence between these two *positions*, and the incoherencies *contrapositions*. Self-appropriation allows us to develop positions and reverse contrapositions.[64] This is how

[63] Large, *Levinas' Totality and Infinity*, 4.

[64] Lonergan, *Insight*, 412.

the crisis of foundations pointed out by Husserl is dissolved.[65] Husserl suggested the dominance of naturalism results from the success of the natural sciences. It is assumed that their explanatory power will eventually account for the being of phenomena. So, it is supposed that neuroscience will eventually explain the subjective states of physicists as they understand physics, thereby bringing natural explanation full circle. But this leads to an infinite regress because neuroscientists would then have to account for their own subjective experience of understanding whatever iteration of understanding they happen to be understanding. So, the key incoherence that phenomenology sets out to reverse is the one that has concerned us all along: aliveness remains largely beyond the horizon of contemporary biology. As Large pointed out, "the successes of natural science have led to its forgetting its own experiential basis."[66] Attending to this forgetting led Lonergan to his *General Empirical Method* that includes data of sense and data of consciousness. Thus, developing positions and reversing contrapositions is not a question of attending to descriptions or understanding explanations. Rather, it is about verifying pronouncements against cognitional activity—our own. Kant, for instance, pronounced that knowing is like taking a good look. But we, and Kant, can observe ourselves doing far more than looking when we are knowing. For instance, we have insights, and these are verified by judgments. This oversight of insight permeates Western thought, not least thanks to Kant.

Following Lonergan, this investigation proposes to dissolve the problem of aliveness excluded from biology by adverting to the nature of distinction, as it results from the process of distinguishing. Paradoxically, this calls for a non-naturalized science of nature, a rigorous discipline with an ideal of knowledge that does not hinge on physiochemical cause-effect but rather on an understanding of patterns of event-response. Moreover, in realizing—again, through self-appropriation—that understanding patterns of event-response is grounded on relationships. This, and no less, is implied by Marlow and Braakman when they say:

> Individual cells perform certain chemical conversions and pass the product down the chain to other cells for subsequent reactions. Sharing cellular building blocks and energetic resources in this way both requires and enables cohabitation among diverse organisms.[67]

What is this, if not relationships?

The entire community has the capacity for a certain function, like nitrogen cycling.[68]

Later, we shall see why this is to embrace the historical. As Large pointed out, to treat the world as a natural object subject to the laws of cause and effect is itself a projection:

[65] Husserl, *The Crisis of European Sciences.*

[66] Large, *Levinas' Totality and Infinity*, 5.

[67] Marlow and Braakman, 36.

[68] Marlow and Braakman, *Team Players*, 36, quoting Laura Hug, professor of environmental microbiology at the University of Waterloo.

> This projection must have an origin. For the world to be a natural object, it first of all must appear. Such an appearing of the world is not subject to the laws of cause and effect, but it is originally present. The aim of phenomenology is to describe this original presentation.[69]

This is ב. We examined at length how ב made living being possible. Crucially, we found that the condition for the possibility for the emergence of ב was plurality, joinings.

This is why it is helpful to follow Levinas' footsteps. For him ethics begins, not with the self, but with the experience of the other. This, again, is an upper scheme, this time about @@A. The ground scheme is that [ethics] begins with relationships, relationships bring about self and other, and this is what unfolds experience. Recall Emeran Mayer: this is what bacteria did for 3 billion years. They set the conditions for the evolution of multicellular life by inventing the most efficient "language" to run complex and robust feedback systems that to this day supports all life on earth. What Levinas says for ethics can be reframed for [ethics]. Here is a statement about Levinas quoted above, now reframed for [ethics]:

> [Ethics] begins not with the self or the other but in their relationships. It is not single organisms that relate, rather, they are forced into relationships to self-maintain. [Ethics] has its source not in will or freedom, but in the demand of the other. . . .

Levinas' ethics is liminalist and not substantialist because it begins with the experience of the Other, rather than starting with the self and working outwards. Likewise, [ethics] begins with relationships that give rise to experience, rather than starting with the single organisms and working outwards.

Levinas accepts Husserl's critique of scientific and common-sense empiricism for its failure to account for phenomena by presupposing an idealism that is not clarified.[70] But he goes directly against Husserl's idea that the other is given as a 'like me.' This is Husserl's approach:

> I treat the other differently because I empathize with them. I see they are a subject just like me, though I cannot literally occupy their minds so as to see the world as they do. But it is I who determine the other, and not the other me. What we notice in this little story is that it is I who determined the other, and not the other me. The other is only other because I relate to them in that way. I of course could fail to empathize with them and treat them as I do any other object.[71]

It must be acknowledged that, for Levinas, 'the Other' referred mostly to humans and, with reservations, for other primates and even dogs. Beyond that, his position converges with Husserl's because most nonhuman organisms, not being Other, appear as merely objects. Still, his framing of ethics serves as a guide for the framing of [ethics] by analogy.

---

[69] Large, 5.

[70] Levinas' is not a return to naturalism, even if it is a kind of "radical empiricism" (Large, *Levinas' Totality and Infinity*, 5). This converges with Lonergan's General Empirical Method.

[71] Large, *Levinas' Totality and Infinity*, 6 (slightly modified).

In another key step, Levinas also broke with Heidegger. In *Being and Time*, Heidegger claimed that Descartes' theoretical subject was founded rather than founding because there is a deeper level of engagement than the theoretical one. This "being-in-the-world" is neither knowing nor even intuiting but what takes place through use.[72] By 'use' is meant something along the lines of affordances. This is forness. It is inherent to ב, to living being as such. It is the founding engagement with the world. With no distinguishing distinctions to engage, there would be no such engagement. Heidegger, however, contended that the human environment differs in kind from the environment of other animals. This bears directly on the discussion of Levinas and beyond. Engelland clarified this point about Heidegger:

> His position rests on an informed understanding of biology that finds confirmation in the recent biological theory of niche construction. Environments exist only when their corresponding organisms exist, because the biology of the organism specifies which features of the physical world will constitute the organism's environment.[73]

This is Uexkull's *Umwelt*, the play biomorphism-aliveness/biomorphism-world (Ba-Bw). The theoretic indeed pertains to the human Ba-Bw and Heidegger rightly criticized Descartes on this account. For Heidegger, our Ba-Bw cannot be as tightly coupled as the Ba-Bw of other organisms, if only because ours includes language and truth, making the theoretic possible. A tight coupling would imply that pursuits such as biology, mathematics, and philosophy are a biological function of *homo sapiens*. So, according to Heidegger, our Ba-Bw involves a historical dimension that is not present in other species. Throughout his works, Heidegger referred to this difference as "an abyss" that separates humans from animals to the point of denying that the human body is an organism.[74] For Engelland, "to deny the human difference amounts to species relativism and leaves the normativity of truth unexplained."[75]

But this is the general problem of continuity-discontinuity that is inherent to biology. In *The Descent of Man*, Darwin wrote, "The difference in mind between man and the higher animals, great as it is, certainly is one of degree and not of kind."[76] The same must apply to the differences between higher and lower animals, between lower and even lower animals, and so forth, all the way down the line. Problems of continuity-discontinuity are typical of biology, and they are never solved by declaring an abyss. Rather, what is required are appropriate terms to make comparisons. For instance, this is the point of identifying the difference between proper and borrowed content of distinctions. So, while Engelland correctly argued on Heidegger's behalf for avoiding species relativism, still, as our examination of @@L has shown, the normativity of truth is the proper content of

[72] Large, 7.

[73] Engelland, "Heidegger and the Human Difference," 176.

[74] Quoted by Engelland.

[75] Engelland, 176.

[76] Adler, *Great Books of the Western World*, 319. Cited in Engelland, "Heidegger and the Human Difference."

distinction—whether it is in organisms as embodied distinctions (Ba) or in their unfolding ("Logic is nothing more than the properties of the act of distinction.") as Bw. So, it obtains in the full range, from the most elementary to the most theoretical distinctions. This proper content accounts for continuity, and borrowed content for discontinuities. For an empirical argument against Heidegger's position, simply recall *Animal Tool Behavior* by Shumaker, Walkup and Beck. What was once assumed to be the exclusive provenance of humans today requires a systematic compendium.

To Heidegger's point, ב is always already *qua* historical, merely because of its arbitrary neglect of concrete residue. This is why anticipation unfolds and why proper contents continue to differentiate—until eventually normative truth is made explicit. This is even more so once it is understood that this unfolding occurs in the context of joinings, relationships. Just how tightly coupled does Heidegger suppose Ba-Bw to be? Or does he have in mind a mechanical coupling, which is not only naturalistic but flatly deterministic? But this is not the living kind of being. This is a conflation of the kinds of things.

The historical dimension Heidegger claimed for humans only—what is it rooted in? In fact, it is part of the proper content of @@A. Historicity is the very nature of the living kind of being, as each organism in a relationship self-maintains to remain itself, and as both together contribute to maintain the relationship. To be alive is to select, in both senses, and not only to be naturally selected for if you happen to make the right selection. The sun is ahistorical because the nonliving in itself is devoid of time. The germ of historicity is inherent to living being, if only because time is a consequence of our being distinguishing distinctions: "due to the limitations of our [embodiment] we [advert] to only a blurred [apprehension] of the world and live in time."[77] The very limitations that allow living beings to be alive and the world to appear make us inherently [historical].

The conundrums of quantum physics result from our hymenomorphic nature, skinned beings in the We that we are. Levinas takes up this We as *the face-to-face*. Thus, historicity, inherent to @@A, is inherently [social] and [ethical]. Being and time pertain to skin-to-skin and their unfolding to face-to-face.

[Ethics] is not ethics. While historicity is intrinsic to [ethics], there is a specificity to the unfolded historicity intrinsic to ethics, to which corresponds an unfolded decoupling of Ba-Bw. Yet even in this narrow sense Heidegger is wrong. Not only humans, but Darwin's "higher animals"—primates, elephants, cetaceans—already systematically show a kind of historicity that cannot be accounted for merely naturalistically. Nor, as it turns out, could parrots or octopi.[78] Nor, it seems, could other

---

[77] Rovelli, *Seven Brief Lessons*, 62–63 (modification in brackets). This quote is explained in Chap. 7.

[78] The literature in primatology on this point is now extensive. A good starting point is Wrangham et al., *Chimpanzee Cultures*. On the last point, see Pepperberg, *Alex & Me*, and the recent film, *My Octopus Teacher*, starring Craig Foster and directed by Pippa Ehrlich and James Reed, and judge for yourself.

animals, like the coyote and badger who play and hunt together, or the chicken and dog who chase each other playfully.[79]

This brings us back to Levinas. In *Totality and Infinity*, he dug down to discover the fundamental principles that constitute a self. Whereas Heidegger argued in *Being and Time* that this ground is ontological, Levinas argued it is ethical. The difference between a foundational ethics and a foundational ontology, hinges on the account of the relation to the other. Heidegger rejected Husserl's intuition as a basis for understanding the relation of the self to the world, but his description of being with others, repeated Husserl's. For both

> the other is an element belonging to the self-understanding of the self, rather than someone who radically calls into question the limits of this self-understanding. In Husserl's description, I realize that the other is like me, whereas in Heidegger's the other belongs to the way in which I relate to my own being.[80]

A fundamental issue of limits is at stake. For Levinas, the other intervenes radically in the self's relation to itself. It does so in a way that is beyond the control of the self. This is true of all joinings. To state it in terms of identity-unity-whole, the other intervenes at the level of *unity*, which mediates *whole*. So, at *identity* we discover ourselves responding to others. Mirror neurons provide a good example:

> A type of sensory-motor cell located in the brain that is activated when an individual performs an action or observes another individual performing the same action. Thus, the neurons "mirror" others' actions. Mirror neurons are of interest in the study of certain social behaviours, such as empathy and imitation, and may provide a mechanistic explanation for social cognition.[81]

According to neuroscientist Giacomo Rizzolatti, who first identified them with his colleagues at the University of Parma in the 1990s while observing macaque monkeys, mirror neurons "could help explain how and why we 'read' other people's minds and feel empathy for them."[82] So, they may explain how birds learn songs and how humans learn dancing. Capoeira dancers, for instance, showed more activation of these neurons when observing capoeira than when observing ballet. Predicting others' actions has been proposed as a function of mirroring, how we anticipate what another will do next.[83] Well-known neuroscientist Ramachandran predicted that "mirror neurons will do for psychology what DNA did for biology; they will provide a unifying framework and help explain a host of mental abilities that have hitherto remained mysterious and inaccessible to experiments."[84]

Neither Husserl nor Heidegger knew about mirror neurons; nor is it likely that Levinas did. For the former two, the relation to the other belongs to *my* existential

[79] Videos can be found online with these phrases.

[80] Large, *Levinas' Totality and Infinity*, 7.

[81] Knoblich and Sebanz, "Mirror Neuron."

[82] Winerman, "The Mind's Mirror."

[83] Knoblich and Sebanz, "Mirror Neuron."

[84] Ramachandran, 1.

project. Even if I do not treat others as I do things, the other is internal to the being of a self, a notion of autonomy that obtains for rights, duties, and virtues. Levinas' view, however, converges with contemporary neurology and its evolutionary origins:

> It is not I who constitutes the other, but the other me, and it is not the other who is internal to my consciousness or understanding, but I who am interrogated. My being is usurped by the other, as though the beginning of my existence were already for the other, before any act of comprehension or understanding.[85]

This is joinings, an @@A model of the full range of [subjectivity]. The subject "is already split from within (as though the other occupied my place despite me)."[86] This is the primordial sense of *A and B*. Levinas pointed out that, with language, "the other 'present' themselves in speech in a manner that is different from vision."[87] The other does not 'disclose' herself, but 'reveals' herself through speech. It is not the words or the objects they refer to, but the way the other attends that results in the ethical inversion of intentionality. Listening, I respond to the demand of the other. I do not initially comprehend the other as being or not being like me or in terms of my own projects. Rather, "I respond before I comprehend or understand."[88] This is how mirror neurons function. We discover the other in ourselves, despite ourselves. Levinas referred to this presence of the other as 'transcendence.' This is not an ontological commitment to a hidden world but an ethical commitment, one that results from the irreducible relationship of the face-to-face, a demand prior to what one can express, know, affirm, or deny. Instantly recognized, this demand is heteronomy, and, therefore, the transcendence of/to the other.

> On this demand is built not only the rest of the world, but the world of justice and morals, which is how we normally understand ethics, but also the world of science, reason and objectivity. It is not ontology that determines ethics but rather ethics that determines ontology.[89]

Levinas' general notion of subjectivity and ethics as an account of human joinings may be extended to the joinings of all [body-subjects]. It goes right back to biogenesis. In this strict sense, [ethics] determines Ba-Bw—what living beings *are* and what they *know*.

Let the term skin-to-skin—antecedent to [face-to-face]—refer to the joint demands that organisms always already place on each other, their mutual it-is-not-I-who-constitutes-the-other-but-the-other-me. [Skin-to-skin] made life possible in the first place. This is the *con-crēscere* primordial condition for the extraction of energy from the environment that underlies the capacity of organisms to be-at-work-staying-themselves. We have direct evidence for the foundational nature of joinings in the cell genomes mentioned above that "often lack the ability to make all the amino acids needed to build their proteins or the nucleotides for constructing

[85] Large, *Levinas' Totality and Infinity*, 8 (slightly modified).

[86] Large, 8.

[87] Large, 8.

[88] Large, 8–9.

[89] Large, 9 (slightly altered).

their DNA, suggesting they acquire these building blocks from neighboring cells with a surplus."[90]

Con-crēscere, we living beings collectively build our selves (Ba, who), and our world (Bw, what). Who and What co-create in the We. Joinings give birth to intensionality and extensionality. As @@L is grounded on @@A, so metaphysics is founded on ethics. This is Levinas' *ethics as first philosophy*. Ethics is more *important*—inner, prior to, deeper than—metaphysics.[91] Aliveness, @@A [skin-to-skin] founded on ב, is inherently [ethical] and [historical]. For [ethics], We founds Who and What. It follows that a non-naturalistic inquiry into life begins from the We. And inquiry into the We—[ethics], skin-to-skin, [face-to-face], A and B—is inquiry into relationships.

[90] Marlow and Braakman, *Team Players*, 36.

[91] This is Levinas' central thesis in *Totality and Infinity*. Recall the technical sense of "important" as inner, prior to, deeper (see section "The unity of mathematics and living being").

Chapter 10
The Paradoxical Nature of Aliveness

Dis-eliding the Dynamics of Joinings

Having established *relationship* as an ~~ontological~~ category—this erasure indicates Levinas' point that ethics antecedes metaphysics—we take up the topic of relationship as foundational.

It was already pointed out that two forces define groups, attraction and repulsion. This is the nature of the We, of joinings, collectives, plurality. [Ethics], then, is intelligible in terms of these two forces. Our own phylogenetic and ontogenetic development as a We begins [skin-to-skin] and eventually unfolds in the face-to-face of our family of origin. This brings us to Bowen's *family systems* theory. It will be recalled that in the section "Liminalism and substantialism", Chap. 6, it was said that liminalism may be understood through systems thinking, whose focus is on relationships. In her primer on Bowen, Roberta Gilbert explains that, regardless of our denying it or cutting off from it, we are greatly determined by our family, both in our weaknesses and strengths. To that family, what we have learned there, and our reactions to it, we owe much, both what is good about ourselves and what get us into trouble, what we wish would be different. The same forces went into creating both.[1] At work in families are the same two forces of attraction and repulsion that define groups. Murray Bowen refers to them as the *togetherness* and the *individuality* forces. There are striking echoes of Levinas in the account of two forces in Bowen's systemic approach to families:

> Much of human existence is dictated by the strong "*togetherness*" force originating in that most basic of social organizations, the family. That force, opposite from the "individuality force," *automatically and instinctually pulls us together* in a family and emotionally defines what that family is. It sticks us together, *absorbing part of each self, demanding* we be there for the group. *Togetherness is the force that tries to pull one back* to the group when one is being "too much" a self. It says, "*Be like us,*" "*Be for us more than for yourself,*" or "*Think as we think.*"

[1] Gilbert, *The Eight Concepts*, 9.

© The Author(s), under exclusive license to Springer Nature Switzerland AG 2023

D. C. Mayer-Foulkes, *The Nature of Living Being*, Biosemiotics 26,
https://doi.org/10.1007/978-3-031-24789-7_10

It can feel good and bad, protective and restrictive.

> In that way, that organism, so much bigger than self, of which we are all a part, the family, is made up of *"donations" of self* from its members—parts of selves that are *there more for the family than for the individual.* In that way, the members *"fuse" together, or "lose self"* into the family in *automatic reactions that are as old as time itself.*[2]

Through these two forces, wholes (/W) both organic wholes (WW), and social wholes (wW), self-regulate, regulate each other, and are regulated by larger self-regulating wholes (/W) they belong in, and regulate wholes (/W) within them. In short, this is how an identity-unity-whole self-regulates. Conflict, trade-offs, division of labor, are at work throughout. In the context of families, Bowen referred to this as "borrowing and lending" self. The repulsive and attractive forces, proper content of @@A, are represented by this very icon. Recall how the two icons @@L and @@A were arrived at. @@A is and icon for the primordial *A and B*. Bowen's starting point is identical to Levinas', as was quoted above:

> The big difference between Levinas' approach to ethics and the traditional way to think about morality is that the former begins with the experience of the other, whereas the latter starts with the self and works outwards.[3]

We are gestated as identity-unity-wholes in our mother's womb, the originary @@A and deepest template of our lived experience. Onto this template we will graft all experience. Our identity is then formed in a We that soon becomes overt, our family of origin. There, our Ba-Bw unfolds in a We that makes sense of the world. Throughout our life, this will inevitably be our primary reference. So, for instance, it will be the root of our over functioning or under functioning in the face of peer pressure. Through affect, the paradoxical yearning to *both* belong *and* be ourselves—the two forces—will underly how we subsequently manage the tension generated by "be like us more than yourself," or "think as we think." This kind of *both/and* is the hallmark of @@A. Likewise, it is the origin of the play of our transparency or opaqueness in affect as we learn to maintain a sense of ourselves as one. This is the founding paradox of joinings, belonging to them while differentiating [I]/[not-I].

A key aim of family systems therapy follows from this—to facilitate a shift in the notions of [I] and [not-I], both in family members and in the family as a whole. So, families learn about the dynamics of borrowing and lending of self that underlies the homeostasis of the family system. This is, in fact, accomplished through dis-elision, not by description, but by promoting actions that reframe experience. That is, by injunction. Following guided injunctions allows family members to grasp the *both/and* or paradoxical nature of @@A. An example is provided below. The dynamics of family systems and small groups are intelligible to us primarily through affect. The notion of the two forces serves to dis-elide and articulate these dynamics explicitly. Reframing follows the Hegel-Lonergan nest of terms: what is at first

---

[2] Gilbert, *The Eight Concepts,* 9. For his collected papers, see Bowen, *Family Therapy in Clinical Practice*. Emphasis added.

[3] Large, *Levinas' Totality and Infinity*, 2.

implicit is made explicit, which at first seems abstract and then becomes alien. Mediated by practice, shifts become possible that alter the system's dynamics which bring reconciliation, in the deeper sense of the term.

Grappling with the dynamics of joinings engages deep biases, not least because of the difficulty in admitting that we are not the fully autonomous individuals we like to imagine we are. Our very notion of relationship must be reframed. This systemic reframing by Bowen points to *relationship* as the proper content of @@A in the family, and, by isomorphism, suggests how *relationship,* as such, may reframe biology as inquiry into aliveness.

Families learn to understand joinings for therapeutic reasons, to learn to live with whom one loves; a similar therapeutic is called for to understand joinings in biology, the love of learning about what lives.

The Paradox of Relationships

Relationships are the dynamics within and between joinings. Bowen identifies five patterns in which the togetherness and individuality forces are expressed: *conflict, distancing, cut-off, triangling* and *overfunctioning/underfunctioning* (or *reciprocity*)[4] Bowen uses the term *anxiety* to refer to any heightened state that results from the two forces, regardless of coloring. Interfere with an ant and the result expressed in movement and physiologically is anxiety. Anxiety is contagious. Others pick it up. A cow takes fright and anxiety ripples through a herd. A tree whose leaves are chewed releases chemicals that raise the alarm for trees downwind. As we now know, some tree signals are species specific while others are for all trees.[5] A single white blood cell sets off a collective response to a wound. In any living system, anxiety expresses the opposition and circular causation between the individuality and the togetherness forces. Our own emotions are heightened states, physiological reactions, which, on becoming conscious, are experienced as feelings. The anxiety of a social system is the tangible yet mostly unspoken paradoxical dynamics that constitute an organization. The infectiousness of anxiety between group members defines the group's limits. This converges precisely with Kraus and Ruxton's definition of a group. These dynamics are systemic principles that underly identity-unity-wholes and their interrelationships. These paradoxical dynamics are the realm of [ethics].

The five patterns, then, are expressions of anxiety. In *conflict*, two sides split a paradox, each giving voice to a single pole. Take a couple that shares the value of frugality when facing a dilemma over an expense. She is for it; he is against it. Their shared value is expressed in polarization. A heated argument serves to vent anxiety but obscures the need to think through a tradeoff. *Distancing* is the mutually

[4] Gilbert, *The Eight Concepts*, 11–20.

[5] Wohlleben, *The Hidden Life of Trees*, 7.

regulated oscillation between attraction and repulsion expressed through increased or lessened distance or interaction. Following an argument, both members of a couple feel hurt and misunderstood, so they avoid contact. A shift from conflict to distancing may even lead to temporary, even permanent *cut-off*. And yet, this only reinforces the systemic bonds.

Triangling is redirection, either conscious or unconscious. Aggravated, a husband communicates anxiety to his wife who, in turn, redirects it at a child. If this pattern is habitual it may result in the child's sickness, an expression of the system's chronic anxiety. Redirection of anxiety of two members mitigates their conflict or distancing, bringing relief to the whole system. Triangling underlies intra- and intergroup dynamics such as gossiping, 'external danger', scapegoating, xenophobia, and so forth. It redirects anxiety towards a present, absent, and even an imaginary third. As shown by circadian clocks (see Fig. 4.3), this is inherent to the forness of two-element regulating systems whose function is referred to systems beyond them.

At the Monkey Sanctuary, tension between Danny, the leader of the group, and Django, the subdominant male, would sometimes reach a very high pitch. Determined to prevent a fight between them, particularly because distancing between them was limited by the size of the territory, it was time for one of us to surreptitiously leave the grounds and return well disguised as External Danger. The alarm would immediately be raised and the whole monkey group would shriek with fear. Danny and Django would huddle together with loud squeaks assuring each other lifelong loyalty as they joined the battle demonstrating excitedly side by side. When the humans joined the fray, the monkeys took courage and cheered ever more loudly as the intruder was chased off. We truly put on a show for the monkeys. For months afterwards, the slightest unfamiliar sound or movement at the front gate would send Danny and Django into excited reassuring huddles.

Overfunctioning/underfunctioning or *reciprocity* is the system self-regulating by reciprocal heightening and lowering activity of its subsystems. Consider the blistering topic of punctuality in a couple. Jane is anxious and gets ready early; John is laid back and gets ready late. Heightened anxiety leads them to blaming, defending, diagnosing, and so on. Yet, as a single system, this is how Jane and John manage to arrive (just) in time. Neither is at ease, and both arrive stressed, she because they are late, and he because they are early. But, to those around them, they are mostly punctual, highlighting the paradox and circular causation by which couples self-regulate. She is early *because* he is late; he is late *because* she is early.[6]

[6] Independently of Bowen, John Bowlby's *Attachment Theory* revolutionized the field of child psychology with the fundamental idea that parent-child and couples' attachment bonds are the natural state that results from our evolutionary past, in contraposition to the long-held assumption that interdependence is fundamentally unhealthy. Thus, starting from joinings, Bowlby classified partners according to three styles of participation in relationships: *anxious*, *avoidant*, or *secure*. The *anxious* over-functions and the *avoidant* under-functions. To become *secure* is to reduce our borrowing and lending of self—using Bowen's terms. Bowen suggests the best way to accomplish this is to spend time in one-to-one relationships with every possible member of our family of origin. The most successful couples therapy today has resulted from Dr. Sue Johnson's extension of Attachment Theory. She has built precisely on the fundamental premise that dependence and inter-

Following injunctions can aid Jane and John in breaking their habit of *reciprocity*. This is the basis of Brief Therapy.[7] Clients are asked to define operationally what they wish to achieve. Having pinned it down, they are given a precise instruction, often one that elicits an amused smile. So, for instance, it may be suggested to Jane to slow down and take her time getting ready, and at all costs avoid hurrying John. This will allow him to mobilize himself, as he claims he is perfectly capable of doing. But, of course, he is now in the habit of depending on Jane for this. For his part, John might receive the suggestion to get ready earlier, thereby removing the cause of Jane's nagging, which he claims bothers him so. Following these injunctions makes explicit their dynamics as a system, dissolves the issue, and starts them on their way to gain greater awareness of their dynamics as a couple. This is the play of [I] and [not-I] inherent to @@A that is beyond the horizon of @@L. Concretely, as it is said in Brief Therapy, the paradox of change in dynamic systems is that very often the solution is the problem, and the problem is the solution. As a system, both partners cause the problem by solving it and both can solve it by doing what they believe causes it.

Returning to Bowen's five patterns, together they make up relationships, and are therefore expressions of the proper content of joinings. These are the dynamics of relationships in the context of the two forces. One would surmise, for instance, that in the *Prochlorococcus-Pelagibacter* partnership, the joinings naturally selected for are those that optimize the two forces through the five patterns. Likewise, in the instance of the transition to multicellularity, the best candidates are those that most optimally respond to the six questions that were quoted in Chap. 9.[8] Cooperation is possible in viable relationships that persist over time because they prevent *cutoff* by minimizing *conflict* through a good balance of *distancing*, *triangling*, and *reciprocity*. This at once enhances the autonomy of individuals and the benefit from their interrelation.

This is all about the *what* underlying the biochemical *how*, the proper content of @@A (joinings, relationships). We humans intuitively grasp this about relationships. This is what relationships are. We all know and share tacit knowledge of these dynamics and the tradeoffs they involve. If we grant that cells are relatively autonomous, and that they thereby engage in relationships, we implicitly grant that these relationships are intelligible through these patterns. This tacit knowledge is accessed by self-appropriation. Just as sound cannot be described to the deaf, nor color to the

---

dependence, far from being unhealthy, are what we are biologically designed for (see in particular *Love Sense: The Revolutionary Science of Romantic Relationships*). Predating the current meaning of the term 'system,' which follows Bertanfly's General Systems Theory, the meaning in Greek is "a whole compounded of parts." Plato on occasion uses the term as "a composition" and Aristotle as "an organized government, constitution" (Liddell and Scott, *An Intermediate Greek-English Lexicon*).

[7] See for instance *Change: Principles of Problem Formation and Problem Resolution* by Richard Fisch, Paul Watzlawick, John H. Weakland.

[8] See section "Ethics."

blind, the paradoxical character and circular causation of interrelationships can only be grasped experientially.

The marker of paradox is *both/and.* I *both* love *and* hate Christmas; he *both* resents *and* appreciates being corrected; she is *both* sad *and* happy to be graduating. Consider again the statement "This statement is false," which is *both* true *and* false. It is so by circular causation and levels. And it allows us to access the nature of paradox. It interlaces four sets of repetitions: true, true, true, . . .; false, false, false, . . .; if-then, if-then, if-then, . . .; value switch, value switch, value switch . . . . Further, it draws awareness that these sets of repetitions go on and mix with each other, in me, who is the hinge between the mathematics and phenomenology, and whose condition of possibility is that I am an identity-unity-whole. This is how I, who parses the world in terms of parts and wholes, or Wholes within Wholes, can understand it. Manifolds of bet structures established at the level of *whole* affect *unity* and become available at *identity*, the level of supervening awareness, leading to my understanding and judgment "it oscillates."

This kind of sense-making is intrinsic to any *both/and.* Both/and is implicitly at work in the five patterns. *Conflict* always involves the underlying desire not to be in conflict; the same is true of *distancing* and *cut-off. Triangling* involves a non-relating stand-in for relating, which boils down to relating without relating, while *reciprocity* is independence through interdependence. The opposite of each of these is also true—non-conflict desires conflict; non-distancing desires distance; non-cutoff desires cut-off. Triangling involves non-relating by relating, and reciprocity interdependence through independence. Relationship is further founded on both/and in three principal ways: (a) the paradox of limits (belongs or not to the group), (b) the paradox of becoming (ceasing to be what one is while not yet being what one will be), and (c) the paradox of labels, the appearance of stasis where there is always movement.

It takes an organic whole (WW) to understand organization (/W). Indeed, it takes a WW to understand. @@L—logic—derives from @@A—joinings, relationship. This is itself paradoxical because, to grasp @@L, we must elide that paradoxical nature of @@A, the condition for its possibility. And yet, only by this elision can we be in @@L and grasp the nature of @@A. In short, we require *a and b* and *A and B* to grasp both, and their difference.

It follows from this nature of @@A that our immersion in relationships is, in principle, partially beyond our grasp. So, their effect on us always remains partially beyond our horizon. This goes far deeper than we imagine. Steve Bierman, for instance, accounts for placebo and nocebo as effects of relationships, in his case between a patient in a vulnerable state and a healer's suggestions. He notes: "The proposition 2 + 2 = 4 holds, whether asserted by a genius or a jerk. However, when a patient regresses to the primal state of helplessness and dependency, *the source matters far more than the suggestion.*"[9] This empirical biological data is challenging in @@L terms but is immediately intelligible as @@A—relationship.

[9] Bierman, *Healing Beyond Pills and Potions*, 45 (emphasis in original). Marchant provides an account of rigorous empirical study of placebo and nocebo effects in *Cure: A Journey into the Science of Mind and Body.*

Paradox in Groups

We have already touched on some of the paradoxes of joinings, for instance those of group membership mentioned in the section "Boundaries", in Chap. 9. Yet, as we have just seen, additional, deeper paradoxes are at work in family dynamics and, in general, group membership. These dynamics, all part of the proper content of @@A, are clearly available for observation in social wholes. These dynamics can be extrapolated to organic wholes. In contrast to family dynamics, which we are born into, the following account pertains to the dynamics of group formation. That is, the formation of joinings.

According to Berg and Smith, three key paradoxes operate in groups: the paradox of dependency, the paradox of boundaries, and the paradox of regression.

Paradox of Dependency

"The metaphor for the paradox of dependency is ecological. For any part of a system to be able to act independently it must accept is dependency on the other parts together with which it makes up a whole."[10] This statement could stand as a definition of @@A. The paradox obtains for both kinds of organized wholes: organic wholes and social wholes. In both cases, the condition for the possibility of organic wholes is a larger whole, whether social or organic. So, the organized whole (/W) calls for a larger organized whole (/W). To be a /W is to depend on other /Ws. The two forces of fusion and individuation are inherent to /Ws. The paradox of dependency is that, to individuate, an individual must fuse with the group; and to fuse with the group, the individual must individuate. Cleary, this only makes sense for organized wholes (/W) and not for mechanical wholes (w/p). Thinking of groups and individuals mechanically leads to a fundamental assumption of independence, and of dependency as intrinsically against their nature—the key fallacy addressed by Bowlby's Attachment Theory, when he pointed out that interdependence is the natural state of social individuals.[11] These difficulties vanish as soon as we think in terms of organized wholes (/W). To avoid thinking of groups in @@L terms it is indispensable to advert to our experience of groups and our interdependence. This is to self-appropriate our nature as @@A.

[10] Berg and Smith, "Paradox and Groups," in Gillette and McCollom, *Groups in Context*, 114.

[11] See the reference to Bowlby in long footnote in the previous section.

Paradox of Boundaries

> Without boundaries, there is fusion. In this regard, boundaries are at the base of everything in [...] life. ... The paradox is that boundaries simultaneously make it possible for a [living entity] to take actions and at the same time limit those actions by what the boundaries define...[12]

This statement goes to the heart of our topic. Boundaries are at once separations and joinings. The authors touch on the issue of labeling already alluded to:

> The paradox of simultaneous possibility and limitation is most evident in the boundary delineation associated with labeling. If we have labeled an experience a particular way, that is the frame within which we will think about it. To develop thinking beyond the limits defined by the symbol requires us to break the frame, but even then, the breaking out is shaped by that which is being broken out of. The deep paradox is that were it not for the symbols, experience could not be sorted and would not be an ongoing part of our experience. At the same time, those very symbols constrain the ways we are able to experience both the past and the future.[13]

In short, A=A; and further, [A=A]➔[A = A]. That organisms organize their experience through symbols points directly to EI. Moreover, it points to distinguishing and EI as integrator-operator. The profound interconnection between labels and boundaries has been discussed in relation to naming. The discussion of naming in Daoism is well illustrated by the koan: "It is not the wind that moves, it is not the flag that moves; it is your mind that moves." There are no boundaries except those established by living beings, and these boundaries are the very means by which they touch each other and the rest of the world.

The paradox of boundaries is at work in identity. As already explained, when we are identity on behalf of an organization we belong to, we both are, and are not, the organization. These dynamics are intimately linked to the paradoxes of dependency and regression.

Paradox of Regression

> Individuals joining a group do so as whole and separate entities, but in so doing they are in the process of becoming parts of something larger than themselves. ... Allowing oneself to be a set of "parts" rather than an integrated "whole" feels regressive, as if one were returning to a time of less personal integration and wholeness. Such fragmentation, and the attendant loss of wholeness, is invariably threatening and runs counter to many of the developmental struggles that individuals go through. Resisting this process of inner partitioning, while perhaps making it easier for the members, makes it very hard for the group as a whole to do what it must in order to come together as an integrated entity made up of wholesome parts."[14]

---

[12] Berg and Smith, "Paradox and Groups," 116.

[13] Ronnie Stenquist, student. Cites Berg and Smith, "Paradox and Groups," 116.

[14] Berg and Smith, "Paradox and Groups," 117–118

To join as social whole, organic wholes must accommodate other organic wholes. This is, yet again, an account of @@A. There is an inverse reciprocal relation between the integration of one and the integration of the other. Integration into a social whole calls for a movement towards dis-integration of an organic whole, and resistance to this dis-integration delays integration into the social whole. This results in a regressive-progressive movement in the process of dis-integration and re-integration as the forces of fusion and individuation work together on two levels, for each organic whole and for the social whole. This paradox is consistent with reciprocal relationship at work in the paradox of dependency: to individuate, an individual must fuse with the group; and to fuse with the group, the individual must individuate. So, both fusion and individuation can be both regressive and progressive, depending on the stage at which the social whole is in.

This process may be characterized using Bruce Tuckman's sequence of group development: forming, storming, norming, performing, adjourning.[15] In the initial, forming stage of a group, participants are polite and accommodating with each other, carefully avoiding conflict. This is fusion, regressive de-individuation in which the togetherness force has the upper hand. It allows the group to come into being. Taken too far, as it must, a point is reached when the opposite happens. Individuation, with its progressive re-individuation, gives the upper hand to the individual force. This can threaten the group's demise. This is storming, the second stage, the result of individuals starting to resist regression. In contrast to the first stage, individuals progress while the group regresses. To preserve both, a balance must be reached, and this is norming, the third stage. Once norms are in place, the relation between group progression and individual regression ceases to be inversely reciprocal. Instead, the individuation of individuals now results in greater integration of the group, while fusion now brings regression. Arriving at this synergy energizes both the individuals and the group. This is the performing stage, when both the individuals and the group truly begin to function as a social whole. It is when interdependence leads to intense ongoing development. The *entelecheia* of a social group always includes a meta level, the process of a group constantly reinventing for itself what it means to be a group. This metalevel, inherent to social wholes, allows them to constantly reinvent themselves—the topic of the following section. All groups eventually come to an end, the state of adjourning. This brings about a process of mourning that can be more or less openly acknowledge and worked through by the group.

Some key points follow from the foregoing. First, the deep similarity in the dynamics at work in both organized kinds of wholes, social and organic. Though they are different kinds of things, the deep isomorphism between them is crucial for inquiry into @@A—joinings, organization. Clearly, these dynamics are utterly unlike anything that occurs in mechanical wholes and understood in @@L terms. The mere fact that the explanations of these paradoxes may stand as accounts of @@A show that paradox is inherent to joinings and organization. Second, in the

[15] Bruce Tuckman coined the phrase "forming, storming, norming, and performing" in his 1965 paper, "Developmental Sequence in Small Groups."

prevailing ideal of knowledge, @@L is predominant and determines what is to be understood by 'understanding', the practical consequence of which is difficulty making sense of these paradoxes. Third, this points to the importance of methods of inquiry proportional to living being such as isomorphism and self-appropriation. Adverting to our own experience as organic wholes in social wholes provides us with direct access to @@A and its inherent paradoxes. We can become aware that our participation in social organizations takes us well beyond the horizon of @@L and deeply into @@A. In the words of Michael Polanyi, about @@A, *we know far more than we can say*. For instance, this was written by a student from one of my own classes:

> Moving out of my mental models made me understand the paradox that I am a whole within a whole. Interestingly this conclusion was not a product of only my experiences, but a collaboration of everyone sharing their feelings during the last live [online] chat. While members were discussing their expectation to apply systems thinking in their lives outside of class, I was reflecting on whether or not systems thinking stops here or continues as part of my approach to the world. I conclude that systems thinking does not stop when I depart from our current system… but rather it stays with me because I myself am a system.[16]

Systematization of the paradoxical nature of @@A is well beyond the scope of this account. Clearly, though, far from being and obstacle, inquiry into our experience of organization in terms of paradox is not only feasible but indispensable. As the student makes evident, it is quite possible to shift from @@L to @@A and thereby get beyond the mechanistic horizon. Moreover, as the current reader can observe in her or himself, it is further possible to learn about organic and social wholes by analogy to each other. This alone points to the two fundamental strategies of inquiry into living being by self-appropriation and isomorphism.

Paradox of Repetition

ב is the heart of the proper content of @@A. In a speech for an inauguration, Jacques Derrida reflected on the paradox of ב, in this case in relation to social wholes. Doing so, he illustrated deconstruction and threw light on some of the issues at play between @@L and @@A. Referring to his essay on Joyce and the word "yes" in Ulysses, he takes up the paradoxes linked to affirmation. "Yes" is the final word in Ulysses in a monologue in which Molly affirms her marriage to herself. A "yes" as a promise or an agreement or an oath is absolutely originary, absolutely inaugural:

> Inauguration is a "yes," I say "yes" as a starting point. Nothing precedes the "yes." The "yes" is the moment of institution, of the origin; it is absolutely originary. But when you say "yes," you imply that in the next moment you will have to confirm the "yes" by a second "yes." When I say "yes," I immediately say "yes, yes." I commit myself to confirm my commitment in the next second, and then tomorrow, and then the day after tomorrow. That

[16] Group essay "Team Two's Journey Through The CIP," (*Final Paper for class LED603*) National University, February 2014, San Diego, CA.

> means that a "yes" immediately duplicates itself, doubles itself. You cannot say "yes" without saying "yes, yes." That implies memory in that promise. I promise to keep the memory of the first "yes." In a wedding, for instance, or in a promise, when you say "yes, I agree," "I will," you imply "I will say 'I will' tomorrow," and "I will confirm my promise;" otherwise there is no promise. That means that the "yes" keeps in advance the memory of its own beginning, and that is the way traditions work. If, tomorrow, you do not confirm that today you have founded a new program, there will not have been any inauguration. Tomorrow, perhaps next year, perhaps twenty years from now, you will know whether there has been an inauguration.[17]

We do not know that; we pretend we do. One must keep repeating the "yes". Derrida names this iterability. This repetition is threatening because it could be mechanical, a parody, a record, the "yes, yes" of a parrot. Such a technical reproduction from the outset threatens the living origin of the "yes." Derrida points out that from the start, "yes" is haunted by its own mechanical ghost even as the second "yes" can re-inaugurate and reinvent the first "yes". To be alive, yesterday's inauguration must be reinvented today, otherwise it is dead. An inauguration must be reinvented daily. This, then, is the paradox of the ongoing foundation of social wholes. Without it, they cease to be. If this paradox is fundamental 'logically' it is so because it is fundamental existentially. This is all about renewal, in the context of the stress imposed by the other paradoxes. "Yes, yes" is the paradox of iteration, repetition, affirmation, confirmation, it is the spirallic nature of aliveness. Not surprisingly, it is the heart of all disciplines, and science, of course, whose watchword is verification. A "yes"—a promise, an agreement, an oath, Amen—all are at once absolutely inaugural and tied to previous inaugurations that attest to our fidelity.

Derrida's "yes, yes" states the second axiom of *LoF*, but in reverse: Γ = Γ Γ. Moreover, he shows why Γ on its own is indeterminate. From the two axioms, stated in terms of *again*, are derived memory ("yes, yes") and oscillation ("yes, no, yes, no..."), thereby making explicit what is implicit in @@A that is at the outer edge of @@L. He even touches on the mechanical ΓΓ = Γ as a parody, a ghost of the living "yes, yes." Unlike mechanisms like an automatic heating system, a living "yes, yes" confirms, re-inaugurates, reinvents, because repetition involves an irreducible element of arbitrariness. This is []→[ ], and it unfolds the proper content of @@A.

Perhaps no topic is more frequently touched on in today's rapidly changing economic, social, and political environment than the indispensable, even urgent need, for institutions not to repeat themselves mechanically but to cultivate the capacity to reinvent themselves.

Monkeys, and other animals, hold you to a second "yes". Institutions, so deeply entwined to our spoken and written promises are latecomers in a long lineage of ב, the reason why "yes" immediately duplicates itself into "yes, yes."

---

[17] In Caputo, John D., Ed. and commentary, *Deconstruction in a Nutshell: A Conversation with Jacques Derrida (Perspectives in Continental Philosophy)*, The Fordham University Press New York 1997, 27–28.

Inner Conflict, Transparency/Opaqueness and the Unfolding of Anticipation

Michael de Montaigne wrote: "We are, I know not how, double in ourselves, so that what we believe we disbelieve, and cannot rid ourselves of what we condemn."[18] According to Sigmund Freud, "We long ago discovered that a thing which in consciousness makes its appearance as two contraries is often in the unconscious a united whole."[19] Whence this sense of inner division? The evidence suggests that our particular capacity for sense-making led to the transition from [ethics] into ethics. Edward O. Wilson, renowned evolutionary biologist, attributed inner conflict—at the level of *identity*, I would say—to multilevel selection over millions of years.[20]

> Probably during the habiline period, a conflict ensued between individual-level selection, with individuals competing with other individuals in the same group, on the one side, and group-level selection, with competition among groups, on the other. The latter force promoted altruism and cooperation among all the group members. It led to innate group-wide morality and a sense of conscience and honor.

How do these two forces compete?

> Within groups selfish individuals beat altruistic individuals, but groups of altruists beat groups of selfish individuals. Or, risking oversimplification, individual selection promoted sin, while group selection promoted virtue.[21]

I shall refer to this interrelating of individuals in their group, and with their group to other groups, as *intersociality*. Intersociality is the evolutionary origin of Bowen's two forces. Clearly, the paradoxical dynamics of groups discussed above operate in these groups. Wilson proposes that the conflict that arose in prehistory left humans suspended between *two extreme forces*, neither of which are we likely to yield to completely.[22] The two forces brought about intersociality and the paradoxical dynamics of relationships. This is the root of our social/ethical Ba-Bw. Again, in Polanyi's words, this is why *we can know more then we can tell.*[23] This is how our paradoxical human *identity* evolved.

Wilson's account of the unfolding of the ethical and its underlying forces involves new levels of transparency/opaqueness (T/O), and of anticipation. Brazilian author Clarice Lispector stated tersely what underlies Montaigne and Freud's statements

---

[18] Hazlitt, *The Complete Works of Michael de Montaigne*, 289.

[19] Quoted in Grunfeld, *Prophets Without Honor*, 40.

[20] Wilson, *The Meaning of Human Existence,* 27–28. I am grateful to my good friend Raul Alba for the gift of this book just as I completed the earliest draft of this chapter.

[21] Wilson 32–33. The "habiline period" refers to *Homo Habilis* or 'handy man'. The name "was given in 1964 because this species was thought to represent the first maker of stone tools. Currently, the oldest stone tools are dated slightly older than the oldest evidence of the genus Homo." (The Smithsonian Institution's Human Origins Program) https://humanorigins.si.edu/evidence/human-fossils/species/homo-habilis (retrieved October 2021).

[22] Wilson, 33 (emphasis added).

[23] Polanyi, *The Tacit Dimension*, 4.

when she reflected: "... perhaps lying is our sharpest mode of thinking."[24] Lying, a very high-level opacity, brings us back to EI and a very specific unfolding of borrowed content. It may be expressed in terms of the emergence of anticipation in five nesting stages that resulted from our own intersociality: anticipation (EI-1); anticipation of anticipation (EI-2); anticipation of non-anticipation (EI-3); non-anticipation of anticipation (EI-4); and non-anticipation of non-anticipation (EI-5). Thus, anticipation unfolds by doubling back on itself. Each subsequent higher stage of anticipation depends on all the previous stages. So, the unfolding of anticipation is recursive in the strict sense in which this term was defined.[25]

Chess is a game of anticipation. Whoever anticipates further, wins. No opaqueness is involved, everything is above board. Simpler board games introduce an element of luck, and thereby guesswork, to make up for their lack of complexity. The EI between the white cell and the bacteria is anticipation pure and simple (EI-1). In contrast, chess players continually anticipate that they must anticipate (EI-2). Indeed, constantly anticipating non-anticipation (EI-3), they seek ways to obscure their own strategy and guard against overlooking their opponent's. In non-anticipation of anticipation (EI-4), play becomes positional, with pieces placed in strong squares to combine for un-anticipated defense or attack. Finally, there is non-anticipation of non-anticipation (EI-5). This would be, for instance, preparation through the general study of chess. Think of expressions of regret and shock following a child's accident: "I cannot forgive myself for not having anticipated this could happen." The remedy to anticipate what we do not know that we do not know is general preparedness and prevention, the principle for the design and creation of emergency response.

This is the unfolding of anticipation as mentioned in Chap. 4.[26] This unfolding depends on the drift [A=A]➔[A = A] as, by natural selection or through social interaction, layers of EI are built one upon another. Anticipation folds back upon itself as EI reaches ever higher levels as an integrator-operator.

Unfolding anticipation drives moral agency. Hand in hand with the decoupling of Ba-Bw, growing individuation and autonomy give way to new forms of organization, deepening paradox, T/O, and intersociality. The horizon of time itself unfolds. Levinas remarked, "The very relationship with the other is the relationship with the future... [T]ime itself refers to [the] situation of the face-to-face with Other."[27]

One could write a lengthy treatise on the unfolding of anticipation. Born of ב, it could touch on every living act. Suffice to say that this unfolding underlies all development and learning. States transition to stages; integrators to operators. Anticipation begets anticipation, the meta structure-process of EI. Eventually, with

[24] Lispector, *La Manzana en la Oscuridad* ("Pero tal vez mentir sea nuestro modo más agudo de pensar").

[25] Chapter 3, section "Autopoiesis with or without cognition I."

[26] See section "EI as integrator-operator."

[27] Levinas, "Time and the Other," 44.

a radically decoupled Ba-Bw, humanity unhinged disgorges the nightmares of history.

The stages of anticipation named, we must wonder how they arose. Inherent to anticipation and relationship is transparency/opaqueness (T/O). In the early 1970s, Jane Goodall related how Figan, a male chimp visiting her camp, spotted a banana that had been overlooked by the other chimps in his group. Sizing up the situation, he deliberately removed himself to avoid glancing at it and giving the game away. Had he reached for it, the dominant male, Goliath, who sat underneath, would have snatched the banana for himself. No sooner did Goliath leave, without a moment's hesitation, Figan collected his snack. On several occasions he led the group away from Goodall's camp when bananas were on offer, only to return alone minutes later to have them all to himself. It became patently obvious that he did this deliberately when, one day, he returned "with his characteristic jaunty walk only to find that the high-ranking male had in the meantime arrived in camp and was sitting eating them." Figan flew into a tantrum and then rushed off to catch up with the group he had led away earlier.[28]

Figan's capacity for opaqueness is linked to high-level anticipation—in this case EI-3, anticipation of non-anticipation—and delay of gratification. Since then, many similar instances such as this have been recorded in many species of animals. Dozens of issues and controversies surround such behaviours, now known as "tactical deception." The topic is controversial because it involves pointed questions about method. How to systematically study events that are inherently occasional? Unsurprisingly, the most sophisticated instances of deception occur in higher primates.[29] But deception runs very deep in nature, bolstering the case for [ethics]. Mitchell and Thompson identified four levels of deception in animals:

1. False markings on animals, such as butterfly markings that indicate their heads are at the back end of their bodies as an aid to escape, or markings to make predators appear safe.
2. False behavior, such as a predator acting in a way to hide its predatory nature around prey.
3. Feigned injury to get or divert attention; for example, a parent bird feigning a broken wing to attract a predator away from its defenseless offspring.
4. Verbal deception, such as a chimp misleading other chimps to hide a food source or a human lying in order to deceive another.[30]

Humans exercise the full range. From camouflage to the pretense that all is well after we stumble—monkeys too feign to forage nonchalantly after a misstep—to deceptions like Fagan's, all the way to Ponzi schemes and systematic political demagogy.

Transparency/opaqueness is a continuum. It may be defined as the gap between what an individual knows and how it acts. Toward the pole of transparency, the gap

[28] Goodall, *In the Shadow of Man*, 97 (modified).

[29] See for instance Taylor Parker and McKinney, *Origins of Intelligence*; Tomasello and Call, *Primate Cognition*; Maestripieri, *Primate Psychology*; and de Waal, *Primates and Philosophers*.

[30] Mitchell and Thompson, "Deception," 21–29.

is small; toward the pole of opaqueness, it is large. Growing sophistication in the face-to-face calls for, and results in, ever more opaqueness. The monitoring of self and other sets up a vector for growing cognitive capacity and self-consciousness. Thus, the vector from [ethics] to ethics accounts for an increasingly sophisticated capacity for anticipation by which organisms make sense of others, of themselves, and of the world.

Unwittingly, we socialize our children to be opaque. Humans certainly do it, as do our fellow apes and monkeys. A mother disciplines her 4-year-old daughter for helping herself to a banana without asking permission. Doing so, she heightens her daughter's awareness of her own transparency. Overtly admonishing her about manners, tacitly the daughter learns she must sharpen her skills at opaqueness. Given our expertise at T/O, we regard animals as naïve. Why does a dominant male gorilla not hide his admiration for the demonstrations of a rival? He simply lacks our degree of malice. A dog presented with a chewed shoe feigns boredom, by our standards a lame denial. "Chimps are very quick to notice and interpret the eye movements of their fellows," said Goodall.[31] This is not restricted to chimps:

> Woolly monkeys are so sensitive and astute in their powers of observation, so skilled in interpreting the idiosyncrasies of human behaviour, that one is made to feel that one's inner thoughts and intentions were known by them long before they were given any outward form of expression.[32]

Caring for the monkeys at the Sanctuary involved matches of anticipation. For instance, scheduling constraints sometimes required an enclosure to be vacated promptly. We would hurriedly shut hatches pretending to wish to trap the monkeys in, and they rushed out in a flash, and *en masse*. This ruse had to be used sparingly—a no-brainer to us, consummate liars.

Theoretically, in a fully transparent social group, members would know what all members are up to. They could not help but communicate, and, therefore, there would be no need for anticipation or intentional expression. But this is not how it is, for a very simple reason: intention is born of opaqueness. As we saw, opaqueness is inherent to identity-unity-whole. The very organization of organisms requires opaqueness. Some states of unity-whole are outwardly evident to others. So, at identity I learn to cover them up. This coverup, birth of expression, seeds intentionality, which is inherently social. The more layered an organism, the greater the opaqueness, and, therefore the flowering of intentionality and expression. For a lovely example, consider *snuffling*, Woolly monkeys' expression of friendliness (Fig. 10.1):

> Snuffling is a highly emotional response that belongs to the appeasement, greeting and friendship behavior. It begins usually with a gesture of covering the mouth with a hand, accompanied by much teeth-chattering and cheek-puffing, and leading finally into an intimate huddle or embrace.[33]

[31] Goodall, *In the Shadow of Man*, 97.

[32] Williams, *Man and Monkey,* 121.

[33] Williams, *The Woolly Monkey*, 23.

Fig. 10.1 Baby Ben finds Ricky's highly emotional snuffling irresistible (Williams, *The Woolly Monkey*)

The full expression conveys: "I am not in the least threatening, my limbs are curled up, my eyes are shut, my mouth covered, and even my teeth are friendly." Intent is stated negatively. Many animals avert their eyes to make themselves look harmless, as an expression of friendliness. Thus, expression and intentionality unfold by T/O in the face-to-face. Intersociality calls for opaqueness to promote group, and group-to-group cohesion. Greater skills at T/O enhance individuation, anticipation, intentionality, and expression, as the face-to-face calls for ever deeper other-monitoring and self-monitoring. Leading in a game of chase a dog zigzag's to throw her playmate off track. Halting abruptly, front legs low, face-to-face ready to spring, the two minutely anticipate each other's next move. Agility aside, this training in T/O, self-monitoring and other-monitoring, inference, and delay, is the core of animal play. It sharpens *identity*, to which at some point we accord the names *consciousness* and *self-consciousness*. Opaqueness involves stillness, inhibition, not doing. A squid learns to hide by donning a costume and keeping still. It even learns evade its predator by riding on him.[34]

Inherent to @@A—joinings—, the sharpness of anticipation can be turned on objects:

> There are two main instinctive drives behind the monkey's investigation of a new object: is it dangerous? Is it edible? Cupboards, boxes, and jars must all be opened; inside may be something very tasty, or very dangerous, or both tasty and dangerous; one never knows. The monkey's chief concern is to find out what is underneath, behind and—most important of all—*inside*.[35]

We continually gauge our surroundings in terms of anticipatory affordances. What level of EI is required by what we are facing? Lower-level EIs are so predictable that

[34] Again, the film *My Octopus Teacher*, is a good instance of the artistic approach to inquiry into aliveness.

[35] Williams, *Samba*, 63.

we can afford to let them be background as higher-level anticipations take precedence in the foreground. To pay attention is to focus on the unexpected, that which is most difficult to anticipate about. To discern the intent of others—that is what is *inside*. Transparency and opaqueness grow each other. This is how intersocial animals sharpen their wits. Only a capacity for loyalty—iteratively keeping promises, both tacit and explicit—ensures long-term group cohesion. So, vampire bats are honor-bound to reciprocate blood loans to lenders on nights when they failed to secure their fill. These [promises] are ratified through social grooming that renders transparent the satisfied state of their bellies. Bats that break their promises risk not obtaining future loans and therefore bankruptcy—starvation.[36] Recurrently keeping [promises] constitutes reciprocity and fairness. Together they are indispensable for social cohesion and individuation.[37] This sets the stage for the eventual emergence of spoken covenants and written contracts. Thus, the T/O at the root of deceit and honor, individuation and expression, fosters the advent of an ever-deepening moral sense.

And so, the face, our sublime capacity for expression and inexpression, is an evolutionary strategy to heighten T/O. Thus, [face-to-face] leads to face-to-face, the full unfolding of relationship in the deeply paradoxical social encounters between fully individuated individuals. Increasingly complex and nested interrelationships unfold ever higher levels of anticipation, intentionality, and morality, making nuanced expression increasingly advantageous.

EIs, from the outset symbolic, are themselves eventually marked by EIs. So, for instance, monkeys can not only read from various cues in the environment that a treat will soon be on its way. The sound of fruit being cut is an EI. Before that, the unique pattern of the back door to monkey kitchen opening and closing when the ingredients for their fruitful are collected and washed. And who knows what other EI precedes it, which is transparent to them and opaque to us. Their delighted welcoming chorus of trills and *eeeolks!* become a shared [name]. Recurrent outward expressions about recurrent events become second order EIs, labels of previous EI sequences. [Named] distinctions eventually lead to language and concepts. Thus, the deeply sedimented T/O of @@A lays the foundation for abstraction and @@L.

The mutual co-creation of anticipation, expression, inhibition, interrelation, all together contribute to a deeper individuation and capacity from sensemaking. Extension—@@L—is thereby the outcome of deepening intension—@@A. So, organisms, identity-unity-wholes emerge sufficiently nested to be capable of

[36] Gerald Wilkinson's research on vampire bats' reciprocal donations of blood is described in Ridley, *The Origins of Virtue*, 62–63.

[37] Based on highly ingenious experiments with many species of animals, primatologist Frans de Waal proposes that the two pillars of morality are reciprocity (fairness) and empathy (compassion). These are shown in his TED talk, "Moral Behavior in Animals." Rabbi Jonathan Sacks contrasts between contracts and covenants. Contracts are zero-sum games about self-interest; covenants are cooperation that involve loyalty and trust and are about identity. See his video, "The Politics of Hope" and discussed in his book, *Morality: Restoring the Common Good in Divided Times*. These give shape to very different kinds of social wholes (wWs), yet in both the twin forces of individuation and togetherness obtain.

making sense through [names]—second order EIs. With names for objects and actions emerges language. This full elision of liminality, the play analog/digital, allows sense-making to take place in terms of [names], fixed concepts. That is digitally, in terms of @@L. It takes a deeply nested and integrated Ba-Bw for Ba-Bw to become decoupled enough for language to emerge, bring about @@L—along with the corresponding expansion of historicity.

Concretely, this is an outcome of the social dynamics of the intersociality described by Wilson. With the unfolding of anticipation and T/O, and the increasing labeling of EI, a sufficiently differentiated @@A emerges to ground the explicitation of @@L. Legislation, codified in writing—clear, distinct, static—brings about history, the transition from [ethics] to ethics, and the explicit differentiation of @@L.

One may obtain a sense of how @@A—joinings, relationship—is precursor to @@L—logic—from the subtle and highly complex nestings of anticipation and T/O at work in the life of a monkey colony. The grandmother monkey, Lulu, for instance, who lived to the ripe old age of 32, was quite unruffled on occasions when she was set upon by an over-dominant young male intent on re-directing at her his anger at Danny, long-time leader of the group and her son. Though this is a highly charged situation at the level of affect, she calmly swung around the room and deliberately withheld ropes as, perched in different spots, she paused to gauge her chaser's next move. If he dashed towards her around the perimeter beams, she would delay until just before he reached her before swinging back to the opposite side of the room. If, instead, he grabbed a rope along the way to cut her short, she anticipated his trajectory and either kept still, swung back to where she had come from, or aimed for a new more distant spot, all the while anticipating her antagonist's choices. Often, she set herself swinging on the center rope in wide circles around the room to keep her options open. One could almost see a glint of mischief in her eyes as she monitored her pursuer, who finally tired of the chase and sat stumped, invariably on the opposite side of the room. Such mastery of extension—the territory and its physical affordances as @@L—are coupled with a mastery of intension—the socio-emotional situations as expressed in terms of movement, or @@A. All monkeys eventually master both these territories and their grammars. Long familiar ourselves with both, their habitual placements around the territory and their activity spoke to us about the overall state and mood of the monkey group. Knowing them all individually, and the group's development, day-by-day over so many years, we could not help but read the room. This was verified time and again through innumerable large and small incidents. Thus, EI1, as it obtains for locations, speeds, and directions of movement, were the background of the higher-level anticipations expressed in relationships through emotion, mood, alliances, and so forth. Their management of extension, thereby served and sharpened their management of intension, together creating an endlessly rich and varied socio-emotional environment.

Nicholas Humphreys proposed that humans, like other primates, are *natural psychologists*. Over 3 months he spent at Dian Fossey's research center in Rwanda, he pondered about the remarkable intelligence of gorillas. Given their placid, almost uneventful daily life, what is it for? In *The Inner Eye,* he discussed the social function of intellect, concluding that, for gorillas just as for humans, social relationships present the greatest cognitive challenges.[38]

If the unfolding of T/O calls for, and makes possible, ever higher levels of anticipation, these eventually lead to [ethical] dilemmas. Carol Gale, an assistant to Jane Goodall, related this story about a chimpanzee that captured a juvenile baboon:

> The chimpanzee was seen hugging the baboon close to his own body, in the same manner as the adult chimpanzee hugs an infant. At the same time his face was opened wide in the characteristic grimace of fear. This strange behavior was baffling for the observer until it was realized that the baboon was still alive. Only when it squirmed or screamed did the chimpanzee express fear and hug the baboon against himself. Between times he gnawed on the baboon's head and arm. When the head eventually hung limp and the animal showed little sign of life, this fear-hugging display ceased.[39]

Alternating between regarding the baboon as prey and as a baby in distress, the chimpanzee vacillated between logic and relationship, low and high-level anticipation, @@L and @@A. A similar incident occurred when a juvenile female jaguar mauled a female baboon. As she sat over her prey, she came face to face with a newborn baby. The young jaguar was immediately moved to protect the infant and did so for hours, even from an aggressive pack of hyenas.[40]

Such dilemmas place huge demands on cognitive ability. So, Wilson stated that "all normal humans are geniuses at reading the intentions of others, whereby they evaluate, proselytize, bond, cooperate, gossip, and control."[41] In his book *Mindsight* on interpersonal neurobiology, Daniel Siegel reminds us that the brain is a social organ.[42] For Humphrey, the key function of art, most obviously the narrative arts such as theater, literature, cinema and even dreams, is to reproduce situations to sharpen our skills as psychologists. All this is, in a word, the function of *mythos*, which Karen Armstrong contrasted with *logos*. Both *logos* and *mythos* respond to a single overarching reality—the endless confrontation between order and chaos in successive higher levels.[43] A key stage on the way to the differentiation of ethics from [ethics] occurred as humans found the way to express profound dilemma by honoring the beasts that laid down their lives for their sake.

> The first great flowering of mythology, therefore, came into being when *homo sapiens* became *homo necans*, 'man the killer,' and found it very difficult to accept the conditions of his existence in a violent world. Mythology often springs from profound anxiety about essentially practical problems, which cannot be assuaged by purely logical arguments. . . . *Logos* is quite different from mythical thinking. Unlike myth, *logos* must correspond accurately to objective facts. It is the mental activity we use when we want to make things happen in the external world: when we organize our society or developed technology. Unlike

[38] Humphrey, *The Inner Eye*. The TV series by the same name was first aired when I lived and worked at the Monkey Sanctuary (Channel 4 Television Co. Ltd., 1996).

[39] Williams, *Challenge to Survival*, 55–56.

[40] Joubert and Joubert, *Eye of the Leopard*, 2006.

[41] Wilson, *The Meaning of Human Existence*, 30.

[42] Siegel, *Mindsight*. Daniel Goleman makes the point in his Forward.

[43] Jordan Peterson's starts here to analyze myth in *Maps of Meaning: The Architecture of Belief*.

> myth, it is essentially pragmatic. . . . A myth could not tell a hunter how to kill his prey or how to organize an expedition efficiently, but it helped him to deal with his complicated emotions about the killing of animals. *Logos* was efficient, practical and rational, but it could not answer questions about the ultimate value of human life nor could it mitigate human pain and sorrow.[44]

Sedimented over millions of years, the complementarity and eventual differentiation between @@L and @@A, logic and relationship, understanding and anxiety, must be reclaimed in biology. Long before the differentiation of logic, there was a keen understanding of ecologies in terms of networks of relationships. We in the West may leave @@A to the artists and the poets, but even this is very recent.[45] Farley Mowat learnt this in a conversation with Ootek, an Inuit who he struck a friendship with during a stint in the Artic while commissioned to unravel the mystery of the disappearing caribou blamed on wolves. Ootek explains that the wolf and the caribou are so closely linked that they almost make up a single entity. He recounts a story that reminds Mowat of the Old Testament:

> "In the beginning there was a Woman and a Man, and nothing else walked or swam or flew in the world until one day the Woman dug a great hole in the ground and began fishing in it. One by one she pulled out all the animals, and the last one she pulled out of the hole was the caribou. Then Kaila, who is the God of the sky, told the Woman that caribou was the greatest gift of all, for the caribou would be the sustenance of man. The Woman set the caribou free and ordered it to go out over the land and multiply, and the caribou did as the Woman said; and in time the land was filled with caribou, so the sons of the Woman hunted well, and they were fed and clothed and had good skin tents to live in, all from the caribou. The sons of the Woman hunted only the big, fat caribou, for they had no wish to kill the weak and the small and the sick, since these were no good to eat, nor were their skins much good. And, after time, it happened that the sick and the weak came to outnumber the fat and the strong, and when the sons saw this, they were dismayed and complained to the Woman. Then the Woman made magic and spoke to Kaila and said: "Your work is no good, for the caribou grow weak and sick, and if we eat them we must grow weak and sick also. Kaila heard, and he said 'My work is good. I shall tell Amorak [the spirit of the Wolf], and he shall tell his children, and they will eat the sick and the weak in the small caribou, so that the land will be left for the fat and the good ones.' And this is what happened, and this is why the caribou and the wolf are one; for the caribou feeds the wolf, but it is the wolf who keeps the caribou strong."

Mowat is astonished to hear a parable that illustrates Darwinian theory.[46] This is inquiry into living being in terms of @@A. Significantly, the story begins with Woman and Man and treats wolf and caribou as "almost a single entity." Well beyond natural selection, it offers a profound insight into the ecology in terms of joinings.

[44] Armstrong, *A Short History of Myth*, 30–32.

[45] Two magnificent instances come to mind where the exact opposite is the case, both about whales, whaling and their impact on civilization. Melville's *Moby Dick* and Heathcote Williams' *Whale Nation* (I never know where find it in my library, is it with zoology or with poetry?).

[46] Mowat, *Never Cry Wolf*, 89–90.

> In the pre-modern world, most people realized that myth and reason were complementary; each had its separate sphere, each its particular area of competence, and human beings needed both these modes of thought.[47]

This complementarity-differentiation is at work in the interrelation between the two hemispheres of our brains. In the 1960s, psychobiologist Roger W. Sperry discovered the left brain, also called the digital brain, specializes in the verbal, the analytical, and the orderly. It is dominant in reading, writing, computations, logic, sequencing, linear thinking, mathematics, facts, and thinking in words. The right brain, also called the analog brain, specializes in the visual, the intuitive, the creative. Less organized than the left brain, it is dominant in imagination, holistic thinking, intuition, the arts, rhythm, nonverbal cues, feelings, visualization, and daydreaming. Needless to say, this differentiation and complementarity of analog and digital is highly suggestive, given that it is the fundamental transition in the process of distinguishing. I would venture to suggests that this may be the general underlying principle in the architecture of brains. Bundles of nerve fibers form an information highway that links the two hemispheres, allowing them to complement each other. Logical and creative functions both require input from both hemispheres. So, the left brain handles language, while the right brain makes sense of context and tone; the left brain handles mathematical equations, while the right brain provides comparisons and rough estimates.[48] Specialized lateralization is inherent to the brain's functioning, even if significant cultural differences occur. For instance, literacy calls for significant neurological rewiring that, among other things, increases collaboration between the hemispheres as it shifts face recognition to the right hemisphere.[49]

@@L and @@A are described by many. Lonergan differentiated them in terms of two forms of understanding that coexist in us—"rational mind" and "emotive psyche." Mind is discursive logical reason, psyche the intensity of imagination and affectivity. The following is my own gloss of Lonergan. Notice that he touches on the key point of linear logic versus paradox:[50]

> A first difference regards contradiction. For mind there is a principle of noncontradiction or excluded middle: either A or not A. Not for psyche: we may love and hate the same person; we may find something at once ugly and strangely beautiful; we may consider an action partly good and partly evil. A second contrast is that each applies its own sets of categories. Whereas mind seeks classes and universals, psyche appeals to representative figures and metaphor. Univocity applies in the first—a one-to-one relationship between terms and referents, while figurative thought thrives on ambiguity. Third, each propounds truth in its own particular way. While rational mind seeks uncontroversial, clear and exact demonstration and encourages others to judge for themselves, psyche repeats and insists with variations,

---

[47] Armstrong, 31.

[48] "Left Brain vs. Right Brain."

[49] Deheane, *Reading in the Brain.*

[50] Mayer, "Interiority and the Challenge of Primatology."

> building by accumulation to a climax for full emotional effect and appeal. A fourth contrast is in the number of themes that each treats; mind converges upon single themes by appraising pertinence in its quest for rigorous distinction. Psyche diverges and touches upon a multiplicity of themes, condensing and combining even the apparently incompatible in its pursuit of emotional effect. In sum, while mind would seek flawless axioms and propositions, psyche thrives in hyperbole and metaphor.[51]

Aliveness—A@@—at once extensional and intensional, differentiates into @@L and @@A. It takes A@@ to know A@@, [body-subjects] to understand [body-subjects] as identity-unity-wholes interrelate to identity-unity-wholes. This is what finds expression through *mythos*. With higher levels of anticipation allowing for deeper inferences about intent, interrelationship and intension come to the foreground. The deeper understanding required of the world is, therefore, founded in animism. From the outset, at the deepest level, to know is to organize. When our social organ is turned to nature we advert to her in terms of relationships, just as Ootek did. One is struck by just how closely Newton's three laws describe the homeostasis of social systems and relationships: "*Every body [system] in a state of uniform motion will remain in that state of motion unless an external force acts on it;*" "*For every action there is an equal and opposite reaction.*" As for the third, *force equals mass times acceleration*, this is well known to monkeys and to any other animal that romps around.

Thus, understanding of nature is born with *mythos*. Personalizing the forces of nature brings them into the horizon of anticipation and the presumption of at least the possibility of prediction. The feedback process Ba-Bw unfolds evolutionarily, developmentally, historically, and biographically. Over billions of years, and in the course of single lifetimes, the proleptic unfolding of [ethical] at long last leads to ethics. We transpose our own nature as identity-unity-wholes onto things—*unity-identity-wholes*. We are primates who, based on a socioemotional foundation, draw social inferences about the world to make it intelligible.

Historically, @@L was abstracted only yesterday as we began to make sense of intersociality through language—a product. If Plato and Aristotle worked the isomorphism between WW and wW, the technological revolution of the seventeenth century led, with its mechanical and conceptual products, to the Cartesian overvaluation of @@L and devaluation of @@A. Descartes rejected Aristotle and with him biology. In the twenty-first century we still refer to "biological mechanisms." Literacy, whether scientific, artistic, or religious, rewires our social organ for rigorous higher levels of anticipation strongly skewed towards @@L.[52] In the late nineteenth century, in response to the hegemony of @@L, represented by science, conceptualism, and technology, Nietzsche attacked Socrates for giving birth to @@L and called for a return to @@A. For Nietzsche, @@L was represented by Apollo—mind, order, harmony, progress, clarity, logic, the principle of

---

[51] Mayer, "Interiority and the Challenge of Primatology," glossing Lonergan, "Exegesis and Dogma."

[52] For a meticulous examination, see Henrich, *The WEIRDdest People in the World.*

individuation—and @@A by Dionysus—passion, chaos, disorder, intoxication, emotion, ecstasy, and unity or joinings.[53]

Logos marginalizes paradox. By contrast, in *mythos*—literature, poetry, religion, mysticism—paradox is ubiquitous. In his early work, Freud pointed out that dreaming requires inhibition of locomotion. Locomotion occurs in the world of extension.[54] This alone shows the deep neurological wherewithal necessary for transparency/opaqueness (T/O). In dreams, with extension and @@L silenced, *identity*, in its aspect of intension, is explored by skipping more freely through *unity*. How are these findings in @@A, to be expressed? By *mythos*. In *Languages of Truth*, Salman Rushdie referred to "tales full of beautiful impossibility, which were not true but by being not true told the truth, often more beautifully and memorably than stories that relied on being true."[55] Consider this story of two jackals:

> Karataka, the good or better guy of the two, and Damanaka, the wicked schemer. At the book's outset they are in the service of the lion king, but Damanaka doesn't like the lion's friendship with another courtier, a bull, and tricks the lion into believing the bull to be an enemy. The lion murders the innocent animal while the jackals watch. The end.[56]

This could be the tale of how a parasite defeats the immune system, or about the lethal effect of an addiction, or about the friend of a child who is no longer welcome because of his bad manners. *Mythos* is all about relationship, as are mysticism and theology.[57] The problem of evil, for instance, is at bottom a problem of Wholes.

The Structure of [Ethics]

Returning to Lonergan's ethics, it provides an elegant, if somewhat abstract, way to verify the isomorphism between [ethics] and ethics. The argument is worth pursuing because of its profound consequences. His theory of ethics is expressed through a tripartite structure because it itself is based on the isomorphism between ethics and cognitional activity: "The three levels of the good correspond to the divisions in our knowing: experience, understanding, and judgment."[58] Just as we discover through self-appropriation the process-structure of our cognitional activity may be specified in terms of experience, understanding, and judgment, so we discover that the process-structure of our ethical activity may be specified in terms of three kinds of goods.

---

[53] "Appolonian and Dionysian."

[54] Freud, "Proyecto de una psicología para neurólogos," 244.

[55] Rushdie, *Languages of Truth*, 15.

[56] Rushdie, 16.

[57] The paradoxical nature of mysticism is treated at length in Walter T. Stace, *Mysticism and Philosophy*, and Gershom Scholem, *Major Trends in Jewish Mysticism*.

[58] Lonergan, *Understanding and Being*, 226.

> Our notion of the possibility of ethics has its ground in the extension of self-appropriation from the subject as knowing to the subject as both knowing and doing. The object of experience ties in with objects of appetite, the good of order ties in with the grasping of intelligibility, value ties in with the judgment that leads to choice.[59]

Call these three kinds of goods G1, G2 and G3, and the whole structure G123:

G3. Goods of value.
G2. Goods of order.
G1. Individual goods.

The first level (G1), then, is the elementary level of the good. An individual good is a spontaneous object of appetite. When obtained, it is experienced as pleasant, enjoyable, satisfying. The second level (G2), the good of order, "is an intelligible unification, something in the way of organization, though it may be much more spontaneous than any deliberately conceived organizing blueprints." The third level (G3), value "lies upon the level of judgment; it is the rational choice."[60]

It follows from Lonergan's isomorphism between cognitional activity and ethics that, just as in cognitional activity the three levels implicitly co-define each other because they form part of a single unified dynamic. This is also the case with the levels of ethics. Here too the three levels require, and therefore co-define each other. For this very reason, the three levels of the good in ethics may be generalized, by isomorphism, to the three levels in [ethics]. Call the three components in [ethics] C1, C2, and C3, and the whole structure C123. If G123 obtains for ethics, then we must demonstrate the isomorphism between G123 and C123.

First, note that Lonergan's differentiates between cognition and ethics by differentiating between the knowing of cognitional activity, on the one hand, and the knowing and doing of ethics, on the other. In a range from the differentiated to the compact, on one end is the high level of differentiation that obtains for human subjects, and on the other the deeply compact nondifferentiation that obtains for elementary [subjects]. Therefore, C123 obtains for the whole range while G123 of ethics, as accounted for by Lonergan, is the highly differentiated instance of C123 that obtains for humans. Since it pertains to the whole range, then, C123 is the proper content of [ethics]. This proper content obtains for any and all instances of @@A—joinings, relationship. Indeed, this suggests that C123 can serve to state the proper content of @@A.

If [ethics] and ethics are isomorphic, several propositions follow that can be verified. First, C123 and G123 must be isomorphic. Second, C123, like G123, must be implicitly defined. Third, G123 must refer to a specific, namely the human segment in the continuum C123. So, it must be possible to determine what characterizes this segment. Fourth, C123 must provide an alternative way of stating the proper content of @@A—relationship, joinings.

---

[59] Lonergan.

[60] Lonergan, *Understanding and Being*, 226.

Start with the isomorphism between C123 and G123. *Prima facie*, the isomorphism between C1 and G1 is unproblematic since it refers to spontaneous [objects of appetite] at the elementary level. For any living being, [objects of appetite] may be defined as any input necessary for *entelecheia*, their being-at-work-to-remain-themselves. Likewise, the isomorphism between C2 and G2 is uncontroversial since both refer to recurrence. This is particularly so since, as we just saw, Lonergan refers to G2 as "intelligible unification, something in the way of organization, though it may be much more spontaneous than any deliberately conceived organizing blueprints." The third level (G3), value, according to Lonergan, "lies upon the level of judgment; it is the rational choice."[61] The isomorphism between C3 and G3 seems less obvious, until it is recalled that it was demonstrated that judgments are a kind of distinction, and that both are verified by recurrence.[62] It was said at the outset of Chap. 1 that verification is similarity displaced in time and accomplished through repetition. "If, for science, the verifiable is the repeatable and to verify is to repeat, for living being in general, to repeat is to verify."[63]

But more must be said about value. [Objects of appetite] (C1), pertain to individual [body-subjects], and therefore so does their recurrence (C2). But these [body-subjects] always already belong to joinings, @@A. In their capacity as partners of such joinings, their [objects of appetite], and particularly their recurrence, is inherently referred to something beyond them. Namely, the partnership itself and its continued success. Recall the bacteria and archaea whose genomes are suspiciously small because "they do not have enough information to build a fully functional cell or to complete the metabolic transformations that convert food into energy."[64] If "the entire community has the capacity for a certain function, like nitrogen cycling" it is because their individual goods (C1) and their recurrence (C2), beyond them, work for the good of the partnership as a whole.[65] Their C1 and C2 are referred to something beyond them and the recurrence is verified in longevity, both of the partnership and of individuals. This verification is C3, it "lies upon the level of judgment." In C123, the binary character of the digital is isomorphic to what Lonergan specifies as "the rational choice."[66] For Lonergan, the rational refers to judgment and stands for the binary yes/no, true/false, guilty/not guilty. Individuals manage themselves in their participation in partnerships, not only for themselves but also for the sake of the partnership. This is value. This is why, in the Inuit myth, the caribou and the wolf are one; it is why the dominant male woolly monkeys Danny and Django work through their conflict; why Jane and Jack manage their punctuality as a couple; and why superpowers contribute to each other's safety through a policy of Mutually

---

[61] Lonergan, *Understanding and Being*, 226.

[62] See Chap. 5, section "Foundations, and proper and borrowed content of distinction."

[63] See Chap. 1, 3.

[64] Marlow and Braakman, 36, citing Laura Hug, professor of environmental microbiology, University of Waterloo.

[65] Marlow and Braakman, 36, citing Laura Hug, professor of environmental microbiology, University of Waterloo.

[66] Lonergan, *Understanding and Being*, 226.

Assured Destruction. In short, recurring interrelationships between partners uphold the recurrence of their interrelationship. This is G3, and prime values, whether spoken or unspoken, are surviving, thriving, flourishing. Thus, there is isomorphism C1 to G1, C2 to G2, and C3 to G3. Therefore, C123 and G123 are isomorphic.

Onto the second point, it follows from this that, if G123 is implicitly defined, so is C123. Since G123 is an instance of C123, it is worth noting that the near circularity of the concluding statement above ("In short, recurring interrelationships between partners uphold the recurrence of their interrelationship") points to the indissoluble relation between partners while they belong to partnerships. The capacity of partners as "parts", and their partnership as "what is more than the sum of the parts", implicitly define each other.

As to the third point—that G123 refers to a specific, namely the human segment in the continuum C123—this points to borrowed content. The structure C123 is the proper content of [ethics], just as G123 is the proper content of ethics. The difference lies in the borrowed content. Specifically, it lies in judgment. How can this segment be characterized? The point was taken up before.[67] Ethics refers to a living being capable of rational choice following a self-reflective judgment that includes the question whether all pertinent questions have been asked. Needless to say, biases may be at work that prevent all pertinent questions from being asked. This is what defines the unethical, which is nonetheless *qua* ethical.

Fourth, as for C123 being an alternative way to state the proper content of @@A—relationship, joining—, the same is true here as in the second point. A "recurring interrelationships between partners upholds the recurrence of their interrelationship" could stand as a definition of @@A—joinings.

Thus, [ethics] and ethics are isomorphic, and ethics is a particular stage in the differentiation of [ethics]. Thus, C123 can stand as the proper content of [ethics], just as it can stand as the definition, and proper content of @@A. In other words, C123, @@A, joinings and [ethics] are all synonymous. Each highlights a different aspect of aliveness, the ground scheme, the process-structure distinguishing distinctions.

Since life emerged and remains in plurality, there is no such thing as single isolated living beings, nor are there acts of single isolated living beings. These are abstractions. Likewise, values; these too are abstractions. On their own, G1 and G3 (and therefore C1 and C3), are abstractions. Their concreteness springs from G2/C2: *con-crēscere*. From joinings of living beings alive because they are in *entelecheia* together. G2/C2, goods of order, refers to organization, organized products such as organizations or material products and material infrastructure. Using Lonergan's phrase again, unlike the abstraction of values (G3) and single isolated acts of living beings (G1), G2, "is an intelligible unification, something in the way of organization." G2 is the concrete synthesis we live in, which embodies the abstract G1 and G3. This is what is expressed by the very helpful tripartite division G123. As identity-unity-wholes, organizations function because we lend them our

[67] See Chap. 5, section "Foundations, and proper and borrowed content of distinction."

identity. In doing so, they in turn in-form our identity and therefore us. We grow and develop in families, schools, teams, churches—all other organizations we belong to. Our learning and development as identity-unity-wholes, then, is in these very *con-crēscere* goods of order. Strictly speaking, the very nature of organization is [ethical]. Since all joinings at all levels are organized, C123 is true of all joinings.

One way to approach this is through liminality. Recall that liminality stands for - -, which in turn stand for continuity-discontinuity. In terms of - - then, C123 would be represented as follows:

C3, goods of value:—

C2, goods of order: - -

C1, individual goods:- -

Continuity and discontinuity on their own are abstract, only together do they make sense concretely. Single acts of individuals are discontinuous and abstract; and values are continuous and abstract. Together, however, single acts of individual vis-à-vis a value brings about a concrete intelligible unification in the way of organization. This converges with the idea that the singe isolated values on/off, or true/false are abstract and derivative from a more primitive and primary concrete oscillation. As was shown, this is another way of saying that @@L derives from @@A.

If continuity and discontinuity on their own are abstract, together they are concrete and paradoxical. We have delved at length on the paradoxes of organization. These obtain at all levels in joinings, all the way down to our cells. Autopoietic units look after their own needs (C1) while also looking after each other's needs (C2) and looking after their joining vis-à-vis their environment (C3). We are back to the fundamental diagram of @@A and Derrida's "yes, yes." Organic wholes, like social wholes, must constantly reinaugurate, reinvent themselves in order to thrive. This has the structure of a promise, of ongoing renewal. This is why relationships between autonomous units are normative. Relationship, as such, is normative. Inherent to the proper content of @@A—joinings—is this normative character: [ethics].

It is worth briefly retracing how we got here. ב is the heart of the proper content of @@A as act of distinguishing, the verb. Insofar as distinguishing involves ב, part of the proper content of @@A is also anticipation, a degree of arbitrariness, and the switch from analog to digital. The proper content of @@A as it occurs in two-element regulatory system and in embodied distinctions is plurality, joinings, *con-crēscere*, intensionality, relationship, organization and thereby their dynamics, which involves the repulsive and attractive forces, paradox, and therefore trade-offs, division of labor, identity-unity-whole, transparency/opaqueness, Ba-Bw and their ongoing decoupling or historicity. These are all normative, and this why all these express the proper content of @@A. In sum, the proper content of @@A is C123, [ethics]. These various terms highlight various aspects of [ethics], or, what is the same, @@A as aliveness, the ground scheme, the process-structure of distinguishing distinctions, as well as []→[ ], their unfolding.

The wherewithal for joinings, organization, or whatever term we choose to use, is identity-unity-whole. Again, inherent to all these terms is liminality. This is how we arrived at hymenomorphism, the ground scheme of distinguishing distinctions. In contrast, the notions of matter and form in hylomorphism pertain to @@L, the upper scheme. The two notions, matter and form, are abstract, far closer to @@L and like the values and acts of single isolated individuals associated with substantialism. In Aristotle's hylomorphism there is no fundamentally concrete category, like C2 or G2, because there is no *con-crēscere*—that which is *for itself* regardless of us.

Recall what was said in Chap. 7 in the section "Products": "The key point is that social structures are products and that *qua* ethical behavior sustains them. To repeat, it is crucial to note that ethics as a whole is grounded in social products. We will return to the significance of this in the final chapter." Here we are. These products, including their attendant material wherewithal, like the ballast of a boat, provide organizations with great stability—the more so given their inherent paradoxical nature. No doubt this is a key reason why it is so tempting, even easy, to think of organizations in mechanical terms. Moreover, this too is why we are tempted to reinforce organizations with formal C2s. Again, this helps, or at least appears to help, overcome the anxiety inherent to their paradoxical nature.

Most strikingly, this is why hylomorphism is not normative. Hylomorphism presents us with is/ought, two spheres that do not touch each other—precisely for lack of C2, that operational, concrete middle term. Indeed, they are abstract precisely because, though they do not touch each other, they require each other. In hymenomorphism, by contrast, C2 is, as has just been shown, about the embodiment of recurring actions-that-instantiate-values. Lonergan shows that the immanent structure of cognitional activity, which is isomorphic to G123 and thereby C123, is normative, ever tending towards increasing intelligibility. Questions lead to answers and answers to further questions—it is that obvious and simple. This is rooted in the distinctional nature of living being. Aliveness is the ongoing interrogating of the environment. And this interrogating is inherently normative. Not surprisingly, this is what we discover by adverting to our own cognitional activity. The cognitional is a species of the distinctional. The C2 inherent to living being is our nature as identity-unity-wholes. The result of this nature is that we assemble the world in term of *things*—Lonergan's *unity-identity-wholes*. Thus, there is a convergence between biomorphism-aliveness and biomorphism-world, Ba-Bw, how we are made and how we make sense of the world. Again, this is why @@A is inherently normative.

The normative, rooted in @@A, is [ethics]. We reach for intelligibility through hymenomorphism. Only liminality—nestings of membraned cells—lead actually and concretely, by way of *con-crēscere* of identity-unity-wholes with one another and within identity-unity-wholes, to C123. To state it differently, living in intersociality, we are organic wholes (WW) made up of organic wholes (WW) interacting with each other in social wholes (wW) and with other social wholes (wW). Given that this is our nature, this is how we construct the world. We do not choose to do so; we do this because this is how we are constituted.

Thus, @@A and [ethics] are synonyms. Together, they point to a fundamental structure-process that is immanently normative. Therefore, the isomorphism between [ethics] and ethics is central to the proper content of biology as well as its method, and to the most fundamental is/ought questions. It follows that self-appropriation is a key resource for inquiry into organization, both within the discipline of biology, and in our actual experience and participation in organizations. We grasp relationships because we are both made up of them and involved in them.

Franz de Waal proposed that reciprocity and fairness are the foundations of morality. Based on a series of experiments now reproduced multiple times with pairs of various species of animals, both participants clearly expect to be rewarded equally for a similar task, particularly if accomplished cooperatively. In the experiment, one member of the pair is deliberately short changed, and this leads to loud complains.[68] The very setup of the experiment is an implicit covenant between the two subjects and the researcher. Thus, the setup, as such, is G2, goods of order that embody values. We all have such covenants with our pets. I take my cats out for a walk every day around 3 pm. The very recurrence embodies this value. Delays are met with long looks, urgent meows, and hops onto the desk to gently paw my face to remind me.

Reciprocity and fairness are implicit in the recurrent process-structure of joinings. This is their nature in ethical terms. These are C3s reflected in what the joined wholes do, whether social or organic, as they recurrently co-determine each other (C2) and insofar as the needs of both are met (C1). C123 makes partnerships viable. This is their proper content. Borrowed content are the concrete differences that make a difference to specific partnerships. The difference between partnerships such as Prochlorococcus-Pelagibacter and de Waal's subjects is that the former are selected for and last over a few millions of years, while the latter is adopted in seconds and lasts a few minutes. The difference between them is in the borrowed content; the proper content is shared by both.

The large variety of animals with similar results in these experiments shows how widely grasped and how central reciprocity and fairness are in the animal kingdom. And the isomorphism in the Prochlorococcus-Pelagibacter partnership shows that aliveness is C123 all the way down the line. Or, properly speaking, it is C123 all the way up the line. This is why from viable partnerships of collectives, higher-level individuals made up of joinings can emerge. At work throughout was expressed by Levinas, the "rise and repetition of the face-to-face encounter, or the intersubjective relation at its precognitive core; viz., being called by another and responding to that other." To paraphrase Levinas:

> If precognitive experience, that is, *living* sensibility, can be characterized conceptually, then it must be described in what is most characteristic to it: a continuum of sensibility and *anxiety*, in other words, sentience and *heightened states that results from the two forces*.[69]

[68] See Franz de Waal, *Moral Behavior in Animals*.

[69] "Emmanuel Levinas." Inserted in italics is my own term *living*, followed by Bowen's term *anxiety* and its definition.

For Levinas, De Waal, Wilson, Bowen, and Bowlby the term *reciprocity* points to the two forces of fusion and individuation. Following Bowen, I have specified five expressions that follow from them. During a long evolutionary process, individuation unfolds from skin-to-skin to face-to-face. The face is a prime organ for anticipation and transparency/opaqueness (T/O). The unfolding of T/O and anticipation is what allows for the gradual transition from C123 to G123. This is the gradual transition from [ethics] to ethics.

Recall the statement cited above:

> The evolution of cooperation requires two conditions. First, there is some ecological or efficiency benefit to cooperation. Second, there is some mechanism that leads to the benefits of cooperation being directed back to the operator and/or their relatives.[70]

This could stand as a biological Golden Rule or [categorical imperative]. Kant framed his ethics in terms of G2. Sociology tends to be in terms of G2. Bowen's terms are articulated in terms of G2, but, incorporating paradox as they do, they express @@A as experienced by individuals, thereby aiding us in the leap from the sociological to the psychological-historical. A key corollary that follows from this is that all reflection on ethics may be repurposed for inquiry into [ethics]. As a treatment of @@A, this is a vast repository on *mythos* for inquiry into biology. This brings us back to the different approaches to ethics as pointed out by Byrne, which could just as well apply to psychology and *mythos* generally. All may be framed in terms of G1, G2, G3 and integrated as G123. For a single example, Leon Kass' book *Founding God's Nation: Reading Exodus,* may be read as a foundational integration of G123 in Western Civilization.[71] A foundational set of *mythos* about nature is crushed by the 10 plagues, and brings about a parting of the ocean that leads up to a systematic legislation set in writing, the inauguration of @@L. The kernel of this legislation, 10 short sentences set in stone, to this day remains an enduring portal from @@A to @@L.

Language, that human product, offers unmatched possibilities for transparency and opaqueness. The possibilities for the deliberate opacity of lying bring about a revolution in the face-to-face. The various stages in the history of language are fundamental to the unfolding of social wholes (wW). Still, the four levels of deception in animals show that opacity goes back a very long way. The picture of monkey Ricky snuffling shows that his gesture carries the intention of avoiding expression, just as Figan when he steps away from the bunch of bananas to avoid giving them away. This is what faces are for, to practice ever higher levels of anticipation. More than once I have been seduced by a monkey's highly emotional snuffling, only to look up from this intimate huddle to catch them with an open eye calculating, "how am I doing?"

---

[70] West et al., "Major Transitions," 10113, emphasis added.

[71] Kass, *Founding God's Nation.*

There was an occasion when Emma and I were walking in the sanctuary grounds along the top bank. As the monkeys always do, she walked on all fours leading me with her tail holding my hand. Slightly ahead I noticed a snail on the ground. A few weeks before the vet had pointed out that this might have been the source of the nasty flu epidemic the monkeys had. So, we decided to stop them eating them. It was a shame because the monkeys relish them, and they are a rich source of protein. After some struggle, they seemed to have resigned themselves. I wondered what Emma would do. The last thing I needed was a temper tantrum in full view of the public with her screaming at the top of her lungs as she had done when we first started seizing them from her. But she had given up on them, so I said nothing. To my delight she walked right past it. I was relieved. At the very last moment, though, she discretely picked it up with her foot. Emma, no! I exclaimed. You…! She glanced at me, reluctantly dropped it, and walked on.

The greater the capacity for opacity, the greater the sense of self. This late stage in the unfolding of *identity* and relationship points to a sharp differentiation of the [social] and the [ethical], while also suggesting that isomorphic use may be made of the stages of moral development as first suggested by Lawrence Kohlberg.[72] *Mythos* expressed T/O and the paradoxes of @@A. This refined expression led to *logos*—and thereby the paradox of labels, the block to un-naming that wedges between the saying and the said. We have seen how labels are both the link and the barrier between distinguishing, the verb and distinctions, the noun. Thus, they are a barrier in the movement between aliveness and logic (@@A←→@@L). This problem, rooted in the historical, is the reason why culture both promotes and becomes an obstacle for inquiry into aliveness.

This opens a vast new topic—*how* to make use of *mythos* for inquiry into aliveness, because we bring historically local concepts to our understanding of evolution. For a single example, consider the differentiation between contracts and covenants introduced by Rabbi Jonathan Sacks to make sense of the state of current society. Both are joinings, but while contracts are zero-sum games, covenants are win-win. This is a differentiation within social wholes (w/W), and therefore an invitation to refine this notion. Marriage is a covenant, not a contract (Sacks 2020b). Sacks points to our tendency to reduce the social and the ethical to the economic. In his own terms, he treats the topic of the reduction of @@A to @@L. Reading his critique closely, and isomorphically, could potentially highly enrich our outlook in biology. So, for instance, Saks' critique of our conception of the social in terms of contracts echoes Marlow's and Braakman's point about our blind spot for microbial partnerships. They thereby show that, whether we are aware of it or not, biology is always already informed by culturally acquired habits of mind. Thus, we implicit transpose habits in ethics to conceptions of [ethics]. We must therefore be historically astute to explore life, the more so if when inquiring into life in terms of @@A.

[72] Kohlberg, *The Development of Modes of Thinking*.

Returning Home

Perhaps the most devastating consequence of the failure to understand aliveness in its own terms is existential. It is a source of angst, a sense of being alien in the world. This has been one of the most important discoveries along this long journey. Consider two expressions of this sense of alienation. The first is Lonergan's:

> Unless there is a moral agent responsible for the world's being and becoming, the world cannot be said to be good in [the] moral sense. If in that sense the world is not good, then goodness in that sense is to be found only in man. If still men would be good, he is alien to the rest of the universe. If on the other hand he renounces authentic living and drifts into the now seductive and now harsh rhythms of his psyche and of nature, then man is alienated from himself.[73]

But if the living world is [ethical], and therefore [good] in the [moral] sense, this has profound transcendental implications. Far from alienating us, our own ethical nature takes us beyond ourselves and embeds us in the [ethics] of nature and, thereby, in the world.

Rabbi Jonathan Sacks made a similar point to Lonergan's:

> There are only two serious possibilities to be entertained by serious minds. Either the one put forward by the Torah that we are here because a Force greater than the universe wanted us to be, or the alternative: that the universe exists because of a random fluctuation in the quantum field, and we are here because of a mindless sequence of genetic mutations blindly sifted by natural selection. Either there is or there is not meaning to the human condition.[74]

If [ethical] meaning making is truly inherent to aliveness, then this is a false dichotomy. There is a third possibility. Our biosphere as a whole, is the locus of meaning. Life is joinings making meaning. The name Gaia stands for the ecology of mind of the planet as a whole, constituted by the distinguishing activity throughout macroenvironments, environments, superorganisms, organisms, organs, tissues, cells, bacteria, viruses. Over eons, their aliveness and their meaning-making have co-emerged. The proper content of aliveness (@@A) that constitutes them all is common to all, as is the proper content of distinction. All are mind-like active participants in the evolution and maintenance of the biosphere. This would be so in any planet with life. And this is a feature of our universe. The evolutionary history of a world with living beings—any such world—is the joint unfolding of intelligibility. Throughout the universe there are pockets of joint meaning making, therefore [ethics], and therefore meaningfulness. Far beyond the human condition, [ethics] is the condition of aliveness in the universe.

This places Sack's two possibilities in perspective. For a start, a "Force greater than the universe" implies a distinction, and these originate with living being. Does this Force draw distinctions? This Force must itself be a distinguishing distinction. Could it be to me as I am to one of my white blood cells? What do my cells know of

[73] Lonergan, "The Subject," 86.

[74] Sacks, *Judaism's Life-Changing Ideas*.

me? Are my cells and I in the same living-knowing horizon? What new understanding is invited by the idea that God must share fundamental features of organisms? By Rovelli's account, the distinctions we draw give form to our universe.[75] Could God's aliveness be at once within and beyond our horizon, and thereby both within and outside our universe? Living *entelecheia* is founded in the arbitrariness intrinsic to ב. Strictly, this condition for the possibility of aliveness is beyond the material world, the world of cause-effect. Whence this arbitrariness?

The third possibility is that the "Force greater than the universe" may be more humble, more local, the force of Gaia as an ecology of mind. This unfathomably massive force of our biosphere is not random fluctuations in the quantum field; these are not sequences of genetic mutations that result in mindless blind sifting by natural selection. Sacks is right, either there is or there is not meaning to the human condition. There is meaning to the human condition, but it is born of the immense and endlessly transforming network of aliveness, even if it is a Force many orders of size smaller than the one he has in mind. And, yet, the universe at large seems to be primed to give rise to humble local forces, indispensable to which is an arbitrariness that transcends cause-effect.

If nothing else, the foregoing reveals the reach of inquiry into aliveness in terms of distinguishing-distinction. This brings us back to the beginning, the proper conjugate form to understand aliveness. We can come out of the bubble. Like the traveler in the Flammarion engraving who stumbles upon a glimpse beyond the veil, we are stumbling back home into the world of nature. The miracles now before us are not the hidden mechanisms of the world, but the astonishing lives of trees, plants, mushroom, animals—our village (Fig. 10.2).

A biology of aliveness shifts from objects and processes to joinings and [ethics], from substantialism to liminalism. We can awaken to a new kind of understanding. Think again of the Hegel-Lonergan nest of terms. *Implicitly*, aliveness is always already deeply intelligible to us. But, assuming our incompetence, we do not even consider making this knowing explicit. In the current ideal of knowledge, the idea that we know about aliveness because we are ourselves alive seems absurd. But Louie and Lulu really are very skilled at math. We excel at navigating our social universe, a world of organizations that provides a superb template to understand organisms. This sounds like an *abstraction*. The formal justification of this idea in this book has certainly called for abstraction. So much so that the dominance of logic (@@L) might be reinforced, rather than provide a scaffold to access aliveness (@@A). The medium is the message. An account in logical terms implicitly affirms logic. The first step to overcome abstraction is to differentiate between distinction and distinguishing, the noun and the verb. That is, to self-appropriate, to advert to our own distinctional-cognitional activity. This itself is *alien*, and all the more so in biology. *Mediation* is to begin to make sense of what we observe in terms of ב, the spirallic, joinings, the proper content of @@A, [ethics]. It is to begin to differentiate the intelligibility specific to aliveness with the aid of differentiations such

[75] See Chap. 7, section "The nonliving."

Fig. 10.2 Flammarion engraving

as scientific approach/artistic approach, mind/psyche, *logos/mythos*. This starts us down a slow and gradual path towards *reconciliation*.

We must stop hiding from the We. We *are* relationships in relationships. We are aliveness (@@A), a We that through and through involves affect. Spontaneously and without realizing it, we grasp the drama of the bacteria chased by the white blood cell. It makes sense through a deep and subtle awareness of the levels of our own anticipation. This is the solid foundation of our skills at intelligibility. Reasoning comes later and is deliberate.

In a world teeming with life, it is shocking how little access children of Western cultures have to a disciplined and rigorous course of learning about nature that is not under the aegis of science. I, for one, yearn to be surrounded by animal sanctuaries, environmental and restoration projects run by artists, musicians, leaders from other cultures, to help us regain our intimacy with nature. This book has been the work of a lifetime. It too stands as a testimony to the dominance of science and mechanical thinking over nature. I too had to submit to the imperatives of our scientific lobsided culture to make myself heard. The perspective of our youth ought to be formed by full immersion experiences with nature, like my own decade at the Monkey Sanctuary. The planet would be an infinitely better place if we all enjoyed friendships with a Charlie, a Kate, a Laura and a Louie.

Realizing that we have always participated in joinings reenchants the world. We invite ourselves back to aliveness. The split between *logos* and *mythos* reflects what C. P. Snow called *the two cultures*.[76] If this split between the sciences and the humanities is, as he argues, a major handicap to solving the world's problems, it is also a major handicap to understanding aliveness. Troves of understanding will remain marginal to learning about aliveness so long as *logos* and *mythos* remain split. To bring them back together, aliveness must be understood in terms of its proper conjugate form. This is to shift from hylomorphism to hymenomorphism.

The monkeys and we understood each other. For years we trained each other's mirrors neurons to exquisitely anticipate each other. We share a couple billion-year history of signaling process-structures. This, too, is transcendental; it took us beyond ourselves. We are a We, free to regard each other as infinites. We are outside the mechanical bubble together. We share an inner sense of subjectivity thanks to a colossally extensive shared evolutionary history. It follows from the proper content of aliveness (@@A) that all living beings share a consensus. We were all built together and together we build the world. To be outside the bubble is not to reject the outer shared objectivity. Rather, it is to realize that its mechanical framing has obscured our inner shared subjectivity, the very subjectivity that is the condition for the possibility of logic (@@L) that obscured aliveness (@@A) in the first place. Outside the bubble, beyond an ideal of knowledge of clear and distinct ideas with no room for paradox, vast resources await.

We have in myth, in historiography, in literature, in treatises on ethics, vast reservoirs of insights into the nature of living being. What might Spinoza's *Ethics* teach us about [ethics]? This question may be generalized to any thinker. Indeed, to any venture in sense-making—painting, poetry, music. By isomorphism, all are treasures about the play of paradox. Societies that value *mythos* value elders for the same reason that elephants do. They have been there before; they know the way. *Mythos* is not merely an expression of yearning; it discloses our deep embeddedness in the world.

The emergence of ethics is an evolutionary blip in the chronicle of [ethics]. With ethics, a new language of truth emerges—*mythos* and its explanatory power. It is time to reclaim its original intent, to understand aliveness in the ongoing struggle between order and chaos. *Logos*—practical, clear, distinct, and indispensable—has trapped us in a bubble of transparency. In this bubble, stories are dismissed for not belonging to a knowing that is not scientific. This is the paradox of labels, language, conceptual structures—to become obstacles to their own ideal, the flow back and forth between distinguishing and distinctions. Science elucidates proper content from borrowed content; art celebrates borrowed content for its own sake. Doing so, it celebrates our intimacy with life.

ב, again, foundation of aliveness as foundation of biology, brings us back home.

[76] Snow, *The Two Cultures,* 3.

A Final Monkey Story: How Charlie Became the Leader of the Group

With Danny's passing, Django had recently become the leader of the group. From one day to the next his jowls had grown, and the crown of his head had puffed-up, a lithe young male turned mature warrior.

One evening the monkeys were upset. We heard those shrill anxious calls one typically hears when a baby is on the wrong monkey and its upset mother consoles it at a distance and is trying to get him back. Sure enough, we arrived to find little Ricky on Django's back while a distraught Sarah tried to approach to retrieve him. Her daughters Sophie and Sally rallied in her support. But with three rather belligerent female monkeys intent upon him, and Ricky tightly gripping onto his back, Django paced about looking conspicuously uncomfortable. He kept well away from them while Ricky clung for all he was worth—the more so because of the general sense of alarm around him. He could see and hear the anguish of his mother and his sisters.

Many times in the past we had seen situations in which the alarm is raised and a baby monkey climbs onto the nearest monkey. If it happens to be a female who its mother is not close to, it can lead to a tense situation. The mother insists on going after her baby and the other female keeps trying to get away. For a moment one wonders just how this impossible situation can be resolved. But there is an elegant solution. The leader of the group will first discipline the mother to keep her away from the other female. Once this female settles down, he will sit by her and encourage the baby onto his own back. Only then will the baby's mother be allowed to approach for the baby to climb back onto her.

If anything, our arrival made things worse. Sarah's pleading intensifies. Django redirects his tension and frustration at us. He demonstrates and rushes about with greater excitement. At intervals he attempts to wriggle himself free from under the baby, clearly hoping to bring the whole unfortunate misunderstanding to an end. Several times he reaches behind him for Ricky's arm or leg and gives them impression bites. This is what mother monkeys will do to persuade insistent toddlers to get off their backs. But they would never do this in such a fraught situation. At the best of time toddlers scream indignantly when their mother does it, and it now terrifies Ricky. He momentarily recoils and lets go with that limb, but it only makes him cling on all the tighter. Django does not settle for even a second. With his violent movement and the momentary loss of grip, Ricky slides down Django's back. He is now sitting on the base of Django's tail, which must feel even stranger to Django, who has now lost all equanimity and self-control. Standing on a beam and demonstrating wildly, furiously shaking a rope in each hand, with one final whipping movement he sends Ricky flying, bouncing off a wall, against a beam, and landing with a dull thud on the ground.

It takes Sarah several seconds to realize what has happened. Then she and her daughters dive down to the baby. Croaking with concern, Sarah is upon Ricky. He climbs on her and holds tightly to her back. The relief of having him securely back on her frees Sarah to express her rage. In a fleeting but remarkable moment of defiance, she looks up and cough-barks indignantly at Django, though in her voice still mingles shock and concern. It is a single glare; a single broken sound soon washed over by eeolks from all directions as monkeys look on from shelves and ropes and beams. It is the reassuring expression that brings the resurgence of gentler sounds and movements at the end of all such incidents.

We could only guess whether Ricky was injured or not. Although a horrific moment for him, and dreadful to watch, he had neither fallen far nor hit himself hard. Monkeys, particularly the babies, are astonishingly resilient. Who knows what falls they occasionally endure in the rainforest. From the few falls we have seen at the sanctuary, it is quite remarkable what they can get away with.

We were appalled by Django's behavior. We had never seen a male act so recklessly with a baby. We empathized fully with Sarah and remarked upon the look she had given him and the indignation we all agreed had been expressed by the very unusual sound she had made.

A few weeks later, as we all sat in the living room, we faintly heard some nyook-nyooking. This is the sound of bulling the monkeys typically make to egg each other on to attack. To the astonishment of some guests, six of us jumped to our feet as one and scrambled out the door.

To everyone's excitement, Django rushed about demonstrating anxiously, as he was hounded by Sarah, Sophie and Sally, while they and others egged on Charlie into battle with his uncle. Charlie—the very Charlie I had babysat when Django was already a full-grown male—now challenged his uncle for dominance with the support from the Sarah family and from the rest of the Jessy clan. Then, in the middle of the commotion, Django and Charlie rush into an appeasement huddle with loud squeaks. They have always been close to each other.

We managed to get Django isolated before anyone got hurt. Next morning, Django seemed rather dejected but was his old lanky self. When we approached the territory, Charlie, sporting new-fangled large jowls and a preened-up crown, demonstrated at us nonstop, leaving us with no doubt as to who was now in charge.

Postscript

I wake up on the top in Monkey Room 1. All is quiet except for the buzzing of the dimmed lights. All around me, still balls of dark fur are soundly sleep. Laura is by my side. I had unsettling dreams. In one, it had been decades since I had left the Sanctuary. My hair and beard were white. Many faces in tiny screens listened as I defended a dissertation about monkeys. Something about distinctions. I felt deeply relieved. Vicky came over and held my hand. She glowed. I am so very glad, she said. You are no Edward Casaubon after all.

Appendix: Internal Critique of *Laws of Form*

Again is the key term used to state the two axioms of *Laws of Form* (*LoF*). It is, in Russell's words, one of those terms that we "must be content to accept as intelligible without definition." This is the cornerstone of the problem of circularity in the definition of distinction. As has been shown, not only is 'again' explicitly used in both axioms; the explanation that follows each is also stated in terms of repetition.

Axioms must be self-evidently true, intuitively obvious, and as elementary as possible. So, there can be no objection to repetition in stating them. However, the formalism explicitly says that this must be formally allowed:

First Canon, Convention of Intention.
Let the intent of a signal be likened to the use allowed to it.
Call this the convention of intention. In general, what is not allowed is forbidden.[1]

Spencer-Brown explicitly emphasized the outstanding importance of this:

> It is part of the mathematical discipline that what is not allowed is forbidden. That is to say, what you don't introduce, you can't use. And until you have introduced shape, size, duration, whatever, distance, you can't use it.[2]

Per force, this must apply to repetition. Indeed, *LoF* does explicitly allow repetition, but not until Chap. 2, under the title *Forms taken out of the form.*[3] A preceding section entitled *Knowledge* defines the notion of *form*:

Let a state distinguished by the distinction be marked with a mark ⌐ of distinction.

and

[1] Spencer-Brown, 3.

[2] Spencer-Brown, *Transcript*, 25.

[3] Spencer-Brown, *Laws of Form*, 3.

© The Editor(s) (if applicable) and The Author(s), under exclusive license to Springer Nature Switzerland AG 2023
D. C. Mayer-Foulkes, *The Nature of Living Being*, Biosemiotics 26,
https://doi.org/10.1007/978-3-031-24789-7

Call a space cloven by any distinction, together with the entire content of that space, the form of distinction. Call the form of distinction the form.[4]
With this, the ground is set to explicitly allow repetition:

Name
Let there be a form distinct from the form.
Let the mark of distinction be copied out of the form into such another form.
Call any such copy of the mark a token of the mark.

Note the use of term "distinct" to grant the possibility of repetition. Are the first and second sentences not equivalent? Either they are, or there is a slight circularity here, because the first statement in effect already calls for copying, which is only granted in the second. An alternative way of making this set of statements might have been "let bet structure be established." This becomes obvious through paraphrasing:

Let there be a form A distinct from the form A.
Let the mark A be copied out of the form A into such another form A.
Let bet structure be established between A and A.
Call any such copy of the mark A a token of the mark A.

The term 'token' is used in the very sense it was used in Chap. 1 regarding ב, which amounts to treating instances as tokens of a type. It was shown there that this is inherently anticipative, that it calls for arbitrariness, and that it involves forness. Therefore, the use of repetition must precede the axioms. ב would need to be an axiom or at least part of one.[5] It could be argued that this is Axiom 1: *The value of a call made again is the value of the call.* But ב is the condition for this possibility because this statement uses of the notion of 'again' just as it is used in the other axiom: both calls and crosses are specified by *again*. Moreover, if Axiom 1 is about repetition, why then is it formally allowed in Chap. 2? An argument could be made that there is a difference between the injunctions by which a formalism is built ("let there be…") and the formalism once completed. This would make the formalism analogous to a finished computer program, not the instructions to build it. So, the not-allowed repetition would belong to these injunctions, presumably with no effect on the finished software.

The problem with this argument is that it contradicts the stated intent of this kind of formalism, the more so this one. Spencer-Brown's acknowledges this when he says that "the system pulls itself up from its own bootstraps."[6] He clearly had in mind the construction of the formalism and not the finished product. Still, this

[4] Spencer-Brown, 4.

[5] In his paper "Extension…," Louis H. Kauffman demonstrated that the two initials of *LoF* may be reduced to a single initial. Also mentioned are two other similar demonstrations, not unlike the Huntington Axiom. However, the opening of this demonstration states, "This note assumes the conventions of *LoF* by G. Spencer-Brown." Presumably, included amongst these is the convention 'Name' that is under discussion. Just as in *LoF*, repetition is taken for granted since, without prior justification, multiple instances of single symbols are taken as being *the same*.

[6] Spencer-Brown, *Transcript*, 8–9.

argument is helpful because it points to something crucial—the formalism is a product. This is obvious enough, but it is an essential point treated in Chap. 4.

In fact, ב could never be an axiom.

Recall the point made in Chap. 2 about the definition of distinction as *perfect continence*. This assumes that a boundary crossed on the outbound journey from A to Ā is the same as the boundary crossed on the return. It is assumed in both axioms that two crossings and two calls are the same, indeed *exactly* the same. But the condition for this is a living subject establishing bet structure, by arbitrary neglect of concrete residue. Indeed, the condition to construct a formalism is a double elision: the elision of the knowing subject and then the elision of this elision.[7] This point is treated at length in Chap. 6.

We are back at the unavoidable problem of performative circularity in the definition of distinction. In Spencer-Brown's sense of the term, ב is more *important* than the axioms because it is inside, deeper, prior to them. Only the elision of the subject can justify him saying, "The system actually comes from nothing and pulls itself up from its own bootstraps, and there it is." Why, as Russell puts it, are we "content to accept as intelligible without definition" the term 'again'? As discussed at length in Chap. 2, because prior to any notion of 'again' is the spirallic nature of aliveness. Taking this for granted may be reasonable in most circumstances but not when we seek the foundations of mathematics. And this is the key point: *this elision itself* is central to the "game one is playing when one plays the game of mathematics." Spencer-Brown made Russell's point as follows:

> In all mathematics it becomes apparent, at some stage, that we have for some time been following a rule without being consciously aware of the fact. This might be described as the use of a *covert* convention. A recognizable aspect of the advancement of mathematics consists in the advancement of the consciousness of what we are doing, whereby the covert becomes overt . . . . The nearer we are to the beginning of what we set out to achieve, the more likely we are to find, there, procedures which have been adopted without comment. Their use can be considered as the presence of an arrangement in the absence of an arrangement.[8]

In this light, becoming conscious of the subject's ב as condition for the possibility of mathematics, *is an advancement of mathematics*. The final line just quoted converges with Fig. 2.4 in Chap. 2 reproduced here, illustrating the elision of the subject in the notion distinction as perfect continence ("We see the first three but must suppose the boundary to be like the fourth"):

Figure A.1 Distinction as circumference

[7] See Chap. 2, section "Definitions of distinction."

[8] Spencer-Brown, *Laws of Form*, 85.

Spencer-Brown noted that we take for granted that proofs are written on a plane surface and not, say, on the surface of a torus, in which case they could be false. Recall that the formalism in *LoF* uses graphic or topological notation and not typographical notation, which in any case comes with its own assumptions, such as direction of writing and reading.[9] Mathematical formalisms are deeply embedded in tradition, which is to say culture, and culture itself is deeply embedded in our primate nature, animal nature, and thence in the nature of living being.[10] These multiple layers all involve ב as the condition for their possibility. Just as it is not obvious that repetition is inherent to distinction as *perfect continence* and imagined as a circle, so it is not obvious that ב is anticipative and that it underlies repetition.

If 'again' is implicitly defined by the axioms, why would its use be formally and explicitly granted later? This would suggest that Spencer-Brown is not aware that 'again' is implicitly defined. Whether he is or not, or if 'again' is assumed, both options point to just how deeply repetition is inherent to the axioms. Again, making obvert what is covert *is an advancement of mathematics*. And what about the larger point, what does this say about mathematics, concretely about its inadequacy as an ideal of knowledge for inquiry into aliveness?

Assume that 'again' is implicitly defined by the formal statement of the axioms presented in Chap. 5, Fig. 5.1 and reproduced here:

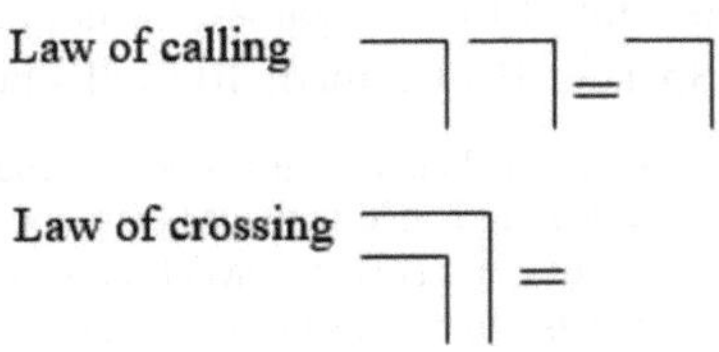

Figure A.2 The axioms of *Laws of Form*

Both equations repeat the sign Γ. The law of calling says that 'again' is equivalent to Γ and the law of crossing that it is equivalent to . So, together they state that distinction consists of two aspects of 'again.' The first axiom says that Γ repeated any number of times is equivalent to Γ. No matter how many repetitions, they all amount to Γ. The second axiom says that two Γ are equal to . As was stated before, this is why the former is a principle of continuity and the latter a principle of discontinuity. As the arithmetic unfolds, it shows how a Γ that pervades the space of another Γ allows them to be cancelled, that is, together they are equivalent to . Three of these repetitions are equal to Γ, four are equal to , five are equal to Γ, and so on. So, the axioms take us back to distinction as A=A, A≠A and continuity-discontinuity. Already in germ here are schemes of recurrence, oscillation, and even counting. Defining distinction through repetition, the axioms appear, at least covertly, to define distinction as ב. That being so, the equivalence between distinction and ב is in the axioms. But it is us who are reading this into the axioms. It is we who are importing the notion of ב. Nothing else indicates this is the case.

[9] "Topological" is used by Schiltz, "Space is the Place," 161. "Graphic" is used by Engstrom, "C. S. Peirce's Precursors to Laws of Form," 25–66.

[10] The notion of animal cultures is now widely accepted. For an overview, see https://en.wikipedia.org/wiki/Animal_culture.

The calculus of indications could be named *the calculus of distinctions* or *the calculus of repetition*. To repeat here what is said in Chap. 4, the mathematician Louis Kaufmann summarized *LoF* saying "Logic is nothing more than the properties of the act of distinction."[11] It could also be said "Logic is nothing more than the properties of ב," or "Logic is nothing more than the properties of the act of repetition." That ב is implied, says that *LoF* is about the compact and *important* foundational level of mathematics at which *indication*, *distinction*, and *repetition* are undifferentiated.

Enough has been said about the axioms to make it at least intuitively obvious why the second-degree algebraic equations derived in LoF oscillate, named by Spencer-Brown "imaginary Boolean values." What could be closer to 'again'? This oscillation makes overt what was covert, and *LoF* is about repetition as always already established by [subjects].

While the argument that 'again' is implicitly defined in the axioms is gratuitous, it does show that 'again' belongs in the gray zone between the implied and the assumed. Is the meaning of the sign '=' implied or assumed? Certainly, both axioms suggest, though they do not strictly specify, the meaning of this sign. But then, is it implied or assumed that these are two equations? Most conspicuously, is it implied or assumed that all instances of the sign ⌐ are *the same*? Is it implied or assumed that ⌐ is repeated?

Any number of questions like these could be asked about the axioms, and about their enunciation in words, if only because many more terms are involved. This all points to the ineluctable limitations of axiomatic systems, even those implicitly defined, because formal systems are grounded in subjects.

Lewis Carroll provides a lovely demonstration of the limits of axiomatic systems that speaks to both Russell's point about what we must be content to accept as intelligible without definition, and Spencer-Brown's point on the advancement of mathematics. In his logic fable "What the Tortoise said to Achilles," trouble starts when the laws of logic are treated as axioms.[12] Traditionally the three fundamental laws of logic are: (a) the law of contradiction, (b) the law of excluded middle, and (c) the principle of identity. The tortoise requires Achilles to write down, as an axiom and additional law, the rule *modus ponens*. This is the rule: *if p, then q*, where *p* and *q* stand for propositions. However, Achilles can only apply this axiom if he has a previous one, which in turn requires a previous one, and so on, *ad infinitum*. So, the laws of logic cannot be axioms. It is noteworthy that the statement of *modus ponens* in terms of the variables *p* and *q* shows it to be algebraic. This is Boolean algebra, the very one for which *LoF* supplies the underlying arithmetic. In fact, *modus ponens* is proved in *LoF* as a theorem in the primary arithmetic.

Nonetheless, Carroll's general point stands. In our present case, the tortoise would require Achilles to grant 'again' as an additional axiom, which could only be applied if he had a previous one, and so on, *ad infinitum*. No statement can be made without the repeated use of signs, and this implies 'again.' So, 'again' or 'ב' could not be stated as axioms, indispensable as they are to any formalism.

---

[11] Kauffman, "Peirce's Existential Graphs," 61 (emphasis added).

[12] Carroll, "What the Tortoise Said," 278–280.

Glossary

- - See **Liminalism.**

[] See **[Subject].**

[]>[ ] Abbreviation for [A=A]➔[A = A].

[A=A]➔[A = A] (abbr. []>[ ]) Given its intrinsic arbitrariness, drift is inherent to A=A. This allows the gradual transformation of EI.

@@A (aliveness) Joinings, the form of organisms as distinctions that is the cond. for the poss. of them being distinguishing distinctions. The ground scheme from which @@L (logic) differentiates.

@@A proper content (This term unfolds throughout the essay.) ב is the heart of the proper content of @@A as act of distinguishing, the verb. Insofar as these acts involve ב, their proper content includes anticipation, a degree of arbitrariness, the switch from analog to digital. As two-element regulatory systems and embodied distinctions, the proper content of @@A is plurality, joinings, *concrēscere* (growing together), intensionality, relationship, organization. Thereby dynamics involving the repulsive and attractive forces, paradoxes, trade-offs, division of labor, identity-unity-whole, transparency/opaqueness, Ba-Bw, historicity. Since these are normative, this too is proper content of @@A, and therefore, C123, [ethics]. Each of these terms highlights aspects of @@A (aliveness), the ground scheme, the process-structure of distinguishing distinctions, and []➔[ ], their unfolding.

@@L (logic) The form of logic as depicted in the diagram of the four logical conjunctions in *LoF*. @@L is the upper scheme that differentiates from @@A.

@@L, proper content Logic, as the properties that follow from distinction, and the irreducible fact that some borrowed content is required for differences to make a difference to a [subject]. Relations (not relationships), therefore the logical conjunctions, express the proper content of @@L.

A=A See **A=A, A≠A.**

A=A, A≠A Full expression of A=A. To be the same (A=A), two things, events, etc. must be numerically different (A≠A).

A/Ā, A=A Full expression of A/A. Distinguishing (A/Ā) calls for sameness (A=A), homogeneity *for* a [subject] for whom some differences made a difference and not others.

Ā→A→/→Ā Functional closure. An outside component (Ā) crosses a semi permeable boundary to become a component (A) integrated into a system/process in functional closure (/), from which components again cross the boundary to cease being part of the system (Ā)

© The Editor(s) (if applicable) and The Author(s), under exclusive license to Springer Nature Switzerland AG 2023

D. C. Mayer-Foulkes, *The Nature of Living Being*, Biosemiotics 26,
https://doi.org/10.1007/978-3-031-24789-7

Aliveness (@@A) The nature of life. The meaning of this term unfolds throughout the essay, showing that distinguishing-distinction is the proper conjugate form for inquiry into aliveness.

Anticipation, unfolding See **Unfolding of Anticipation.**

Arbitrariness See **Neglect of concrete residue.**

Artistic approach In contraposition to *scientific approach*, learning to care rigorously for monkeys, for example, by practical necessity setting aside the limitations of science to play things by ear. As one works on musicianship, so one can work on one's 'primateship.'

ב Hebrew letter bet (B). To establish bet structure, to distinguish. See **Bet structure, establish.**

Ba Biomorphism-a, biomorphism of aliveness, the form of aliveness. Organisms as *distinctions* (that distinguish). See also **Bp** and **Bw.**

Bet structure The relation between the vertical and the horizontal sequences inherent to schemes of recurrence.

Bet structure, establish, ב To distinguish (A/Ā, A=A) by arbitrary neglect of concrete residue to differentiate between differences that make a difference and differences that do not. To construe events as members of a set, as tokens of a type. This is repetition, it calls for the arbitrariness inherent to forness. With ב, concrete particulars are treated as abstract generals. Through anticipation, future events are treated as earlier events.

Bp Biomorphism-p, biomorphism of products, the (form of) products produced by living beings.

Bw Biomorphism-w, biomorphism of the world, the (form of) the world as formed by living beings. Organisms as *distinguishing* (distinctions).

Ba-Bw The necessary relation between Ba and Bw. Their differentiation is abstract. See **Ba** and **Bw.**

[Body-subject] A [subject] as embodied. At once a *unity-identity-whole* being-at-work-staying-itself as subject ("on the side of" the subject), and an identity-unity-whole being-at-work-staying-itself as object ("on the side of the object").

Borrowed content See **Distinction, proper content** and **Distinction, borrowed content.**

Byproducts (See **Four kinds of things**).

C123 (Abbreviation of C1, C2, C3). Generalization for [ethics] of G123; C1, individual components; C2, components of order; C3, components of value. See **G123.**

[Cognitive] See **[Subject].**

Con-crēscere Latin etymology of *concrete*, "growing together". The 'concrete' proper content of @@A. See **@@A, proper content.**

Concrete residue Differences that do not make a difference for a [subject].

Content, proper and borrowed See **Distinction proper content** and **Distinction borrowed content**

Distinction, abstract definition Distinction is perfect continence, often represented as a circle on a flat plane.

Distinction, as continuity/discontinuity The concatenation of continuity/discontinuity or simply continuity/discontinuity.

Distinction, borrowed content Contingent values that are the differences that make a difference for a [subject]. See **Distinction, proper content.**

Distinction, concrete A limen, a membrane or set of membranes that constitute a [subject].May sometimes refer to a scheme of recurrence that a [subject] biomorphically construes as a distinction, for instance a fence. The former is Ba the latter Bw.

Distinctions, lineage The lineage of synthetic acts that result from the unfolding of (the borrowed content of) distinction. All involve a [subject's] anticipation and forness in relation to an [object]. Both [subjects] and [objects] may be from the most compact to the most differentiated. These synthetic acts include *umwelten*, affordances, human judgments, things, concepts, etc.

Distinction, noun The result of distinguishing activity (the verb). See **Distinguishing, verb.**

Distinction, proper A distinction for itself, a living being.

Distinction, proper content The necessary structure of distinction, including the irreducible fact that distinction involves some, any, given value (borrowed content). See **Distinction, borrowed content.**

Distinctional Distinctional is to distinction as cognitional is to cognition. The distinctional comprises distinction (noun) and distinguishing (verb).

Distinguishing To establish A/Ā, A=A, to establish bet structure, ב, by arbitrary neglect of concrete residue; to differentiate between differences that make a difference and differences that do not.

Distinguishing, verb Synthetic act from which results a distinction (noun). See **Distinction, noun.**

Diff-diff Abbr. of differences that make a difference and differences that do not.

Double elision To forget the knowing subject and then to forget that it has been forgotten.

Establish bet structure See **Bet structure, establish.**

[Ethics] Inquiry into the skin-to-skin and/or face-to-face encounter inherent to joinings, that is, in both social and organic Wholes (/Ws). The actual acts of individuals, at the level of *identity*, and as counterpart to the [social]. See **[Subject]** and **@@A, proper content.**

EI Abbreviation for Earliest Indication. See **Earliest Indication.**

EI-1, E-2, EI-3, EI-4, EI-5 See **Unfolding of Anticipation.**

Earliest Indication Recursive process that allows a chain of earlier events to be identified by an organism as indicating that a certain event is imminent. EI refers both to the set of earliest events and to the process or heuristic structure by which these are arrived at.

Entelecheia Being-at-work-staying-itself or holding together actively as a whole.

Establishing bet structure, ב See **bet structure, establishing.**

Forness The relation between a living being and the differences that make a difference to it. The synthesis of subject and object is always already *for* a subject. The foundation of signification.

Four kinds of things Things, or kinds of being, either involve distinguishing activity or they do not. If they do, or did, this cannot be left out of the account of a thing. A proper account of a thing calls for its highest conjugate form. There are three *distinguishing kinds of things*: ***living beings***, ***products***, and ***byproducts***. Distinguishing-distinction is, or was, at work in all. Ba-Bw (biomorphism-aliveness and biomorphism-world) are at work, or there are traces of their having been at work. Of the four kinds of thing, to ***the nonliving***, distinguishing only obtains as our own Bw. ***Living beings*** establish bet structure by arbitrary neglect of concrete residue to differentiate between differences that make a difference and differences that do not. This arbitrariness makes them nodes of causation. ***Products*** are produced by living beings to establish bet structure recurrently. ***Byproducts*** are highly improbable molecular distributions that extend beyond the boundaries of living beings (expressed by the second Ā in Ā→A→/→Ā). Sectors of ***the nonliving*** are real descriptively and/or explanatorily (galaxies, stars, and planets, for instance). These result from the spontaneous optimization of flow that result from the second law of thermodynamics. But it is we living beings who, by misplaced concreteness, regard these sectors as 'bodies' already out there now. See **Thing.**

Functional closure A living cell is structurally open but functionally closed, open to the input of matter but closed with regard to the dynamics of the relations that generate it. See Ā→A→/→Ā.

G123 (Abbreviation of G1, G2, G3). G1, individual goods; G2, goods of order; G3, goods of value.A G1 obtained is experienced as pleasant, enjoyable, satisfying. A G2 is an intelligible unification, a pattern of organization for the recurrence of G1s. G3 are values instantiated by G1 and G2. See **C123.**

Ground scheme See **Two schemes.**

Highest conjugate form The viewpoint that allows for the verification of correlations of correlations of all intelligible relations that occur systematically leaving out only what occurs coincidentally.

Hymenomorphism From Greek *morphē*, form, shape; beauty, outward appearance; and *hymen* membrane (especially 'virginal membrane,' the membrane par excellence). ב is done by membranes. Echoing Aristotle's *hylomorphism*, it refers to the fundamental metaphysical commitments that follow from **- -**, **liminalism**.

Identity-unity-whole Organisms function as an identity-unity-whole. Whole refers to the distinguishing activity of an organism in its totality; unity, to the subdivision of the whole available intermittently to the highest-level system, which is identity. Identity is the subdivision of unity, the sequence available in the here and now to the highest-level system.

[Inquiry] See **[Subject].**

Integrator-operator The unfolding of biological processes (i.e., growth, development, learning) must be accounted for through a genetic or developmental heuristic structure. Concretely, through successive integrations in terms of an operator. The operator accounts for the sequence of prior integrations being replaced by more developed successor integrations.

Intension In contraposition to Descartes' *extension*, the stretch or tendency inwards and towards 'not'.

Intersociality The interrelating of individuals in their group and with their group to other groups, evolutionary origin of the two forces, fusion and individuation.

Joinings Living beings do not originate from severance but from joinings. The condition for the possibility of distinguishing distinctions, as entailed in @@. Joinings originate nesting and therefore identity-unity-whole.

Kinds of goods See **G123.**

Kinds of things See **Four kinds of things.**

Liminalism and Substantialism Liminalism grasps things intelligently and affirms them rationally in relational terms that may be fuzzy, ambiguous, paradoxical; Substantialism grasps things as 'bodies' already out there now, concrete, bounded clearly and distinctly.

Liminality, - - The layered play of continuity/discontinuity that constitutes identity-unity-wholes, in the multifaceted synchronic and diachronic sense instantiated by membranes of living beings. Makes explicit the biological nature of ב. See **Hymenomorphism.**

Lineage See **Distinctions, lineage.**

Living being See **Four kinds of things**.

LoF *Laws of Form* by G. Spencer Brown. See bibliography.

Logic (@@L) See **@@L.**

Lower scheme See **Two schemes.**

[Mind] See **[Subject].**

[Name] See **[Subject].**

Neglect of concrete residue, Arbitrariness Neglect of differences that do not make a difference.

Nesting Successive joinings of distinguishing distinctions.

Nodes of causation Cells are [subjects] with agency, nodes of event-response and therefore nodes of causation, because their ב involves anticipation and thereby arbitrary neglect of concrete residue.

Nonliving See **Four kinds of things.**

[Object] See **[Subject].**

Opaqueness See **Transparency/Opaqueness.**

Organic wholes See **WW.**

Plurality The specific instance of multiplicity that refers to multiple individuals, actual distinguishing distinctions for themselves by functional closure.

Products See **Four kinds of things, wp.**

Proper conjugate form In inquiry, the highest viewpoint possible. In the case of inquiry into living beings, product and byproducts, the proper conjugate form is distinguishing-distinction.

Proper content See **Distinction, proper content** and **Distinction, borrowed content. Proper content of @@A,** see **@@A, proper content. Proper content of @@L,** see **@@L, proper content.**

Relation/relationship Relations are abstract and pertain to @@L, Bw; relationships, recurring two-way and reciprocally anticipative acts between organisms, pertain to @@A, Ba.

Scientific approach See **Artistic approach**.

Social wholes See **wW**.

[Social] See **[Subject].**

Spirallic time The ongoing process of establishing bet-structure at the heart of living experiencing, or of living being, or of aliveness.

[Subject] To sidestep debate about levels or qualitative degrees of subjecthood, and what might constitute them, the full range of these is referred to by enclosing the term in square brackets. The same applies to all similarly bracketed terms.

Substantialism see **Liminalism and Substantialism.**

Synthetic acts See **Distinction, lineage.**

The nonliving See **Four kinds of things.**

Thing A unity-identity-whole.

T/O Abbreviation for Transparency/Opaqueness.

Transparency and opaqueness (T/O) The gap between how an organism acts and what it knows, constitutive of identity-unity-whole. T/O is a continuum. At the pole of transparency, the gap is small, and at the pole of opaqueness the gap is large. Higher levels of opaqueness and anticipation correspond to each other. Opaqueness is a *sine qua non* for higher levels of sociality, anticipation, distinguishing, distinctions, knowing.

Two Forces Tendencies or pulls in a group towards both fusion and individuation.

Two schemes, upper and lower Origin of what we are content to accept as intelligible without definition in formal systems. The ground scheme is the form of organisms as distinctions, the upper scheme is the form of their distinguishing: ב: →Repetition, *Again*→Axioms of *LoF*→Calculus of distinction or "Boolean Arithmetic"→ "various models"→Boolean Algebra or Logic.ב→Repetition, *Again*→Axioms of *LoF*→Calculus of distinction or "Boolean Arithmetic"→**Cells as distinctions**→Boolean Algebra or Logic.

Unfolding of anticipation EI-1, anticipation; EI-2, anticipation of anticipation; EI-3, anticipation of non-anticipation; EI-4, non-anticipation of anticipation; EI-5, non-anticipation of non-anticipation.

Unity-identity-whole A thing.

Upper scheme See **Two schemes.**

Wholes See **WW, Ww, wp.**

wp (whole made up of parts) Material products are wholes made up of parts.

WW (organic wholes) A living whole that is a joining of living wholes. A multicellular organism is relatively autonomous and made up of relatively autonomous cells. Even single celled organisms might be considered organic wholes, given the relative autonomy of their constituents. See **Ww**.

Ww (social wholes) A joining made up of living wholes. A social product. Any organization of living beings. The key difference between WWs and wWs is at the level of Identity (in the sense of identity-unity-whole).

Bibliography

Abiogenesis. 2020. Wikimedia Foundation, last modified on February 9, 2020, at 23:57. https://en.wikipedia.org/wiki/Abiogenesis.

Adler, Mortimer, ed. 1952. *Great Books of the Western World by Encyclopedia Britannica,* vol. 49 (Darwin). Chicago: Encyclopedia Britannica.

Affordance. 2019. Wikimedia Foundation, last modified on December 19, 2019, at 20:27. https://en.wikipedia.org/wiki/Affordance.

Alter, Robert. 1996. *Genesis, Translation and Commentary*. New York: Norton.

Apollonian and Dionysian. 2021. Wikimedia Foundation, last modified on April 1, 2021, at 10:06. https://en.wikipedia.org/wiki/Apollonian_and_Dionysian.

Aristotle. 2002. *Metaphysics*. Trans. Joe Sachs. Santa Fe: Green Lion Press.

———. 2004. *On the Soul*. Trans. Joe Sachs. Santa Fe: Green Lion Press.

Armstrong, Karen. 2005. *A Short History of Myth*. Edinburgh: Canongate.

Atkins, Peter. 2007. *Four Laws that Drive the Universe*. Oxford: Oxford University Press.

Atteberry, Jonathan. How Geysers Work. *HowStuffWorks*. https://science.howstuffworks.com/environmental/earth/geophysics/geyser2.htm.

Axelrod, Robert, and William D. Hamilton. 1981. The Evolution of Cooperation. *Science, New Series* 211 (4489): 1390–1396.

Bahadur, Krishna. 1974. Photochemical Formation of Self-Sustaining Coacervates. *Procedures of the Indian National Academy of Science* 39 (4): 455–467.

Bakken, Tore. 2014. George Spencer-Brown. In *The Oxford Handbook of Processed Philosophy and Organization Studies*, ed. Jenny Helin, Tor Hernes, Daniel Hjorth, and Robin Holt, 481–498. Oxford: Oxford University Press.

Barbieri, Marcello. 2003. *The Organic Codes: An Introduction to Semantic Biology*. Cambridge: Cambridge University Press.

Bartolomé y Lupercio Leonardo de Argensola. 2021. Wikimedia Foundation, last modified on April 12, 2021, at 21:04. On the difficulty of attributing authorship see https://es.wikipedia.org/wiki/Bartolom%C3%A9_y_Lupercio_Leonardo_de_Argensola.

Bateson, Gregory. 1980. *Mind and Nature: A Necessary Unity*. Glasgow: Fontana.

———. 2000. *Steps to an Ecology of Mind*. Chicago: University of Chicago Press.

Bateson, Gregory, and Mary Catherine Bateson. 1987. *Angel's Fear: Towards and Epistemology of the Sacred*. New York: Bantam.

BBC Home Making: Weaver Bird. 2009. Video uploaded April 10, 2009. Available at https://www.youtube.com/watch?v=6svAIgEnFvw.

© The Editor(s) (if applicable) and The Author(s), under exclusive license to Springer Nature Switzerland AG 2023

D. C. Mayer-Foulkes, *The Nature of Living Being*, Biosemiotics 26,
https://doi.org/10.1007/978-3-031-24789-7

Behan, Adrian, and J. Peder Zane. 2012. *Design in Nature: How the Constructal Law Governs the Evolution in Biology, Physics, Technology, and Social Organization*. New York: Doubleday.

Berg, David N., and Kenwyn K. Smith. 1995. Paradox and Groups. In *Groups in Context*, ed. Jonathon Gillette and Marion McCollom. Lanham: University Press of America.

Bernal, John Desmond. 1949. The Physical Basis of Life. *Proceedings of the Physical Society, Series A* 62 (9): 537–538.

———. 1967. *The Origin of Life*. London: Weidenfield and Nicholson.

Bianconi, Eva, Allison Piovesan, Federica Facchin, Alina Beraudi, Raffaella Casadei, Flavia Frabetti, Lorenza Vitale, Maria Chiara Pelleri, Simone Tassani, Francesco Piva, Soledad Perez-Amodio, Pierluigi Strippoli, and Silvia Canaider. 2013. An Estimation of the Number of Cells in the Human Body. *Annals of Human Biology* 40 (6): 463–471.

Bierman, Steve. 2020. *Healing Beyond Pills and Potions: Core Principles for Healers and Helpers*. California: Gyros Press International.

Biological Altruism. 2013. *Stanford Encyclopedia of Philosophy*. July 21, 2013. https://plato.stanford.edu/entries/altruism-biological/.

Biology Dictionary. 2021. Ectoderm. Retrieved January 2021. https://biologydictionary.net/ectoderm/#:~:text=The%20main%20function%20of%20the%20ectoderm%20is%20to,neural%20tube%20in%20a%20process%20known%20as%20neurulation.

Biosemiotics. 2019. Wikimedia Foundation, last modified on November 28, 2019, at 19:50. http://en.wikipedia.org/wiki/Biosemiotics.

Bitbol, Michel, and Pier Luigi Luisi. 2004. Autopoiesis With or Without Cognition: Defining Life at Its Edge. *Journal of the Royal Society Interface* 1: 99–107.

Blaser, Martin J. 2014. *Missing Microbes: How the Overuse of Antibiotics is Fueling Modern Plagues*. New York: Henry Holt.

Blumberg, Hans. 2013. *Teoría del Mundo de la Vida*. Mexico City: Fondo de Cultura Económica.

Boswell, James. *Life of Samuel Johnson*. Buckland

Bourgine, P., and J. Stewart. 2004. Autopoiesis and Cognition. *Artificial Life* 10 (3): 327.

Bourke, Andrew F.G. 2011. *Principles of Social Evolution*. Oxford: Oxford University Press.

Bowen, Murray. 2004 [1985]. *Family Therapy in Clinical Practice*. Lanham: Rowman & Littlefield.

Bretz, Michael. 2008. Emergent Probability. In *Unifying Themes in Complex Systems IV: Proceedings of the Fourth International Conference on Complex Systems*, ed. Ali A. Minai and Yaneer Bar-Yam, 123–132. Heidelberg: Springer.

Byrne, Patrick H. 2016. *The Ethics of Discernment: Lonergan's Foundations for Ethics*. Toronto: University of Toronto Press.

Capra, Fritjof, and Pier Luigi Luisi. 2014. *The Systems View of Life: A Unifying Vision*. Cambridge: Cambridge University Press.

Carroll, Lewis. 1895. What the Tortoise Said to Achilles. *Mind* 4 (14): 278–280.

Castelfranco, Ann, and Daniel Hartline. 2015, 2015. The Evolution of Vertebrate and Invertebrate Myelin: A Theoretical Computational Study. *Journal of Computational Neuroscience* 38 (3).

Challoner, John. 2015. *The Cell: A Visual Tour of the Building Block of Life*. Chicago: The University of Chicago Press.

Chen, Justin, Laura A. Jacox, Francesca Saldanha, and Hazel Sive. Mouth Development. *WIREs Developmental Biology* 6: e275: 1–16.

Coffey, Jerry. 2008, July 16. How Old is the Universe? *Universe Today*. http://www.universetoday.com/15575/how-old-is-the-solar-system/.

Cooper, Guy A., and Stuart A. West. 2018. Division of Labour and the Evolution of Extreme Specialization. *Nature Ecology and Evolution* 2: 1161–1167.

Copi, Irving M., and Carl Cohen. 2005. *Introduction to Logic*. 12th ed. Upper Saddle River: Pearson.

Corballis, Michael C. 2011. *The Recursive Mind: The Origins of Human Language, Thought, and Civilization*. Princeton: Princeton University Press.

Corbey, Raymond. 2005. *The Metaphysics of Apes: Negotiating the Animal-Human Boundary*. Cambridge: Cambridge University Press.

Coyle, Daniel. 2012. *The Little Book of Talent: 52 Tips for Improving your Skills*. New York: Bantam.
Darwin, Charles. 1987. Notebook M. In *Charles Darwin's Notebooks, 1836–1844*, ed. Paul H. Barrett, Peter J. Gautrey, Sandra Herbert, David Kohn, and Sydney Smith. Cambridge: Cambridge University Press.
de Waal, Frans. 2006. *Primates and Philosophers: How Morality Evolved*. Princeton: Princeton University Press.
———. 2012, April 10. Moral Behavior in Animals. *TED video*, 16:52. https://www.youtube.com/watch?v=GcJxRqTs5nk.
Derrida, Jacques. 1997. *Deconstruction in a Nutshell: A Conversation with Jacques Derrida (Perspectives in Continental Philosophy)*. Edited and with commentary by John D. Caputo. New York: The Fordham University Press.
Descombes, Vincent. 1980. *Lo Mismo y lo Otro: cuarenta y cinco años de filosofía francesa (1933-1978)*. Translated into Spanish by Elena Benarroch. Madrid: Ediciones Cátedra, 1982. Translated into English by L. Scott-Fox and J. M. Harding as *Modern French Philosophy*. Cambridge: Cambridge University Press. This translation does not include the paragraph cited here.
Devlin, Keith. 2000. *The Language of Mathematics: Making the Invisible Visible*. New York: Holt Paperbacks.
Dictionary.com. 2006. Website based on the *Random House Unabridged Dictionary*. New York: Random House. https://www.dictionary.com/.
Douglas, Mary. 1966. *Purity and Danger*. London: Routledge and Kegan Paul.
Egbert, Matthew D., Xabier E. Barandiaran, and Ezequiel A. Di Paolo. 2010. A Minimal Model of Metabolism-Based Chemotaxis. *PLOS Computational Biology* 6 (12).
Ehrlich, Pippa, and James Reed, dirs. 2020. *My Octopus Teacher*. A Netflix Original Documentary.
Eliot, George. 2002. *Daniel Deronda*. New York: The Modern Library.
Emmanuel Levinas. 2019. *Stanford Encyclopedia of Philosophy*. Last modified on August 7, 2019. https://plato.stanford.edu/entries/levinas/.
Endosymbiosis: Lynn Margulis. n.d. Understanding Evolution. https://evolution.berkeley.edu/evolibrary/article/history_24.
Endosymbiosis: The Appearance of the Eukaryotes. n.d. The Virtual Fossil Museum. Retrieved April 2019 from http://www.fossilmuseum.net/Evolution/Endosymbiosis.htm.
Engelland, Chad. 2015. Heidegger and the Human Difference. *Journal of the American Philosophical Association* 1: 175–193.
Engstrom, Jack. 2001. C. S. Peirce's Precursors to Laws of Form. *Cybernetics and Human Knowing* 8 (1–2): 25–66.
Ethology. 2021. Wikimedia Foundation, last modified on July 4, 2021, at 13:21. https://en.wikipedia.org/wiki/Ethology.
Favareau, Donald. 2010. *Essential Readings in Biosemiotics: Anthology and Commentary*. Dordrecht: Springer.
Fikes, Bradley. 2014, December 30. Does Red Meat Increase Your Cancer Risk? *San Diego Union Tribune*. http://www.utsandiego.com/news/2014/dec/29/red-meat-cancer-varki-neu5Gc-sialic/.
Finkelstein, Israel, and Neil Asher Silberman. 2001. *The Bible Unearthed: Archaeology's New Vision of Ancient Israel and the Origin of Its Sacred Texts*. New York: Free Press.
Fischer, Lars. 2021. Slime Memory. *Scientific American* 324 (6): 19.
Fischer-Schreiber, Ingrid, Franz-Karl Ehrhard, and Michael S. Diener. 1991. *Shambala Dictionary of Buddhism and Zen*. Boston: Shambala Dragon Editions.
Follmann, H., and C. Brownson. 2009. Darwin's Warm Little Pond Revisited: From Molecules to the Origin of Life. *Naturwissenschaften* 96 (11): 1265–1292.
Freedman, Richard Elliott. 2003. *Commentary on the Torah*. New York: Harper Collins.
Freud, Sigmund. 1981. Proyecto de una psicología para neurólogos. *Biblioteca Nueva,* Tomo 1. Madrid, 1981, trad. López Ballesteros.
Geological History of Oxygen. 2022. Wikimedia Foundation. Last modified on 11 July, 2022, at 21:49. https://en.wikipedia.org/wiki/Geological_history_of_oxygen.

Gilbert, Roberta M. 2006. *The Eight Concepts of Bowen Theory: A New Way To Think About The Individual and The Group*. Virginia Falls: Leading System Press.

Gill, Steven R., Milhai Pop, Robert T. Deboy, Paul B. Eckburg, Peter J. Turnbaugh, Buck S. Samuel, Jeffrey I. Gordon, David A. Relman, Clair M. Fraser-Liggett, and Karen E. Nelson. 2006. Metagenomic Analysis of the Human Distal Gut Microbiome. *Science* 312 (5778): 1355–1359.

Gombrich, Ernst Hans. 1995. *The Story of Art*. 15th ed. New York: Phaidon.

Goodall, Jane. 2000. *In the Shadow of Man*. Rev. ed. New York: Houghton Mifflin.

Gorham, Geoffrey, Benjamin Hill, Edward Slowik, and Kenneth C. Waters, eds 2016. *The Language of Nature: Reassessing the Mathematization of Natural Philosophy in the Seventeenth Century*. Minneapolis Studies in the Philosophy of Science, 20. Minneapolis: University of Minnesota Press.

Gray, Peter. *Psychology*. 6th ed. New York: Worth.

Grunfeld, Frederic V. 1979. *Prophets Without Honour: A Background to Freud, Kafka, Einstein and Their World*. London: Hutchinson.

Haeckel, E. 1866. *Generalle Morphologie der Organismen*. Berlin: Georg Reimer.

Hadhazy, Adam. 2010, June 14. Fact or Fiction: The Days (and Nights) Are Getting Longer. *Scientific American*. http://www.scientificamerican.com/article/earth-rotation-summer-solstice/.

Hamilton, Jon. 2011, December 23. Myth Busting: The Truth About Animals and Tools. *National Public Radio*. https://www.npr.org/2011/12/23/143833929/myth-busting-the-truth-about-animals-and-tools.

Hanschen, Erik, Dinah Davison, Zachariah Grochau-Wright, and Richard Michod. 2018. Individuality and the Major Evolutionary Transitions. In *Landscapes of Collectivity in the Life Sciences*, ed. Snait B. Gissis, Ehud Lamm, and Ayelet Shavit, 255–268. Cambridge: MIT Press.

Harari, Yuval Noah. 2016. *The Future of Humanity*. Address at the Royal Institution, London. Video uploaded September 28, 2016. https://www.youtube.com/watch?v=XOmQqBX6Dn4.

Hawking, Stephen. 2005. *God Created the Integers: The Mathematical Breakthroughs that Changed History*. Philadelphia: Punning Press.

Hazlitt, William. 1842. *The Complete Works of Michael de Montaigne*, 289. Retrieved January 2020, https://quotes.yourdictionary.com/condemn

Health. 2015. *San Diego Union Tribune*. Thursday April 21, 2015, 4.

Henrich, Joseph. 2020. *The WEIRDdest People in the World: How the West Became Psychologically Peculiar and Particularly Prosperous*. New York: Farrar, Straus and Giroux.

Hilbert, David. 1997. *The Foundations of Geometry*. Trans. Leo Unger and Paul Bernays. La Salle: Open Court.

Hoffmeyer, Jesper. 1996. *Signs of Meaning in the Universe*. Trans. Barbara J. Haveland. Bloomington: Indiana University Press.

———. 2008. *Biosemiotics: An Examination into the Signs of Life and the Life of Signs*. Edited and translated by Donald Favareau. Scranton: Scranton University Press.

Wumen Huihai. 2000. Mumokan, Case 29. In *Two Zen Classics*: *Mumonkan (The Gateless Gate) and Hekiganroku (The Blue Cliff Records)*. Trans. Katsuki Sekida. New York: Wheatherhill.

Humphrey, Nicholas. 2013. *The Inner Eye*. London: Faber and Faber.

Husserl, Edmund. 1970. *The Crisis of European Sciences and Transcendental Phenomenology: An Introduction to Phenomonological Philosophy*. Trans. David Carr. Evanston: Northwestern University Press.

Hut, Roelof A., and Domien G.M. Beersma. 2011. Evolution of Time-Keeping Mechanisms: Early Emergence and Adaptation to Photoperiod. *Philosophical Transactions of the Royal Society, B* 366: 2141–2154.

International Astronomical Association Definition of Planet. 2020. Wikimedia Foundation. Last modified on January 5, 2020, at 21:18. https://en.wikipedia.org/wiki/IAU_definition_of_planet.

Introduction to Cyanbacteria: Architects of Earth's Atmosphere. n.d. University of California Museum of Paleontology. https://ucmp.berkeley.edu/bacteria/cyanointro.html.

Jablonka, Eva, and Marion J. Lamb. 2014. *Evolution in Four Dimensions: Genetic, Epigenetic, Behavioral, and Symbolic Variation in the History of Life.* Rev. ed. Cambridge: MIT Press.

Jablonsky, Nina G. 2013. *Skin: A Natural History*. Berkeley: University of California Press.

James Boswell. 2019. Wikimedia Foundation. Last modified on November 26, 2019, at 03:44. https://en.wikiquote.org/wiki/James_Boswell.

Janke, Paul R. 2017, June 15. *A Correlated History of Earth.* 3rd ed. Chart. The Worldwide Museum of Natural History, Hill City.

Johnson, Sue. 2013. *Love Sense: The Revolutionary Science of Romantic Relationships*. New York: Little, Brown Spark.

Johnson, David W., and P. Johnson Frank. 2013. *Joining Together: Group Theory and Group Skills*. 11th ed. Upper Saddle River: Pearson.

Jonas, Hans. 2001. *The Phenomenon of Life: Toward a Philosophical Biology*. Evanston: Northwestern University Press.

Joubert, Dereck, and Beverly Joubert. 2006. *Eye of the Leopard.* Film produced by the National Geographic Society. Video uploaded April 23, 2014. Available at https://www.youtube.com/watch?v=ugi4x8kZJzk.

Joyce, James. 1992. *Ulysses*. London: Penguin Books.

Joye Research Group. 2021. Hydrothermal Vents. Retrieved April 5, 2021 from, https://joyelab.org/hydrothermal-vents.

Kant, Immanuel. 1998. *Critique of Pure Reason*. Trans. Paul Guyer. Cambridge: Cambridge University Press

Kass, Leon R. 2021. *Founding God's Nation: Reading Exodus*. New Haven: Yale University Press.

Kauffman, Stuart. 1995. *At Home in the Universe: The Search for the Laws of Complexity*. London: Penguin.

Kaufman, Louise H. 2011. Peirce's Existential Graphs. *Cybernetics of Human Knowing* 18 (1–2). Exeter: Imprint Academic.

Kaufman, Louise H. n.d.-a Laws of Form: An Exploration in Mathematics and Foundations. Unpublished manuscript. Available at http://www.math.uic.edu/~kauffman/Laws.pdf.

———. n.d.-b Extension is a Single Initial of the Primary Algebra. Unpublished manuscript. Available at http://homepages.math.uic.edu/~kauffman/Extension.pdf.

Kawai, Masao. 1965. Newly-acquired Pre-cultural Behavior of the Natural Troop of Japanese Monkeys on Koshima Islet. *Primates* 6 (1). Available at https://www.j-monkey.jp/lng/pdf/sweet-potato-washing/Kawai-(1965)-Monkey-culture-PRIMATES.pdf

Kelly, Kevin. 1994. *Out of Control: The New Biology of Machines, Social Systems and the Economic World.* Boston: Addison-Wesley.

Keys, James. *See* Spencer-Brown.

Kimmerer, Robin. 2015. *Braiding Sweetgrass*. Minneapolis: Milkweed Editions.

Knoblich, Günther, and Natalie Sebanz. 2014, May 28. Mirror Neuron. *Encyclopædia Britannica Online*. https://www.britannica.com/science/mirror-neuron,

Kohlberg, Lawrence. 1958. *The Development of Modes of Thinking and Choices in Years 10 to 16.* PhD diss., University of Chicago.

Kotler, Stephen, and Jamie Wheal. 2017. *Stealing Fire*. New York: Dey Street Books.

Kranendonk, Van, J. Martin, David W. Deamer, and Tara Djokic. 2017. Life on Earth Came from a Hot Volcanic Pool, Not the Sea. *Scientific American* 317 (2).

Krause, Jens, and Graeme D. Ruxton. 2002. *Living in Groups*. Oxford: Oxford University Press.

Kull, Kalevi. 2010. Umwelt. In *The Routledge Companion to Semiotics*, ed. Paul Cobley, 43–56. London: Routledge.

Large, William. 2015. *Levinas' Totality and Infinity*. London: Bloomsbury.

Left Brain vs. Right Brain: What Does This Mean for Me? *Healthline*. https://www.healthline.com/health/left-brain-vs-right-brain#_noHeaderPrefixedContent.

Levinas, Emmanuel. 1989. Time and the Other. In *The Levinas Reader*, ed. Seán Hand. Malden: Blackwell.

Lewin, Sarah. 2016, January 12. Life on Earth Can Thank Its Lucky Stars for Jupiter and Saturn. *Space.com*. https://www.space.com/31577-earth-life-jupiter-saturn-giant-impacts.html.

Lezama Lima, José. 2010a. *Refutación de los Espejos: Poesía, Poética, Prosa*. Edited by Victor Toledo. Ciudad de Puebla, México: Benemérita Universidad Autónoma de Puebla.

———. 2010b. Danza de la Jerigonza. *Refutación de los Espejos: Poesía, Poética, Prosa*. Antología por Víctor Toledo, 90. Puebla, México: Universidad de Puebla.

Liddell, Henry George, and Robert Scott. 1997. *An Intermediate Greek-English Lexicon*. Oxford: Oxford University Press.

Lipton, Bruce. 2005. *The Biology of Belief: Unleashing the Power of Consciousness, Matter & Miracles*. Santa Rosa: Mountain of Love/Elite Books.

Lispector, Clarice. 1974. *La Manzana en la Oscuridad*. Translated into Spanish by Juan Garcia Gayo. Buenos Aires: Editorial Sudamericana.

Lonergan, Bernard. 1971. *Method in Theology*. Toronto: University of Toronto Press.

———. 1988. The Subject. In *Collected Works of Bernard Lonergan: A Second Collection*, ed. Robert M. Duran and John D. Dadosky, 60–74. Toronto: University of Toronto Press.

———. 1990. *Understanding and Being, Collected Works of Bernard Lonergan*. Vol. 5. Toronto: University of Toronto Press.

———. 1992. *Insight: A Study of Human Understanding, Collected Works of Bernard Lonergan*. Vol. 3. Toronto: University of Toronto Press.

———. 1993. Cognitional Structure. In *Collected Works of Bernard Lonergan*, ed. Frederick E. Crowe and Robert M. Doran, vol. 4, 2nd ed. Toronto: University of Toronto Press.

———. 1996. Exegesis and Dogma. In *Philosophical and Theological Papers 1958-1964, Collected Works of Bernard Lonergan*, vol. 6. Toronto: University of Toronto Press.

Lorino, Philippe. 2014. Charles Sanders Peirce (1839-1914). In *The Oxford Handbook of Process Philosophy and Organizational Studies*, ed. Jenny Helin, Tor Hernes, Daniel Hjorth, and Robin Holt. Oxford: Oxford University Press.

Lovelock, James E. 1982. The Regulation of Carbon Dioxide and Climate: Gaia or Geochemistry. *Planetary and Space Science* 30 (8): 795–802. http://www.jameslovelock.org/the-regulation-of-carbon-dioxide-and-climate-gaia-or-geochemistry/.

Loza-Correa, Maria, Laura Gomez-Valero, and Carmen Buchrieser. 2010. Circadian Clock Proteins in Prokaryotes: Hidden Rhythms? *Frontiers in Microbiology* 1: 130.

Luisi, Pier Luigi. 2006. *The Emergence of Life: From Chemical Origins to Synthetic Biology*. Cambridge: Cambridge University Press.

Macy, Joanna. 1991. *Mutual Causality in Buddhism and General Systems Theory*. New York: SUNY Press.

Maestripieri, Dario, ed. 2003. *Primate Psychology*. Cambridge: Harvard University Press.

Mancuso, Stephano, and Alessandra Viola. 2015. *Brilliant Green*. Trans. Joan Benham. Washington: Island Press.

Marchant, Jo. 2016. *Cure: A Journey into the Science of Mind and Body*. New York: Crown Publishers.

Margulis, Lynn, and Karlene V. Schwartz. *Five Kingdoms: An Illustrated Guide to the Phyla of Life on Earth*. 3rd ed, 9–10. New York: W. H. Freeman.

Marlow, Jeffrey, and Rogier Braakman. 2018. Team Players: Microbial Partnerships That Rule the Planet. *Scientific American* 319 (5): 1118–1132.

Marx, Karl, and Friedrich Engels. 1956. *The Holy Family*. Trans. Richard Dixon and Clement Dutts. Moscow: Foreign Languages Publishing House.

Maslow, Abraham H. 1962. *Toward a Psychology of Being*. New York: Wiley.

———. 1966. *The Psychology of Science: A Reconnaissance*. New York: Harper & Row.

Maturana, Humberto. 2004. The Knowledge of Knowledge Entails Responsibility. In *The Certainty of Uncertainty, Dialogues Introducing Constructivism*, ed. Bernhard Poerksen, 47–84. Translated by Alison Rosemary Koeck and Wolfram Karl Koeck. Exeter: Imprint Academic.

Mayer, Daniel. Fall 2003. Interiority and the Challenge of Primatology: Anthropomorphism as an Instance of High-level Cognition about High-level Cognition. *Method: Journal of Lonergan Studies* 21 (2): 145–160.

———. 2011, June 24. From Distinction to Anticipation. Presentation at the Eleventh Annual International Gathering in Biosemiotics, New York. https://www.academia.edu/723995/From_distinction_to_anticipation_the_advent_of_the_semiosic.

———. 2012. Significance. In *A More Developed Sign: Interpreting the Work of Jesper Hoffmeyer*, ed. Donald Favareau, Paul Cobley, and Kalevi Kull, 259–262. Tartu, Estonia: Tartu University Press.

Mayer, Emeran. 2016. *The Mind-Gut Connection: How the Hidden Conversation Within Our Bodies Impacts Our Mood, Our Choices and Our Overall Health*. New York: HarperCollins.

Mayer, Daniel. 2022. McShane on the Transformation of Botany: Outline for a Dissertation. *Divyadaan: Journal of Philosophy and Education* 33 (1): 95–104. Nashik: Salesian Institute of Philosophy

Mazur, James E. 2015. *Learning and Behavior*. 7th ed. New York: Pearson.

Meigs, James. 2021, June. Thank God for Big Pharma. *Commentary*. https://www.commentary-magazine.com/articles/james-meigs/big-pharma-covid-vaccines/.

Mikulecky, Peter J., Michelle R. Gilman, and Brian Peterson. 2008. *AP Biology for Dummies*. Indianapolis: Wiley.

———. n.d. Endosymbiotic Theory. *Dummies.com*. https://www.dummies.com/test-prep/endosymbiotic-theory/.

Mitchell, Robert W., and Nicholas S. Thompson. 1986. *Deception: Perspectives on Human and Nonhuman Deceit*. New York: SUNY Press.

Mowat, Farley. 1969. *Never Cry Wolf*. New York: Dell Publishing.

Nizri, Vicky. 2002. *Lilith, La Otra Carta de Dios*. México City: Miguel Ángel Porrúa.

Offord, Catherine. 2016, December. Elephant Footprints Create Habitat for Tiny Aquatic Creatures. *The Scientist*. https://www.the-scientist.com/notebook/elephant-footprints-create-habitat-for-tiny-aquatic-creatures-32437.

Okumura, Shohaku. 2010. *Realizing Genjokoan: The Key to Dōgen's* Shobogenzo. Somerville, MA: Wisdom Publications.

Parker, Sue Taylor, and Michael L. McKinney. 1999. *Origins of Intelligence: The Evolution of Cognitive Development in Monkeys, Apes and Humans*. Baltimore: John's Hopkins University Press.

Partridge, Eric. 1958. *Origins: A Short Etymological Dictionary of Modern English*. London: Routledge and Kegan Paul.

Patterson, Penny, and Eugene Linden. 1981. *The Education of Koko*. New York: Holt, Rinehart and Winston.

Peirce, Charles Sander. 2010. The Logic of Signs. In *Essential Readings in Biosemiotics: Anthology and Commentary*, ed. Donald Favareau. Dordrecht: Springer.

———. n.d. Reflections on Real and Unreal Objects. MS (R) 966. http://www.helsinki.fi/science/commens/terms/object.html.

Pennisi, Elizabeth. 2018, June 28. The Momentous Transition to Multicellular Life May Not Have Been so Hard After All. *Science*. https://www.sciencemag.org/news/2018/06/momentous-transition-multicellular-life-may-not-have-been-so-hard-after-all#.

Pepperberg, Irene. 2009. *Alex & Me: How a Scientist and a Parrot Discovered a Hidden World of Animal Intelligence—and Formed a Deep Bond in the Process*. New York: Harper-Collins.

Perry, Jon. 2018, November 25. Irreducible Complexity? What Organ Evolved First: Brain, Heart, Stomach? *Streamed Live*. https://www.youtube.com/watch?v=S4I8YjMgV6E.

Peterson, Jordan B. 1999. *Maps of Meaning: The Architecture of Belief*. New York: Routledge.

Petrovich, Douglas. 2016. *The World's Oldest Alphabet: Hebrew as the Proto-Consonantal Script*. Jerusalem: CARTA Jerusalem.

Photographing Smoke Patterns. 2012, May 12. ePhotozine. https://www.ephotozine.com/article/photographing-smoke-patterns-13055.

Plessner, Helmuth. 2019. *Levels of Organic Life and the Human: An Introduction to Philosophical Anthropology*. Trans. Millay Hyatt. New York: Fordham University Press.

Polanyi, Michael. 1983. *The Tacit Dimension*. Gloucester: Peter Smith.

Popa, Radu. 2004. *Between Necessity and Probability: Searching for the Definition and Origin of Life*. Heidelberg: Springer.

Powers, Richard. 2018. *The Overstory*. New York: W. W. Norton.

Prokaryote. 2020. Wikimedia Foundation, last modified on February 8, 2020, at 12:25. https://en.wikipedia.org/wiki/Prokaryote.

Protocell. 2020. Wikimedia Foundation, last modified on January 10, 2020, at 14:00. https://en.wikipedia.org/wiki/Protocell.

Pullen, Lara C. 2014, November 19. Bacterial Microbiome Move Over: The Gut Virome Makes Its Debut. *Medscape*. https://www.medscape.com/viewarticle/835193.

Ramachandran V. S. 2000. Mirror Neurons and Imitation Learning as the Driving Force Behind The Great Leap Forward in Human Evolution. *Edge* 69. https://www.edge.org/conversation/vilayanur_ramachandran-mirror-neurons-and-imitation-learning-as-the-driving-force.

Recio, Gonzáles, and José Luis. 2004. *Teorías de la Vida*. Madrid: Editorial Síntesis.

Ridley, Matt. 1996. *The Origins of Virtue: Human Instincts and the Evolution of Cooperation*. London: Penguin Books.

Robbins, Gary. 2015, October 22. Star Biologist Pinpoints Role of Microbes in Disease: Interview with Rob Knight. *San Diego Union Tribune*.

Roberts, Edward A., and Bárbara Pastor. 1997. *Diccionario Etimológico Indoeuropeo de la Lengua Española*. Madrid: Alianza Diccionarios.

Romano, Claude. 2021. *De la Couleur*. Paris: Gallimard.

Roots, Betty I. 2008. The Phylogeny of Invertebrates and the Evolution of Myelin. *Neuron Glia Biology* 4 (2): 101–109.

Rosen, Robert. 1985. *Anticipatory Systems: Philosophical, Mathematical and Methodological Foundations. IFSR International Series on System Science and Engineering*. Vol. 1. New York: Pergamon Press.

———. 1991. *Life Itself: A Comprehensive Inquiry into the Nature, Origin and Fabrication of Life*. New York: Columbia University Press.

Rovelli, Carlo 2016. *Seven Brief Lessons on Physics*. Trans. Simon Carnell and Erica Segre. New York: Riverhead Books.

———. 2018. *The Order of Time*. New York: Riverhead Books.

Rushdie, Salman. 2021. *Languages of Truth, Essays 2003-2020*. New York: Random House.

Russell, Michael. 2009. *Origins, Abiogenesis and the Search for Life*. Cambridge: Cosmology Science Publications.

Sacks, Jonathan. 2018, January 8. The Politics of Hope. *Video*. Available at https://www.youtube.com/watch?v=7BvLp4VdW1A.

———. 2020a. *Judaism's Life-Changing Ideas: A Weekly Reading of the Jewish Bible*. New Milford: Maggid Books.

———. 2020b. *Morality: Restoring the Common Good in Divided Times*. New York: Basic Books.

Sartre, Jean Paul 1968. *Search for a Method*. Trans. Hazel Barnes. New York: Vantage Books

Schacter, Daniel L., Daniel T. Gilbert, and Daniel M. Wegner. *Psychology*. 2nd ed. New York: Worth.

Schidlowski, Manfred. 1992. The Initiation of Biological Processes on Earth: Summary of Empirical Evidence. *Advances in Space Research* 12 (4): 143–156.

Schiltz, Michael. 2009. Space is the Place: The *Laws of Form* and Social Systems. In *Emergence and Embodiment: New Essays on Second-Order Systems Theory*, ed. Bruce Clark and Mark B.N. Hansen, 157–178. Durham: Duke University Press.

Schumaker, Robert W., Kristina R. Walkup, and Benjamin B. Beck. 2011. *Animal Tool Behavior: The Use and Manufacture of Tools by Animals*. Rev. ed. Baltimore: Johns Hopkins University Press.

Schwab, Ivan R. 2012. *Evolution's Witness: How Eyes Evolved*. Oxford: Oxford University Press.

Shapiro, Steward. 2000. *Thinking About Mathematics: The Philosophy of Mathematics*. Oxford: University Press.

Sharma, Vijay Kumar. 2003. Adaptive Significance of Circadian Clocks. *Chronobiology International* 20 (6): 901–919.

Shubin, Neil. 2009. *Your Inner Fish: A Journey into the 3.5 Billion-Year History of the Human Body*. New York: Vintage Books.
Shumaker, Robert W., Kristina R. Walkup, and Benjamin B. Beck. 2011. *Animal Tool Behavior*. Baltimore: John Hopkins University Press.
Siegel, Ronald K. 1989. *Intoxication: The Universal Drive for Mind-Altering Substances*. Rochester: Park Street Press.
Siegel, Daniel J. 2011. *Mindsight: The New Science of Personal Transformation*. New York: Bantam.
Snow, Charles Percy. 1959. *The Two Cultures*. London: Cambridge University Press.
Soho, Takuan. 2010. *Tao Te Ching: Zen Teachings on the Taoist Classic*. Trans. Thomas Clearly. Boston: Shambala.
Spencer-Brown, George. 1969. *Laws of Form*. 1st ed. London: George Allen and Unwin.
———. 1973a. *Laws of Form*. Limited ed. Portland: Cognizer.
———. 1973b, March 19. Transcript of discussion of *Laws of Form* at American University of Masters Conference. Sessions audiotaped by Kurt von Meier and transcribed by Cliff Barney. https://web.archive.org/web/20060821204917/http://lawsofform.org/aum/session1.html. Reproduced in *Cybernetics of Human Knowing* 26 (2–3): 17–34.
——— [James Keys, pseud.]. 1974. *Only Two Can Play This Game*. New York: Bantam Books.
———. 1997. *Gesetze der Form*. Lübeck: Bohmeier Verlag.
Steiner, Roger Y., Michael Doudoroff, and Edward A. Adelberg. 1963. *The Microbial World*. 3rd ed. Englewood Cliffs: Prentice-Hall.
Takahashi, Joseph. 2013, November 21. The Clock in our Genes and in Every Cell of Your Body. TED video, 20:08. https://www.youtube.com/watch?v=ocqn3wYTCRM.
The Endosymbiotic Origin of Domain Eukaryota. n.d. Fossilmuseum.net. http://www.fossilmuseum.net/Evolution/Endosymbiosis.htm.
The Evolution of the Cell. n.d. Learn.Genetics: Genetic Science Learning Center. https://learn.genetics.utah.edu/content/cells/organelles/.
Tomasello, Michael, and Josep Call. 1997. *Primate Cognition*. Oxford: Oxford University Press.
Tuckman, Bruce W. 1965. Developmental sequence in small groups. *Psychological Bulletin* 63 (6): 384–399.
Turek, Fred. 2015, March 10. Circadian Surprise: How Our Body Clocks Help Shape Our Waistlines. Interview on National Public Radio's *Morning Edition*. https://www.npr.org/sections/thesalt/2015/03/10/389596946/circadian-surprise-how-our-body-clocks-help-shape-our-waistlines.
Lao Tzu. 1990. *Tao Te Ching: The Classic Book of Integrity and the Way*. Translated by Victor H. Mair. New York: Bantham.
———. 2001. *Tao Te Ching: The Definitive Edition*. Translated and commentary by Jonathan Star and Jeremy Tarcher. New York: Penguin.
———. 2003. *Dao De Jing: A Philosophical Translation*. Translation and commentary by Roger T. Ames and David L. Hall. New York: Balantine.
———. 2005. *Tao Te Ching: A New Translation*. Trans. Sam Hamill. Boston: Shambala.
U.S. Department of Commerce. 2020, November 5. Chemosynthesis and Hydrothermal Vent Life. National Oceanic and Atmospheric Administration, Office of Ocean Exploration and Research. https://oceanexplorer.noaa.gov/edu/learning/5_chemosynthesis/chemosynthesis.html.
Ursell, Luke K., Jessica L. Metcalf, Laura W. Parfrey, and Rob Knight. 2012. Defining the Human Microbiome. *Nutrition Reviews* 70 (Suppl 1): S38–S44.
Van Dover, Cindy. *The Ecology of Deep-Sea Hydrothermal Vents*. Princeton: Princeton University Press, 2000.
Varela, F. J. (1996). In *Defining Life: The Central Problem in Theoretical Biology*, ed M. Rizzoti, 149. Padua: University of Padua.
Vigotsky, L. 1978. In *Mind in Society: The Development of Higher Psychological Processes*, ed. M. Cole, J. Scribner, V. John- Steiner, and E. Souberman. Cambridge: Harvard University Press.

Waggoner, Ben. 2020. Introduction to the Vendian Period: 650 to 543 Million Years Ago. *Vendian Animals*. Special Exhibit of the University of Californiat Museum of Paleontology. Retrieved August 2020 from Introduction to the Vendian Period (berkeley.edu).

Watzlawick, Paul, John H. Weakland, and Richard Fish. 1974. *Change: Principles of Problem Formation and Problem Resolution*. New York: W. W. Norton.

West, Stuart, Claire El Mouden, and Andy Gardner. 2011. Sixteen Common Misconceptions About the Evolution of Cooperation in Humans. *Evolution and Human Behaviour* 32: 231–262.

West, Stuart A., Roberta M. Fisher, Andy Gardner, and E. Toby Kiers. 2015. Major Evolutionary Transitions in Individuality. *Proceedings of the National Academy of Sciences in the United States of America* 112 (33): 10112–10119.

White Blood Cell Chases Bacteria. 2008, March 7. *YouTube Video* 28. https://www.youtube.com/watch?v=JnlULOjUhSQ.

Whitehead, Alfred North. 1925. *Science and the Modern World*. New York: Simon and Schuster.

Whitman, William B., David C. Coleman, and William J. Wiebe. 1998. Prokaryotes: The Unseen Majority. *Proceedings of the National Academy of Sciences of the United States of America* 95 (12): 6578–6583.

Whittle, Mark. 2008. *Cosmology: The History and Nature of Our Universe, Part 1*. Chantilly: The Teaching Company.

Whyte, David. 2009. *The Three Marriages: Reimagining Work, Self and Relationship*. New York: Riverhead Books.

Williams, Leonard. 1967. *Man and Monkey*. London: Andre Deutsch.

———. 1974. *The Wooly Monkey*. Looe: Monkey Sanctuary Publications.

———. 1978. *Challenge to Survival: A Philosophy of Evolution*. 3rd ed. London: Allison & Busby.

———. 1980. *Samba*. London: Allison and Busby.

Wilson, Edward O. 2014. *The Meaning of Human Existence*. New York: W. W. Norton.

Wilson, Edward O., and Bert Hölldobler. 2009. *The Superorganism: The Beauty, Elegance, and Strangeness of Insect Societies*. New York: W. W. Horton.

Winerman, Lea. (2005, October). The Mind's Mirror. *Monitor on Psychology* 36 (9). http://www.apa.org/monitor/oct05/mirror.

Woese, Carl R., Otto Kandler, and Mark L. Wheelis. 1990. Towards a Natural System of Organisms: Proposal for the Domains Archaea, Bacteria, and Eucarya. *Proceedings of the National Academy of Sciences of the United States of America* 87 (12): 4576–4579.

Wohlleben, Peter. 2015. *The Hidden Life of Trees, What they Feel, How They Communicate*. Trans. Jane Billinghurst. Vancouver: Greystone Books.

Wolpert, Lewis. 2009. *How We Live and Why We Die: The Secret Lives of Cells*. New York: W. W. Norton.

Wrangham, Richard W., W.C. McGrew, Frans B.M. de Waal, and Paul G. Heltne, eds. 1994. *Chimpanzee Cutures*. Cambridge: Harvard University Press.

Wu Ming. 2019. Wikipedia Foundation, last modified on December 29, 2019, at 21:35. https://en.wikipedia.org/wiki/Wu_Ming.

Yang, Joy. 2012, July 16. The Human Microbiome Project: Extending the Definition of What Constitutes a Human. National Human Genome Research Institute. https://www.genome.gov/27549400/the-human-microbiome-project-extending-the-definition-of-what-constitutes-a-human.

Yasutani, Hakuun. 1996. *Flowers Fall: A Commentary on Zen Master's Dōgen's* Genjōkōan. Boston: Shambhala.

Young, Robert M. 1992. *Excursions in Calculus: An Interplay of the Continuous and the Discrete*. Dolciani Mathematica Expositions, Number 13. Mathematical Association of America.

Zeeya, Merali. 2017. *A Big Bang in a Little Room: The Quest to Create New Universes*. New York: Basic Books.

Zimmer, Carl. 2016, April 11. Scientists Unveil New 'Tree of Life.' *New York Times*.

Index

© The Editor(s) (if applicable) and The Author(s), under exclusive license to Springer Nature Switzerland AG 2023

D. C. Mayer-Foulkes, *The Nature of Living Being*, Biosemiotics 26,
https://doi.org/10.1007/978-3-031-24789-7

C

F

N

O

P

MIX
Papier aus verantwortungsvollen Quellen
Paper from responsible sources
FSC® C105338

If you have any concerns about our products,
you can contact us on
ProductSafety@springernature.com

In case Publisher is established outside the EU,
the EU authorized representative is:
Springer Nature Customer Service Center GmbH
Europaplatz 3, 69115 Heidelberg, Germany

Printed by Libri Plureos GmbH
in Hamburg, Germany